Machining Polymer Matrix Composites:

Tools, Techniques, and Sustainability

Francisco Mata Cabrera
Universidad de Castilla-La Mancha, Spain

Issam Hanafi
LSIA, ENSAH, Université Abdelmalek Essaâdi, Morocco

A volume in the Advances in
Chemical and Materials Engineering
(ACME) Book Series

IGI Global
PUBLISHER of TIMELY KNOWLEDGE

Published in the United States of America by
IGI Global
Engineering Science Reference (an imprint of IGI Global)
701 E. Chocolate Avenue
Hershey PA, USA 17033
Tel: 717-533-8845
Fax: 717-533-8661
E-mail: cust@igi-global.com
Web site: http://www.igi-global.com

Library of Congress Cataloging-in-Publication Data

Names: Cabrera, Francisco Mata, editor. | Hanafi, Issam, 1982- editor.
Title: Machining polymer matrix composites : tools, techniques, and sustainability / Francisco Mata Cabrera, Issam Hanafi.
Description: Hershey, PA : Engineering Science Reference, 2024. | Includes bibliographical references and index. | Summary: "The impact of the book is significant in the field of composite materials and manufacturing. It provides a comprehensive understanding of machining composite materials with a polymer matrix and can help researchers, engineers, and students in the selection of machining processes and optimization of cutting parameters"-- Provided by publisher.
Identifiers: LCCN 2023038670 (print) | LCCN 2023038671 (ebook) | ISBN 9781668499276 (hardcover) | ISBN 9781668499283 (ebook)
Subjects: LCSH: Polymeric composites.
Classification: LCC TA455.P55 M25 2023 (print) | LCC TA455.P55 (ebook) | DDC 620.1/92--dc23/eng/20240312
LC record available at https://lccn.loc.gov/2023038670
LC ebook record available at https://lccn.loc.gov/2023038671

This book is published in the IGI Global book series Advances in Chemical and Materials Engineering (ACME) (ISSN: 2327-5448; eISSN: 2327-5456)

British Cataloguing in Publication Data
A Cataloguing in Publication record for this book is available from the British Library.
All work contributed to this book is new, previously-unpublished material.

For electronic access to this publication, please contact: eresources@igi-global.com.

Advances in Chemical and Materials Engineering (ACME) Book Series

ISSN:2327-5448
EISSN:2327-5456

Editor-in-Chief: J. Paulo Davim, University of Aveiro, Portugal

MISSION

The cross disciplinary approach of chemical and materials engineering is rapidly growing as it applies to the study of educational, scientific and industrial research activities by solving complex chemical problems using computational techniques and statistical methods.

The **Advances in Chemical and Materials Engineering (ACME) Book Series** provides research on the recent advances throughout computational and statistical methods of analysis and modeling. This series brings together collaboration between chemists, engineers, statisticians, and computer scientists and offers a wealth of knowledge and useful tools to academics, practitioners, and professionals through high quality publications.

COVERAGE

- Polymers Engineering
- Fracture Mechanics
- Wear of Materials
- Composites
- Coatings and surface treatments
- Heat Treatments
- Electrochemical and Corrosion
- Fatigue and Creep
- Metallic Alloys
- Biomaterials

IGI Global is currently accepting manuscripts for publication within this series. To submit a proposal for a volume in this series, please contact our Acquisition Editors at Acquisitions@igi-global.com or visit: http://www.igi-global.com/publish/.

The Advances in Chemical and Materials Engineering (ACME) Book Series (ISSN 2327-5448) is published by IGI Global, 701 E. Chocolate Avenue, Hershey, PA 17033-1240, USA, www.igi-global.com. This series is composed of titles available for purchase individually; each title is edited to be contextually exclusive from any other title within the series. For pricing and ordering information please visit http://www.igi-global.com/book-series/advances-chemical-materials-engineering/73687. Postmaster: Send all address changes to above address.

Titles in this Series

For a list of additional titles in this series, please visit:
http://www.igi-global.com/book-series/advances-chemical-materials-engineering/73687

Advancements in Renewable Energy and Green Hydrogen
Maria Simona Raboaca (National Research and Development Institute for Cryogenic and Isotopic Technologies, Râmnicu Vâlcea, Romania) Saheb Djohra (Renewable Energy Development Centre, Algeria) Omar Moussaoui (University Mohammed First, Oujda, Morocco) Traian Candin Mihaltan (Technical University of Cluj, Romania) and Mustapha Koussa (Renewable Energy Development Centre,Algeria)
Engineering Science Reference • © 2024 • 327pp • H/C (ISBN: 9798369310144) • US $260.00

Production, Properties, and Applications of Engineered Cementitious Composites
S. Praveenkumar (PSG College of Technology, India) and J. Paulo Davim (University of Aveiro, Portugal)
Engineering Science Reference • © 2024 • 344pp • H/C (ISBN: 9781668481820) • US $250.00

FlexPDE and Finite Element Method Applications in Thermal Energy Storage and Cavities
Khashayar Hosseinzadeh (Sharif University of Technology, Iran) Mehdi Mahboobtosi (Babol Noshirvani University of Technology, Iran) and Davood Domairry Ganji (Babol Noshirvani University of Technology, Iran)
Engineering Science Reference • © 2024 • 380pp • H/C (ISBN: 9798369309322) • US $225.00

Next Generation Materials for Sustainable Engineering
Shrikaant Kulkarni (Padmashri Dr. V.B. Kolte College of Engineering, India) Boris Kharisov (Autonomous University of Nuevo Leon, Mexico) A.K. Haghi (University of Coimbra, Portugal) and Vipul Srivastava (Lovely Professional University, India)
Engineering Science Reference • © 2024 • 442pp • H/C (ISBN: 9798369313060) • US $295.00

For an entire list of titles in this series, please visit:
http://www.igi-global.com/book-series/advances-chemical-materials-engineering/73687

IGI Global
PUBLISHER of TIMELY KNOWLEDGE

701 East Chocolate Avenue, Hershey, PA 17033, USA
Tel: 717-533-8845 x100 • Fax: 717-533-8661
E-Mail: cust@igi-global.com • www.igi-global.com

Table of Contents

Detailed Table of Contents

Francisco Mata Cabrera, Universidad de Castilla-La Mancha, Spain
Issam Hanafi, ENSAH, Université Abdelmalek Essaâdi, Morocco
Wanderson de Oliveira Leite, Institudo Federal de Minas Gerais, Brazil
Khadija Haboubi, ENSAH, Université Abdelmalek Essaâdi, Morocco

This study employs a bibliometric approach to analyze the evolution of highly cited research in machining of reinforced plastics from 1986 to June 2021. The research examines various document formats, such as journal articles, reviews, chapters, and books. Focusing on reinforced plastics and composites, the study evaluates prolific contributors, temporal trends, and citation metrics to assess scholarly prominence. The research uses a curated dataset of 161 sources from the Web of Science Core Collection database, aiming to identify emerging research directions and gaps in the field. The study uncovers quantitative dimensions and interrelationships among research domains, providing insights into scholarly development. The data categorizes document types, indicating that articles form the majority (85.005%) of the dataset, followed by proceeding papers (16.387%) and review articles (2.536%). The dataset comprises various other types, with each representing a smaller portion. The study's empirical foundation is drawn from the Web of Science Core Collection database.

The marine industry is adopting composite materials with a polymer matrix for shipbuilding due to new technologies and high-performance expectations. However, selecting the right material is challenging, especially considering the machining behavior required for manufacturing cylindrical composite structures. This study focused on evaluating the turning operation and mechanical properties of composites made of glass and basalt fibers with a polymer matrix specifically for maritime construction. The composites consisted of two fabric types, six layers, and two fiber orientation lay-ups, using vinyl ester resin and polyester as the matrix. Test specimens were machined and analyzed, revealing that the hardness and mechanical properties had an impact on the machining process. Glass fiber reinforced polymer (GFRP) samples were successfully machined, resulting in well-formed and undamaged FRP bars. In contrast, basalt fiber reinforced polymer (BFRP) specimens faced difficulties during machining, with matrix splitting, interfacial delamination, fiber ripping.

In contrast to traditional materials, polymer matrix-based composites (PMCs) have become increasingly popular due to their versatility in a wide range of industrial applications. These composites possess the necessary properties and offer endless possibilities for construction and composite levels. This chapter aims to provide a comprehensive overview of the machining process for composite materials with a

The chapter underscores the significance of sustainable practices within machining, focusing on waste reduction, energy efficiency, and environmentally conscious methodologies to align with contemporary manufacturing standards.

This review chapter focuses on enhancing the cutting performance of diamond-coated tools when machining CFRP composite materials. The enhancement is achieved by utilizing advanced polycrystalline diamond coatings on WC-Co-based cutting tools. Additionally, a proposed reconditioning method for these diamond-coated tools is discussed. The chapter elaborates on a hot filament chemical vapor deposition (HFCVD) technique for preparing diamond coatings with various film morphologies, achieved through boron doping. It covers the fundamentals, equipment used, deposition parameters, and substrate pretreatment techniques involved in the HFCVD method. Furthermore, it explores the strategy of improving interface adhesion by doping chemical elements into diamond coatings.

Polymer matrix composites (PMCs) represent a class of advanced materials that have gained widespread attention and adoption across various industries due to their exceptional mechanical properties and versatility in applications. Understanding the full spectrum of mechanical properties and their interplay with other material characteristics is essential for optimizing the performance of PMCs in various real-world applications. This chapter offers a comprehensive exploration of the intricate relationship between machining parameters and the mechanical properties of PMCs. The scope of this chapter encompasses an in-depth exploration of various mechanical properties relevant to machining, including tensile strength, modulus, flexural strength, impact resistance, shear strength, fatigue resistance, and creep behavior. It explores the underlying mechanisms behind the alterations in mechanical properties. Furthermore, the chapter highlights strategies to mitigate the adverse effects of machining on the mechanical properties of PMCs.

Preface

In the ever-expanding realm of academic research and industrial innovation, the machining of polymer matrix composites emerges as a formidable challenge, a complex puzzle requiring scholars to navigate through the intricate maze of material property variations, intricate structures, and the relentless pursuit of flawless surface quality. Within this challenging landscape, a beacon of knowledge and practical guidance emerges: *Machining Polymer Matrix Composites: Tools, Techniques, and Sustainability*. This seminal work transcends the boundaries of conventional academic literature, offering a profound exploration of the multifaceted world of machining processes specific to polymer matrix composites.

At its core, this book serves as a comprehensive compendium, meticulously crafted to address the diverse and intricate aspects of machining these advanced materials. It delves deep into the subtle nuances of tool selection, cutting parameters, surface quality, and tool wear, providing scholars with not just theoretical insights but a practical wisdom honed through experience. Through the detailed exploration of turning, milling, drilling, and grinding, alongside the latest advancements in high-speed and ultrasonic machining, the book curates a vast toolbox for scholars eager to optimize their machining techniques. More than just theoretical knowledge, this book offers a hands-on approach, providing readers with the keys to mastering the delicate art and science of machining polymer matrix composites.

In the contemporary landscape of materials science and engineering, polymer matrix composites have become linchpins, finding applications in diverse sectors such as aerospace, automotive, renewable energy, and electronics. The surging demand for lightweight, durable, and high-performance components has elevated the significance of understanding and mastering the machining of polymer matrix composites to unprecedented levels. *Machining Polymer Matrix Composites: Tools, Techniques, and Sustainability* bridges the chasm between theoretical knowledge and practical application, providing a timely and indispensable resource for scholars and professionals alike. This bridge empowers them not only to meet but to surpass the demands of the modern world.

This meticulously crafted book caters to a wide-ranging audience, making it accessible to beginners and seasoned professionals alike. For beginners, it serves as a foundational guide, laying the essential groundwork and offering the fundamental understanding necessary to embark on the intricate journey of machining polymer matrix composites. Simultaneously, for experienced professionals, it serves as an invaluable repository of advanced insights, enabling them to refine their existing techniques and stay at the forefront of the ever-evolving field. Academics, researchers, engineers, and industrial practitioners deeply involved in the intricate world of machining polymer matrix composites will discover in this book an indispensable companion, a reliable source of knowledge and practical wisdom.

The pages of this groundbreaking work house a profound exploration of machining techniques, complemented by real-world case studies that illuminate both challenges and opportunities. However, this book goes beyond the mere technicalities of machining. It embraces a pressing global concern: sustainability. Woven intricately into its fabric, the book serves as a guide, directing scholars toward environmentally conscious machining practices. In an era where environmental awareness is paramount, the book equips academic scholars with the necessary tools not only to achieve high-quality machined components but to do so while minimizing their environmental footprint, contributing significantly to the global drive for eco-friendly practices.

As you embark on this intellectual odyssey through the pages of *Machining Polymer Matrix Composites: Tools, Techniques, and Sustainability*, you are invited to immerse yourself in a rich tapestry of knowledge, practical wisdom, and sustainability considerations. May this book serve as a guiding light, illuminating your path toward mastery in the intricate and challenging domain of machining polymer matrix composites, reshaping the future of academic inquiry, industrial practices, and environmental stewardship.

Within the pages of this chapter lies a meticulous and comprehensive examination that transcends the temporal boundaries of academic research in the machining of reinforced plastics. From 1986 to June 2021, the first chapter embarks on a scholarly odyssey, delving into the annals of knowledge through a detailed bibliometric analysis. Employing a methodological rigor that forms the bedrock of this investigation, the chapter meticulously evaluates a diverse array of document formats and prolific contributors, crafting an intricate tapestry that illuminates the scholarly landscape in this nuanced field. At its essence, this chapter serves as a guiding beacon, offering an invaluable vantage point into the evolutionary trajectory of research in machining reinforced plastics. Through the lens of temporal trends and citation metrics, this analysis unveils not just the surface of academic achievements but the very roots and branches of scholarly prominence. The chapter's insights, drawn from a curated dataset, offer a deep dive into the quantitative dimensions of research domains. Beyond the mere statistical figures, this analysis encapsulates the essence of interrelationships among these domains, forming a cohesive narrative that informs, educates, and inspires. This chapter stands as more than a mere compilation of data; it is a foundational resource shaping the future of research endeavors in the machining of reinforced plastics. Its insights serve as a compass guiding the scholarly community toward unexplored avenues and illuminating emerging research directions. By identifying gaps in existing knowledge, it not only highlights areas ripe for exploration but also empowers researchers to chart a course toward groundbreaking discoveries. As academic inquiry continues to evolve, this chapter stands as a testament to the enduring importance of bibliometric analysis, providing a sturdy scaffold upon which the edifice of future research can be erected. In the grand tapestry of academic exploration, this chapter weaves a narrative of scholarly excellence and intellectual curiosity. Through its meticulous analysis and insightful revelations, it becomes a cornerstone upon which the future of research in machining reinforced plastics is built. As scholars and researchers delve into its depths, they are not merely accessing historical data; they are embracing a legacy of knowledge, a torch passed from one generation of researchers to the next, illuminating the path toward a future where the boundaries of understanding are constantly pushed, and the horizons of knowledge are perpetually expanded.

Within the vast expanse of the maritime industry, Chapter 2 emerges as a beacon of insight, shedding light on the intricate world of machining behavior and mechanical properties inherent in glass and basalt fiber reinforced polymer composites. In the quest for efficiency and precision, this study meticulously investigates the turning operation and delves deep into the mechanical attributes that underpin the very fabric of these advanced materials. As shipbuilding techniques continue to evolve, the significance of understanding the nuances of machining processes becomes paramount. This chapter, nestled within the heart of maritime engineering, offers a profound exploration into the challenges faced during the machining process, providing a wealth of knowledge that resonates far beyond the confines of its pages. At its core, this chapter serves as a linchpin, bridging the gap between theoretical understanding and practical application. Through a lens finely tuned to material properties, particularly hardness, the research illuminates the intricate dance between the composite's composition and the machining outcomes. The turning operation, a cornerstone of manufacturing cylindrical composite structures in the maritime industry, undergoes meticulous scrutiny. This microscopic examination uncovers not just the challenges but also the opportunities inherent in machining glass and basalt fiber reinforced polymer composites. By dissecting composite specimens with precision, the chapter offers more than insights; it provides an indispensable toolkit for engineers and practitioners seeking to navigate the labyrinth of material selection and machining optimization in shipbuilding and its related applications. As readers delve into the depths of this chapter, they embark on a journey into the heart of industrial innovation. The chapter's meticulous analysis serves as a compass, guiding shipbuilders and engineers toward informed decision-making. Every piece of information, every revelation about the intricacies of machining, is a stepping stone toward the creation of stronger, lighter, and more durable maritime components. The knowledge encapsulated within these pages is not static; it is a dynamic force, empowering practitioners to push the boundaries of what is possible in the maritime industry. In an era where technological advancements are revolutionizing the maritime landscape, this chapter stands as a testament to the power of understanding the fundamentals. It becomes a catalyst for change, inspiring engineers to innovate, experiment, and redefine the very essence of shipbuilding. Through its comprehensive exploration of machining behavior and mechanical properties, this chapter becomes more than a scholarly work; it becomes a cornerstone upon which the future of the maritime industry is built. Armed with the insights gleaned from this chapter, engineers embark on a transformative journey, harnessing the potential of glass and basalt fiber reinforced polymer composites to propel the maritime industry into a future defined by precision, efficiency, and unparalleled resilience.

Within the intricate landscape of polymer matrix-based composites, Chapter 3 stands as a lighthouse guiding scholars and practitioners through the complex terrain of the machining process. It serves as more than just a chapter; it is a comprehensive guide, offering an illuminating journey into the challenges and intricacies of working with these versatile materials. At its core, this chapter meticulously dissects the multifaceted aspects of machining, delving deep into the very essence of these composites and the hurdles encountered during the pursuit of high-quality surface finishes. One of the chapter's remarkable strengths lies in its ability to articulate the challenges inherent in machining polymer matrix-based composites. Through a discerning lens, it navigates the maze of complex structural compositions and varying material properties, shedding light on the limitations that often hinder the attainment of flawless surface finishes. This nuanced understanding serves as a cornerstone, not just for theoretical comprehension but as a practical roadmap for practitioners striving for precision in their machining processes. By addressing these challenges head-on, the chapter empowers readers with insights that extend beyond conventional knowledge, offering practical solutions and strategies for overcoming the hurdles encountered in the machining of polymer matrix-based composites. Furthermore, the chapter acts as a synthesis hub, weaving together the threads of various research studies into a cohesive tapestry of knowledge. By amalgamating findings from diverse sources, it paints a holistic picture of material handling and properties in polymer matrix composites milling. This synthesis not only enriches the theoretical framework but also provides a bridge between theoretical understanding and real-world application. Researchers and practitioners alike benefit from this amalgamation of knowledge, gaining access to a repository of insights that fuel further exploration and experimentation in the field. As readers immerse themselves in the depths of this chapter, they embark on a transformative intellectual odyssey. It is not merely a theoretical discourse but a practical compendium of wisdom honed through hands-on experience and rigorous research. Armed with a profound understanding of the factors influencing mechanical properties in polymer matrix composites milling, scholars and practitioners enter the field equipped with a unique perspective. This chapter serves as an indispensable resource, guiding them through the challenges, illuminating the path toward precision, and shaping the future of machining in the realm of polymer matrix-based composites. Through its depth, breadth, and practical applicability, this chapter becomes a beacon, illuminating the way for researchers and practitioners, propelling the field toward new horizons of innovation and excellence.

Within the realm of machining fiber-reinforced polymer (FRP) matrices, Chapter 4 stands as a beacon of knowledge, meticulously illuminating the intricate nuances inherent in this intricate process. It transcends mere theoretical discourse, offering readers a profound journey into the heart of precision engineering. At its core, this chapter is not just a review; it is a compendium of wisdom, carefully curated from acclaimed scientific literature and distilled into insights that shape the future of FRP machining processes. The chapter embarks on a meticulous exploration, dissecting the complexities of machining FRP matrices with surgical precision. Through the establishment of robust physical and statistical models, it becomes a cornerstone upon which the edifice of optimization in polymer matrix composites machining is built. These models, born from a synthesis of empirical data and theoretical frameworks, serve as a guiding light for engineers and researchers, paving the way for the refinement of surface finishes and the profound understanding of cutting force dynamics. By delving deep into the mechanics of FRP machining, the chapter goes beyond the surface, uncovering the subtle interplay of forces, materials, and precision that underpin every successful machining operation. What sets this chapter apart is its ability to draw from a rich tapestry of scientific literature, offering readers an exhaustive overview of the prevailing landscape in FRP machining. The insights gleaned from this review are not merely historical artifacts; they are stepping stones toward future empirical endeavors. In the ever-evolving field of precision engineering, knowledge is not static; it is a dynamic force that propels innovation. The chapter's contribution goes beyond being a reference; it becomes a catalyst for progress, guiding researchers and engineers toward unexplored horizons and inspiring them to push the boundaries of what is achievable in FRP machining. As readers delve into the depths of this chapter, they find themselves not just immersed in technical details but engaged in a transformative experience. The knowledge encapsulated within its pages becomes a toolkit, empowering engineers and researchers to refine precision cutting models with unparalleled efficiency. It is more than an essential reference; it is a roadmap, guiding the way for future endeavors, sparking the curiosity of young researchers, and fueling the passion of seasoned engineers. Through its insights and methodologies, this chapter becomes a cornerstone of innovation, shaping the future of FRP machining processes and paving the way for a new era of precision and efficiency in the realm of fiber-reinforced polymer matrices.

In the realm of modern manufacturing, the utilization of advanced materials has become synonymous with progress and innovation. Among these materials, Polymer Matrix Composites (PMCs) stand out as a remarkable class, offering a unique blend of lightweight properties, high strength, and exceptional durability. As PMCs continue to revolutionize various industries, their machining processes have become a focal point for researchers and practitioners alike.

Chapter 5 aims to provide a comprehensive exploration of machining strategies specifically designed for Polymer Matrix Composites, addressing the intricate challenges inherent in working with these advanced materials. The chapters within delve into a nuanced understanding of machining methodologies, emphasizing optimization techniques that are integral for achieving precision and productivity offering an in-depth analysis of various machining techniques tailored to suit the characteristics of PMCs. From tool selection to cutting parameters, the chapter scrutinizes the fundamental aspects of machining, providing insights into the challenges and opportunities presented by these composite materials. Through a systematic examination, the reader gains a profound understanding of the intricacies involved in achieving optimal machining outcomes.

One of the key focal points of this exploration is the selection of appropriate tools and the establishment of cutting parameters. Chapter 5 investigates the latest developments in tooling technologies, shedding light on innovative advancements that cater to the unique demands of PMCs. As the field continues to evolve, this chapter serves as a guide to navigating the dynamic landscape of machining tools and technologies.

Chapter 6 embarks on a meticulous journey, focusing on the augmentation of cutting performance through the integration of diamond-coated tools adorned with advanced polycrystalline diamond coatings. Within these pages, readers will uncover the intricacies of the hot filament chemical vapor deposition (HFCVD) technique, a veritable cornerstone in the preparation of diamond coatings exhibiting diverse morphologies, fortified by the infusion of boron doping. Delving deeper, the chapter unravels the nuances of substrate pretreatment, deposition parameters, and the strategic doping of chemical elements to bolster interface adhesion, offering a comprehensive roadmap towards enhanced machining efficacy.

Complementing this exploration, Chapter 7 casts its spotlight on the multifaceted realm of Polymer Matrix Composites (PMCs), heralded for their unparalleled mechanical prowess and widespread industrial adoption. Here, readers are beckoned into a realm where the interplay between machining parameters and mechanical properties unveils a tapestry of insight. From tensile strength to fatigue resistance, each facet of mechanical behavior is dissected with precision, elucidating the underlying mechanisms and unveiling strategies to navigate the delicate balance between machining exigencies and material integrity.

In the ever-evolving landscape of materials science and engineering, the machining of polymer matrix composites stands as a critical frontier. As industries increasingly turn to advanced composites for their lightweight, durable, and high-performance properties, the demand for efficient and sustainable machining techniques has never been more pressing. This book, *Machining Polymer Matrix Composites: Tools, Techniques, and Sustainability*, emerges as a seminal contribution to the field, offering a comprehensive and insightful exploration that transcends traditional boundaries. Throughout its pages, the book delves deep into the intricacies of machining processes, providing a wealth of knowledge, practical guidance, and sustainability considerations. In this concluding reflection, we delve into the profound impact of this book on the academic landscape, industrial practices, and environmental consciousness, elucidating its transformative role in shaping the future of polymer matrix composite machining. At its core, this book reshapes the academic discourse surrounding polymer matrix composite machining. Through meticulous bibliometric analysis and comprehensive scientific exegesis, the book elucidates the evolution of research in the field, offering a roadmap for scholars and researchers. The in-depth exploration of highly cited research in machining reinforced plastics, as detailed in Chapter 1, not only provides historical context but also identifies emerging trends, thereby guiding future academic inquiries. Chapter 3's focus on accurate estimation of mechanical properties in PMC milling fills a crucial knowledge gap, offering valuable insights into the challenges faced during the machining process. By synthesizing findings from diverse research studies, the book fosters a holistic understanding of polymer matrix composite machining, creating a robust foundation for further scholarly endeavors. In the realm of industry, the impact of this book is nothing short of transformative. Chapter 2, which examines the effect of fiber orientation and matrix type on machining behavior, serves as a practical guide for manufacturers, especially in sectors like maritime construction. By providing actionable insights into material selection and machining processes, the book equips industry professionals with the knowledge needed to enhance efficiency and precision. Chapter 4, the comprehensive scientific exegesis on fiber-reinforced polymer machining, revolutionizes industrial practices by offering advanced models and methodologies. These tools empower engineers and technicians to optimize machining techniques, leading to the production of high-quality components for diverse applications, from aerospace to automotive sectors. One of the book's most significant contributions lies in its unwavering focus on sustainability. In today's world, where environmental concerns are paramount, Chapter 2's exploration of basalt fiber reinforced polymer machining challenges prompts a reevaluation of materials and processes. By highlighting the difficulties faced during machining, the chapter advocates for the development of eco-friendly alternatives, driving the industry toward sustainable practices. Furthermore, the book's emphasis on the environmental impact of machining processes, showcased in Chapter 3's discussion on the challenges of achieving high-quality surface finishes, fosters a heightened environmental consciousness among practitioners.

This book offering an in-depth analysis of various machining techniques tailored to suit the characteristics of PMCs. From tool selection to cutting parameters, it's scrutinizing the fundamental aspects of machining, providing insights into the challenges and opportunities presented by these composite materials. Through a systematic examination, the reader gains a profound understanding of the intricacies involved in achieving optimal machining outcomes.

One of the key focal points of this exploration is the selection of appropriate tools and the establishment of cutting parameters. This book investigates the latest developments in tooling technologies, shedding light on innovative advancements that cater to the unique demands of PMCs. As the field continues to evolve, this chapter serves as a guide to navigating the dynamic landscape of machining tools and technologies.

Furthermore, the chapter delves into novel approaches aimed at mitigating damage while enhancing overall machining efficacy. Recognizing that PMCs pose challenges distinct from traditional materials, the exploration of adaptive machining processes becomes crucial. This includes the integration of smart technologies and data-driven approaches to optimize machining outcomes in real-time, aligning with the demand for precision in contemporary manufacturing.

In addition to the technical aspects, this chapter places a strong emphasis on sustainability within machining processes. Recognizing the global call for environmentally conscious practices, it explores waste reduction strategies, energy efficiency measures, and other eco-friendly methodologies. The significance of aligning machining practices with contemporary environmental standards is underscored, reinforcing the need for a holistic approach to manufacturing.

It guides scholars and industry professionals alike toward environmentally friendly solutions, aligning the field of polymer matrix composite machining with the broader global imperative of sustainability. In essence, *Machining Polymer Matrix Composites: Tools, Techniques, and Sustainability* transcends the conventional boundaries of academic publications. It is not merely a book but a transformative force, shaping the future of polymer matrix composite machining. Through its meticulous research, practical insights, and unwavering commitment to sustainability, the book empowers scholars, researchers, and industry professionals to navigate the complexities of machining advanced composites with precision, efficiency, and environmental responsibility. As this knowledge permeates academic institutions and industrial sectors, it heralds a new era of innovation, ushering in a future where polymer matrix composite machining is not just a technical challenge but an opportunity for sustainable progress and global impact. This book, with its profound influence, serves as a cornerstone upon which the future of polymer matrix composite machining is built, inspiring generations of scholars and practitioners to redefine what is possible in this dynamic and ever-evolving field.

Francisco Mata Cabrera
Universidad de Castilla-La Mancha, Spain

Issam Hanafi
Université Abdelmalek Essaâdi, Morocco

Chapter 1
Analyzing the Evolution of Highly Cited Research in Machining of Reinforced Plastics

Francisco Mata Cabrera
Universidad de Castilla-La Mancha, Spain

Issam Hanafi
https://orcid.org/0000-0001-8765-2631
ENSAH, Université Abdelmalek Essaâdi, Morocco

Wanderson de Oliveira Leite
https://orcid.org/0000-0002-5723-8525
Institudo Federal de Minas Gerais, Brazil

Khadija Haboubi
ENSAH, Université Abdelmalek Essaâdi, Morocco

ABSTRACT

This study employs a bibliometric approach to analyze the evolution of highly cited research in machining of reinforced plastics from 1986 to June 2021. The research examines various document formats, such as journal articles, reviews, chapters, and books. Focusing on reinforced plastics and composites, the study evaluates prolific contributors, temporal trends, and citation metrics to assess scholarly prominence. The research uses a curated dataset of 161 sources from the Web of Science Core Collection database, aiming to identify emerging research directions and gaps in the field. The study uncovers quantitative dimensions and interrelationships among research domains, providing insights into scholarly development. The data categorizes document types, indicating that articles form the majority (85.005%) of the dataset, followed by proceeding papers (16.387%) and review articles (2.536%). The dataset comprises various other types, with each representing a smaller portion. The study's empirical foundation is drawn from the Web of Science Core Collection database.

DOI: 10.4018/978-1-6684-9927-6.ch001

INTRODUCTION

The continual pursuit of lightweight yet high-strength materials has driven substantial research and innovation in the realm of reinforced plastics and composites(Rajak, Pagar, Menezes, et al., 2019) (Boey et al., 2022). The integration of reinforcing agents within polymer matrices has paved the way for a diverse range of applications across industries, including aerospace, automotive, construction, and electronics (Chohan et al., 2022) (Olonisakin et al., 2022). In recent years, the scientific community has witnessed a dynamic evolution of research trends aimed at enhancing the mechanical, thermal, and multifunctional properties of these materials while addressing pressing sustainability concerns. Reinforced plastics and composites, as multifaceted materials, are characterized by their composite nature, wherein a polymeric matrix is fortified with various reinforcement agents, such as fibers, nanoparticles, or natural fillers. This amalgamation capitalizes on the distinct advantages of both components, resulting in materials that surpass the limitations of individual constituents. Consequently, a comprehensive understanding of the underlying principles governing the interactions between the matrix and reinforcement is essential for designing materials tailored to specific performance criteria (Rajak, Pagar, Kumar, et al., 2019) (Seydibeyoğlu et al., 2023). The emergence of advanced manufacturing techniques, notably additive manufacturing and automated fiber placement, has revolutionized the fabrication of reinforced plastics and composites. These techniques offer unprecedented control over material architecture, enabling the creation of intricate geometries and optimized load-bearing structures(Frketic et al., 2017) (Raspall et al., 2019). Concurrently, the exploration of nanomaterials as reinforcement agents has introduced a paradigm shift in material design. Nanoparticles, such as carbon nanotubes and graphene, imbue composites with exceptional mechanical properties, electrical conductivity, and thermal stability (Zhang et al., 2020) (Kumar et al., 2020). Addressing environmental concerns, researchers have embraced sustainable practices by investigating recyclable and bio-based matrices for reinforced composites. This trend aligns with the broader movement toward eco-friendly materials and manufacturing processes (Hanafi et al., 2012) (Jha et al., 2019). Additionally, the integration of natural fibers as reinforcements has garnered attention due to their renewability, low environmental impact, and potential for weight reduction (Nurazzi et al., 2021) (Fuentes Molina et al., 2021). In response to the demands of high-performance applications, the aerospace and automotive industries have gravitated toward reinforced plastics and composites as substitutes for conventional materials like metals. These materials exhibit a remarkable strength-to-weight ratio, rendering them pivotal for enhancing fuel efficiency, reducing emissions, and ensuring structural integrity. The advancement of reinforced plastics and composites is intricately linked with cutting-edge modeling and simulation techniques (Cabrera et al., 2010). Finite element analysis, multiscale modeling, and computational materials science have contributed significantly to unraveling the complex mechanical behavior of these materials under various loading conditions. This computational insight aids in precision engineering and facilitates the optimal design of materials for specific applications. As researchers delve into the durability and long-term behavior of reinforced plastics and composites, insights into factors such as fatigue, creep, and degradation mechanisms are crucial. Understanding these phenomena guides the formulation of materials resilient to extended service conditions, bolstering their reliability and lifespan (Ahmadifar et al., 2022) (Ribeiro et al., 2023).

In recent years, the research landscape of reinforced plastics and composites has been significantly influenced by the application of bibliometric studies (Hanafi et al., 2023). Bibliometric analysis, a quantitative approach to understanding patterns of publication, collaboration, and impact within a specific field, has provided valuable insights into the evolving trends, influential authors, key research areas, and interdisciplinary collaborations within the domain of reinforced plastics and composites. By conducting a bibliometric study of the literature pertaining to reinforced plastics and composites, researchers have been able to identify the most prolific authors, institutions, and countries contributing to the field. This analysis helps paint a comprehensive picture of the global research network, enabling the identification of key research hubs and collaborative networks. Furthermore, it sheds light on the dissemination of knowledge across different journals and conferences, revealing the most prominent platforms for sharing research findings. Bibliometric techniques have proven invaluable in tracking the evolution of research trends in reinforced plastics and composites. These methods reveal the emergence of new materials, manufacturing techniques, and applications over time, aiding researchers, policymakers, and industry stakeholders in resource allocation and technological advancement. Additionally, bibliometric studies identify impactful works, highly cited articles, and influential journals, allowing researchers to assess the scientific influence of specific publications. These analyses also uncover the intellectual structure of the field through citation patterns and co-citation networks, highlighting interconnected research topics. By revealing cross-disciplinary collaborations, bibliometric studies promote innovation and problem-solving in the field. This synthesis of bibliometric analysis with the study of reinforced plastics and composites offers a comprehensive perspective on progress, trends, and impact, guiding decision-making and fostering collaboration among researchers. As the field evolves, bibliometric studies are poised to play a vital role in shaping research trajectories.

Methods

This study aims to analyze the evolution of highly cited research in the realm of reinforced plastics, covering the period from 1986 to June 2021, using a bibliometric approach. This method involves a comprehensive examination of research content and employs statistical techniques to scrutinize publication data from various formats, including journal articles, reviews, conference proceedings, chapters, and books. By focusing on the thematic area of reinforced plastics and composites, the study explores prolific contributors, temporal trends, citation metrics, and other quantitative markers to assess the scholarly prominence of researchers and journals. The study's empirical foundation is drawn from the Web of Science Core Collection database, a widely used resource that facilitates the analysis of a curated dataset comprising 161 sources related to reinforced plastics and composites. This research approach serves a dual purpose: identifying emerging research trajectories and revealing gaps within the academic landscape. Through this bibliometric framework, the study sheds light on the quantitative dimensions and interrelationships of research domains, offering valuable insights into the field's scholarly development.

Figure 1. A breakdown of document types

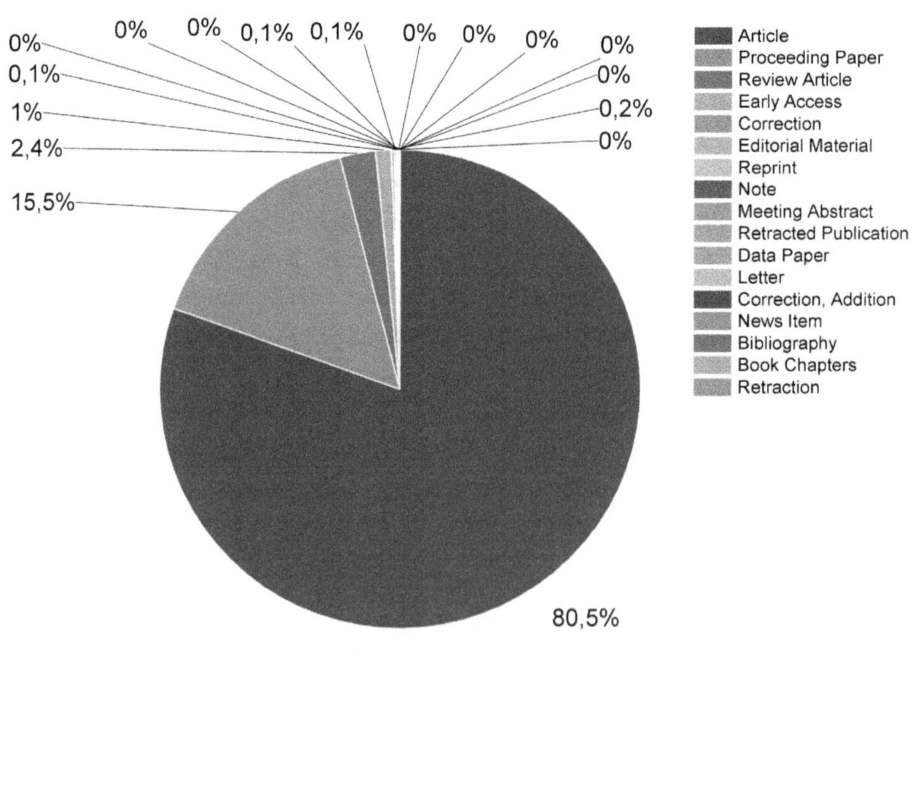

The data appears in Figure 1 to represent a breakdown of document types along with their respective record counts and percentages out of a total of 22,908 records. Here's a summary of the data: There are 19,473 records classified as articles, accounting for approximately 85.005% of the total records. This suggests that articles are the most prevalent type of document in the dataset. There are 3,754 proceeding papers, making up around 16.387% of the total. Proceeding papers are the second most common type of document. Review articles represent 581 records, which is about 2.536% of the total. These articles provide evaluations and summaries of existing research. There are 253 early access documents, constituting approximately 1.104% of the total. Early access typically refers to articles that are available before they have been formally peer-reviewed. Corrections account for 44 records, making up about 0.192% of the total. Corrections are published to rectify errors or mistakes in previous publications. There are 31 editorial materials, making up roughly 0.135% of the total. These might include notes from editors or other supplementary content. Reprints represent 14 records, accounting for about 0.061% of the total. Reprints are usually republications of previously published work. There are 13 notes, constituting approximately 0.057% of the total. Notes could be brief comments or observations. There are 10 meeting abstracts, making up about 0.044% of the total. These are summaries of presentations given at meetings or conferences. There are 9 retracted publications, accounting for around 0.039% of the total. This category likely includes documents that have been withdrawn due to errors or other issues. There are 7 data papers, representing about 0.031% of the total. Data papers typically focus on describing datasets. There are 6 letters, making up approximately 0.026% of the total. Letters might include short communications or opinions. Correction, Addition: There are 2 records falling under this category, accounting for about 0.009% of the total. This might involve corrections with additional information. There are 2 news items, also making up about 0.009% of the total. News items could be updates or announcements. Bibliography, Book Chapters, Retraction: Each of these categories has 1 record, each accounting for about 0.004% of the total. These represent less common document types. Overall, this dataset seems to consist primarily of articles, followed by proceeding papers and review articles. The other types of documents are less prevalent, with each representing a small portion of the dataset.

Table 1. The distribution of research records

Web of Science Categories	Record Count	% of 22 908
Materials Science Composites	10006	43.679
Polymer Science	8383	36.594
Materials Science Multidisciplinary	5131	22.398
Mechanics	2219	9.687
Engineering Mechanical	1870	8.163
Engineering Civil	1582	6.906
Metallurgy Metallurgical Engineering	1517	6.622
Engineering Manufacturing	1329	5.801
Engineering Multidisciplinary	1041	4.544
Physics Applied	1000	4.365
Construction Building Technology	885	3.863
Materials Science Characterization Testing	750	3.274
Engineering Chemical	724	3.160
Nanoscience Nanotechnology	494	2.156
Chemistry Physical	476	2.078
Physics Condensed Matter	427	1.864
Optics	323	1.410
Instruments Instrumentation	310	1.353
Engineering Electrical Electronic	295	1.288
Chemistry Multidisciplinary	287	1.253
Materials Science Textiles	262	1.144
Automation Control Systems	240	1.048
Engineering Industrial	236	1.030
Engineering Environmental	216	0.943
Materials Science Paper Wood	211	0.921

The data related to the distribution of research records across different Web of Science categories, as shown in table 1. The data appears to represent the record count and the percentage of each category out of a total of 22,908 records. Here's a breakdown and analysis of the data: Materials Science Composites: This category has the highest record count with 10,006 records, accounting for approximately 43.679% of the total. Composites are materials composed of two or more distinct constituent materials. Polymer Science: Polymer Science follows closely with 8,383 records, making up about 36.594% of the total. This category focuses on the study of polymers and their properties. Materials Science Multidisciplinary: With 5,131 records, this category represents around 22.398% of the total. It covers research that spans multiple subfields within materials science. Mechanics: This category has 2,219 records, accounting for about 9.687% of the total. Mechanics involves the study of physical forces and their effects on objects and systems. Engineering Mechanical: With 1,870 records, this category makes up around 8.163% of the total. It encompasses research related to mechanical engineering principles and applications. Engineering Civil: There are 1,582 records in this category, constituting about 6.906% of the total. Civil engineering focuses on infrastructure and construction projects. Metallurgy Metallurgical Engineering: This category has 1,517 records, making up approximately 6.622% of the total. It covers the study of metals and their properties. Engineering Manufacturing: With 1,329 records, this category accounts for about 5.801% of the total. It involves research related to manufacturing processes and techniques. Engineering Multidisciplinary: There are 1,041 records in this category, representing around 4.544% of the total. This category encompasses research that spans multiple subfields within engineering. Physics Applied: With 1,000 records, this category makes up about 4.365% of the total. Applied physics involves the application of physical principles to real-world problems. Construction Building Technology: There are 885 records in this category, constituting around 3.863% of the total. It focuses on research related to construction techniques and building technology. Materials Science Characterization Testing: This category has 750 records, making up approximately 3.274% of the total. It involves the study of methods to characterize and test materials. Engineering Chemical: With 724 records, this category accounts for about 3.160% of the total. It encompasses research related to chemical engineering principles and applications. Nanoscience Nanotechnology: There are 494 records in this category, representing around 2.156% of the total. This category covers research related to nanoscale materials and applications. Chemistry Physical: With 476 records, this category makes up about 2.078% of the total. It involves research related to the physical properties of chemicals. Physics Condensed Matter: There are 427 records in this category, constituting around 1.864% of the total. Condensed matter physics studies the properties of solid and liquid matter. Optics: With 323 records, this category accounts for about 1.410% of the total. It covers research related to the behavior and properties of light. Instruments Instrumentation: There are 310 records in this category, representing around 1.353% of the total. It involves research related to instruments and measurement techniques. Engineering Electrical Electronic: With 295 records, this category makes up about 1.288% of the total. It encompasses research related to electrical and electronic engineering. Chemistry Multidisciplinary: There are 287 records in this category, constituting around 1.253% of the total. This category covers research that spans multiple subfields within chemistry. Materials Science Textiles: With 262 records, this category accounts for about 1.144% of the total. It involves research related to textile materials and applications. Automation Control Systems: There are 240 records in this category, representing around 1.048% of the total. It covers research related to automation and control technologies. Engineering Industrial: With 236 records, this category makes up about 1.030% of the total. It involves research related to industrial engineering and operations. Engineering Environmental: There are 216 records in this category, constituting around 0.943% of the total. It focuses on research related to environmental engineering and sustainability. Materials Science Paper Wood: This category has 211 records, making up approximately 0.921% of the total. This distribution highlights the significance of materials science and engineering disciplines in the dataset.

Table 2. Main information about data

Description	Results
Timespan	1986:2021
Sources (Journals, Books, etc)	161
Documents	590
Annual Growth Rate %	2
Document Average Age	16,5
Average citations per doc	219,2
References	23394
DOCUMENT CONTENTS	
Keywords Plus (ID)	1507
Author's Keywords (DE)	1414
AUTHORS	
Authors	1653
Authors of single-authored docs	36
AUTHORS COLLABORATION	
Single-authored docs	40
Co-Authors per Doc	3,48
International co-authorships %	22,88
DOCUMENT TYPES	
article	465
article; proceedings paper	39
review	86

The analyzed dataset spans from 1986 to 2021 and encompasses 161 diverse sources, including journals and books, focusing on highly cited papers (citations > 100), as shown in table 2. The dataset consists of 590 documents, displaying an annual growth rate of 2%, and each document garners an average of 219.2 citations. The content is rich with 1,507 "Keywords Plus" identifiers and 1,414 author-provided keywords, enhancing discoverability. It involves 1,653 unique authors, featuring both single-authored (36 documents) and collaboratively authored papers (average of 3.48 co-authors per document). Approximately 22.88% of the documents entail international co-authorships, reflecting global collaboration. The dataset comprises various document types, including 465 articles, 39 articles that also appear as proceedings papers, and 86 review articles. While the specific most cited document isn't mentioned, the dataset's characteristics underscore its influential impact on scholarly discourse. Collaboration is a prominent theme, encompassing a wide range of document types, from single-authored contributions (40 instances) to collaborative efforts with an international dimension.

DISCUSSION

The dataset presented in figure 2 reveals a chronological progression of articles published over a span of 36 years. Here are some notable observations; Early Years and Growth: The dataset begins in 1986 with 1 article, followed by another article in 1987. The early years show limited activity, which is not uncommon given the nascent stage of many academic disciplines in digital publishing during that time. Emergence and Acceleration: Starting from the 1990s, there's a noticeable increase in the number of articles. In particular, 1991 marks a significant jump with 11 articles, suggesting the emergence of increased research activity. Periods of Variation: The 1990s and early 2000s witness fluctuations in the number of articles published annually, with counts ranging from 7 to 33. These variations could be influenced by changes in research trends, funding availability, and technological advancements. Steady Rise and Plateaus: From the mid-2000s onward, there's a steady upward trend in articles published per year, peaking at 33 articles in 2004 and 2005. This period might signify increased interest in the subject matter or the establishment of research communities around specific topics. Fluctuations in Recent Years: The latter half of the 2010s and early 2020s show some fluctuation, with counts ranging between 9 and 32 articles. Such fluctuations can be attributed to various factors including shifts in research focus, changes in publication policies, or external events like global crises. Recent Years: The years 2020 and 2021 exhibit lower counts of 9 and 2 articles, respectively. The impact of the COVID-19 pandemic could potentially explain the decrease in publications during these years, as disruptions in research and publishing activities were widespread.

The provided data presents the average citations per year for articles published in different years (figure 3). Fluctuating Early Years: In the early years of the dataset (1986-1990), the average citation counts vary. While 1987 stands out with an average of 7.47 citations, other years have lower averages, suggesting fluctuation in the attention received by articles during this period. Steady Increase in the 1990s: Starting from 1991, there's a consistent upward trend in average citations per year. The early 1990s witness a steady increase, and by 1999, the average citation count reaches 17.38. Ups and Downs in the 2000s: The 2000s show fluctuations in average citations. Some years, like 2002 and 2003, have relatively high averages (16.03 and 13.73 respectively), indicating notable attention to articles published in those years.

Increased Attention in the Late 2000s to 2010s: The latter half of the 2000s and the 2010s see a rise in average citation counts. Notably, 2017 and 2019 stand out with average citation counts of 34.90 and 35.55 respectively, suggesting heightened interest in articles published during those years. Remarkable Growth in Recent Years: The most recent years, particularly 2020 and 2021, exhibit remarkable growth in average citations. With average counts of 46.00 and 60.50 respectively, this signifies a substantial increase in the recognition and impact of articles published during these years. Influence of Research Focus: The varying average citation counts across years could be attributed to shifts in research trends, advancements in the field, and the significance of the topics covered by the articles. External Factors: Factors such as technological advancements, societal relevance, and the accessibility of research might have contributed to the observed changes in citation patterns.

Figure 2. Chronological progression of articles published over a span of 36 years

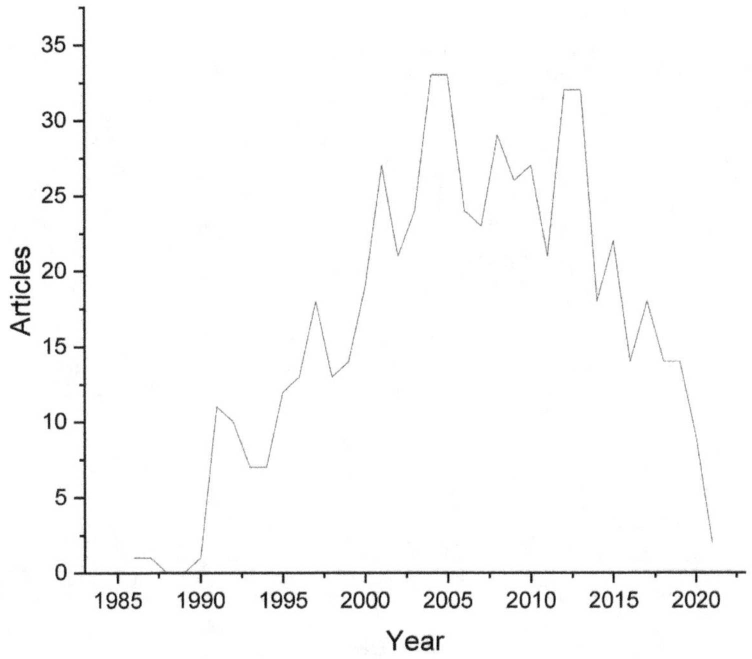

The provided data in figure 4 presents the distribution of articles across different sources in the field of composites science and technology. COMPOSITES SCIENCE AND TECHNOLOGY (46 articles): This source has the highest number of articles, indicating its prominence in the field. It's likely a well-established journal that covers a wide range of topics in composites science and technology. COMPOSITE STRUCTURES (36 articles): With a substantial number of articles, this source is another influential publication in the composites field. The focus might be on the structural aspects and applications of composite materials. JOURNAL OF REINFORCED PLASTICS AND COMPOSITES (36 articles): This journal's title suggests that it specializes in reinforced plastics and composites, likely exploring the combination of materials to enhance properties. COMPOSITES PART A-APPLIED SCIENCE AND MANUFACTURING (29 articles): This source might emphasize the practical applications and manufacturing processes of composite materials, contributing to their real-world utilization.

Figure 3. The average citations per year

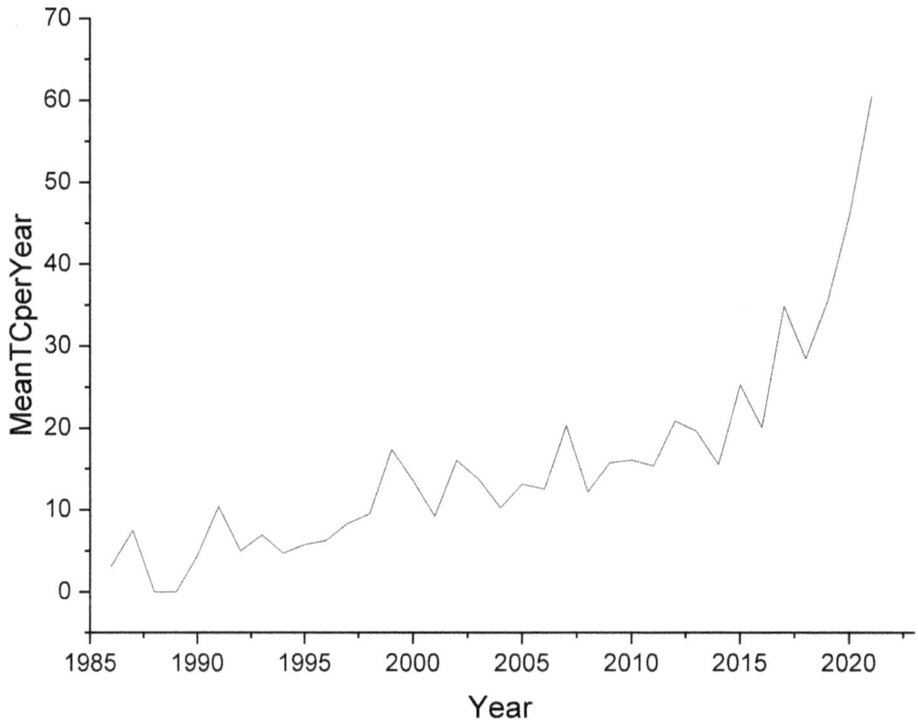

COMPOSITES PART B-ENGINEERING (24 articles): This source may concentrate more on the engineering aspects of composite materials, such as design, analysis, and optimization. MATERIALS & DESIGN (20 articles): This source could explore the design considerations and material selection for composite structures, providing insights into creating efficient and innovative composite products. ACTA MATERIALIA (19 articles): Although not exclusively focused on composites, this source publishes articles on various materials science topics, possibly covering composite materials within a broader context. INTERNATIONAL JOURNAL OF MACHINE TOOLS & MANUFACTURE (16 articles): This source's focus on machine tools and manufacture suggests that it may publish articles that delve into the manufacturing processes and techniques of composite materials. POLYMER COMPOSITES (14 articles): As the name suggests, this journal might specialize in the study of composites using polymer matrices, a crucial component of many composite materials. ACTA METALLURGICA ET MATERIALIA (13 articles): Similar to ACTA MATERIALIA, this source might cover materials science more broadly but could have a subset of articles focused on composites and related topics.

Figure 4. The distribution of articles across different sources

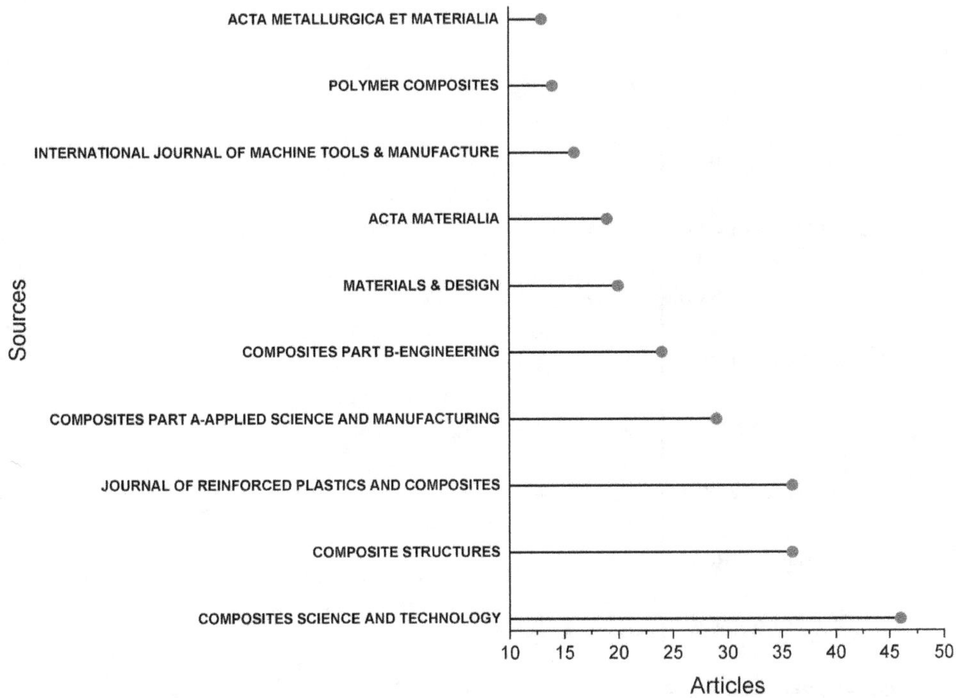

The distribution of articles across these sources reflects the diversity and multidisciplinary nature of composite materials research. Researchers in this field publish their findings in a range of journals that cater to various aspects of composite science, technology, design, manufacturing, and application. The prominence of certain sources suggests key hubs for disseminating new knowledge and advancements in the field.

The list of most cited sources provided in figure 5 seems to focus on the field of composite materials and related areas of material science and engineering. These sources are likely some of the most influential and frequently referenced publications within this domain. COMPOSITES SCIENCE AND TECHNOLOGY (COMPOS SCI TECHNOL) - With 1208 citations, this journal is at the top of the list. It covers a wide range of topics related to composite materials, including their fabrication, characterization, properties, and applications. Given its high citation count, it likely publishes impactful research that significantly advances the field. COMPOSITES PART A: APPLIED SCIENCE AND MANUFACTURING (COMPOS PART A-APPL S) - With 803 citations, this journal appears to focus on the practical applications and manufacturing aspects of composite materials. It likely covers topics such as production methods, process optimization, and innovative applications of composite materials.

Figure 5. The most cited sources

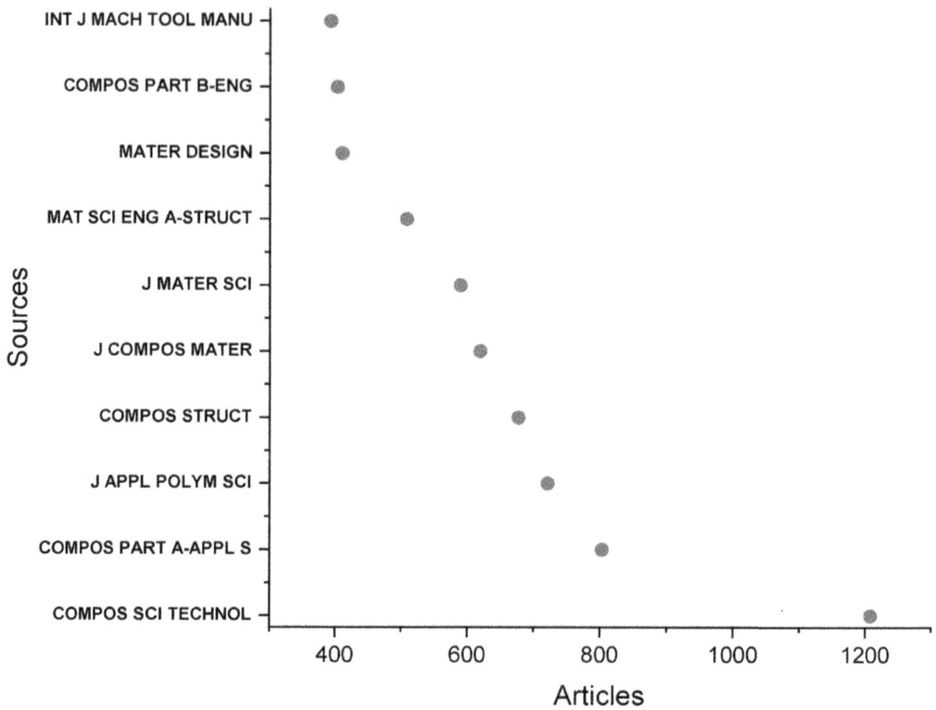

JOURNAL OF APPLIED POLYMER SCIENCE (J APPL POLYM SCI) - This journal, with 721 citations, likely specializes in research related to polymer materials and their applications. Although not exclusively focused on composites, polymers are often key components in composite materials. COMPOSITES STRUCTURES (COMPOS STRUCT) - With 677 citations, this journal is probably dedicated to the structural aspects of composite materials. It likely covers topics like mechanical behavior, analysis, and design of composite structures for various applications. JOURNAL OF COMPOSITE MATERIALS (J COMPOS MATER) - With 619 citations, this journal likely covers a broad spectrum of composite materials research, including their mechanical properties, durability, and novel compositions. It could serve as a comprehensive source for researchers in the field. JOURNAL OF MATERIALS SCIENCE (J MATER SCI) - With 589 citations, this journal covers a wide range of material science topics, including composites. It's likely that the composites-related articles published in this journal are highly influential in the field of material science. MATERIALS SCIENCE AND ENGINEERING A: STRUCTURAL MATERIALS (MAT SCI ENG A-STRUCT) - This journal, with 508 citations, focuses on the structural aspects of materials science, which would include research on composites used in structural applications. MATERIALS & DESIGN (MATERIALS DESIGN) - With 410 citations, this journal likely explores the design and selection of materials for various applications. In the context of composites, it may cover topics such as material selection, optimization, and integration into specific designs. COMPOSITES PART B: ENGINEERING (COMPOS PART B-ENG) - With 403 citations, this journal likely emphasizes the engineering aspects of composite materials. It may cover topics such as manufacturing processes, mechanical properties, and applications in engineering fields. INTERNATIONAL JOURNAL OF MACHINE TOOLS AND MANUFACTURE (INT J MACH TOOL MANU) - With 393 citations, this journal seems to focus on manufacturing technologies, which could include techniques for fabricating composite materials and components.

Bradford's Law is a bibliometric principle that suggests that scientific literature can be divided into a core of highly cited journals or sources, and a broader set of less-cited sources (Figure 6). This principle is particularly applicable to scientific fields where researchers publish their work in a limited number of high-impact journals. By analyzing your list of sources based on Bradford's Law, we can identify the core journals that contribute significantly to the field of composite materials and related disciplines. In Bradford's Law analysis, the data is typically divided into three zones; Zone 1 (Core Journals) - This zone contains a small number of highly cited journals that contribute to a significant portion of the overall citations. These journals are considered the core sources in the field. Zone 2 (Additional Journals) - This zone contains a larger number of moderately cited journals. These journals receive fewer citations compared to the core journals but still contribute to the overall body of research. Zone 3 (Peripheral Journals) - This zone contains a large number of journals that receive very few citations.

These journals contribute the least to the overall citations. Based on your data, we can observe the following:

- The journal "COMPOSITES SCIENCE AND TECHNOLOGY" is in Zone 1, making it a core source in the field.
- "COMPOSITE STRUCTURES" and "JOURNAL OF REINFORCED PLASTICS AND COMPOSITES" are also in Zone 1, indicating their significance in the field.

Based on this analysis, it's clear that there are a few core journals that are highly influential in the field of composite materials, as well as a set of additional journals that contribute to the broader literature. Researchers in this field would likely benefit from regularly referring to the sources in Zone 1 and Zone 2 to stay updated on the latest developments and research trends.

Figure 6. Bradford's law

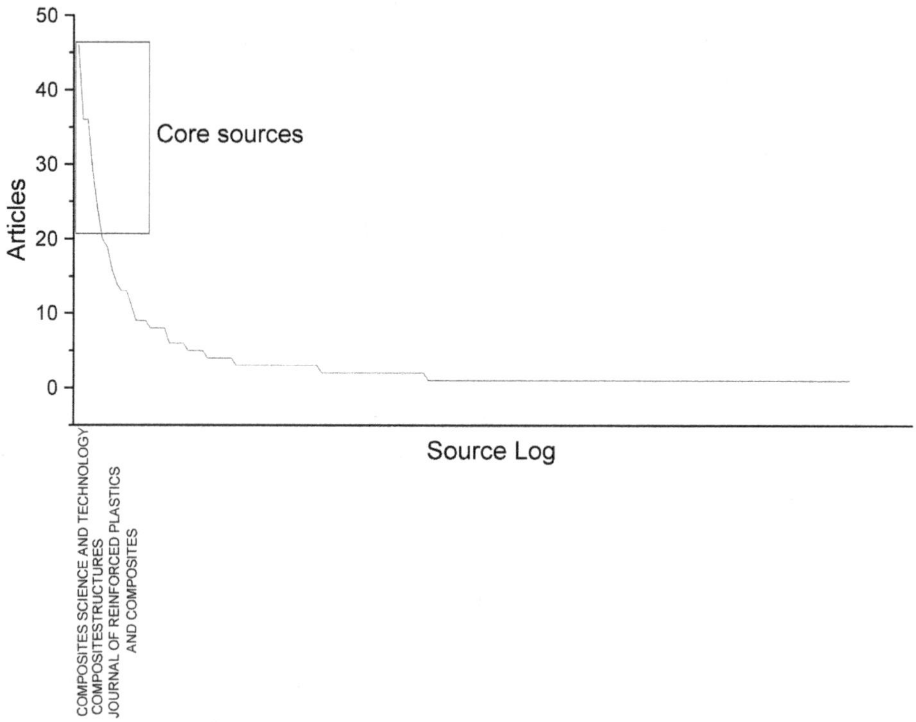

These journals represent research in the field of composite materials, engineering, and related disciplines (Table 3). The provided metrics offer insights into the influence and impact of these journals within the academic community. Here are some key points to consider:

The h-index represents the highest number of papers in a journal that have been cited at least h times. It's an indicator of both the productivity and impact of the journal's research.

Similar to the h-index, the g-index also takes into account the distribution of citations across papers. It aims to measure the impact of the highly cited papers in addition to the overall citation count.

The m-index provides an indicator of the journal's productivity by considering the median number of citations received by its papers. It's an attempt to balance the influence of a few highly cited papers.

Total Citations (TC): This metric represents the total number of times the journal's papers have been cited by other researchers. A higher TC indicates greater recognition and influence in the field.

The number of papers published in the journal. This gives an idea of the journal's overall output and contribution to the literature.

The year in which the journal started publishing. This provides context regarding the journal's longevity and history.

Table 3. Impact of journals

Element	h_ index	g_ index	m_ index	TC	NP	PY_ start
COMPOSITES SCIENCE AND TECHNOLOGY	46	46	1,353	9891	46	1990
COMPOSITE STRUCTURES	36	36	1,5	6226	36	2000
JOURNAL OF REINFORCED PLASTICS AND COMPOSITES	36	36	1,161	6397	36	1993
COMPOSITES PART A-APPLIED SCIENCE AND MANUFACTURING	29	29	1,036	6287	29	1996
COMPOSITES PART B-ENGINEERING	24	24	0,857	4315	24	1996
MATERIALS & DESIGN	20	20	0,769	3502	20	1998
ACTA MATERIALIA	19	19	0,679	6367	19	1996
INTERNATIONAL JOURNAL OF MACHINE TOOLS & MANUFACTURE	16	16	0,552	3525	16	1995
POLYMER COMPOSITES	14	14	0,438	2991	14	1992
ACTA METALLURGICA ET MATERIALIA	13	13	0,394	2438	13	1991

When evaluating the impact of these journals, it's important to consider factors such as the specific field they cover, the quality and significance of the research they publish, and the relevance of their topics in the current academic landscape. Researchers often use these metrics to make informed decisions about where to publish their work and to assess the reputation and impact of different journals within their field.

The provided data in figure 7 represents the publication dynamics of various journals in terms of the number of publications per year over a span of several years.

- **Journal Evolution and Impact:** The data shows the growth in the number of publications for each journal over the years. Journals like "COMPOSITES SCIENCE AND TECHNOLOGY," "COMPOSITE STRUCTURES," and "JOURNAL OF REINFORCED PLASTICS AND COMPOSITES" have experienced consistent growth in the number of publications, indicating their continued relevance and impact. The fact that these journals have maintained their publication output suggests that they are likely contributing valuable research and attracting submissions from researchers.

Figure 7. The publication dynamics of various journals in terms of the number of publications per year

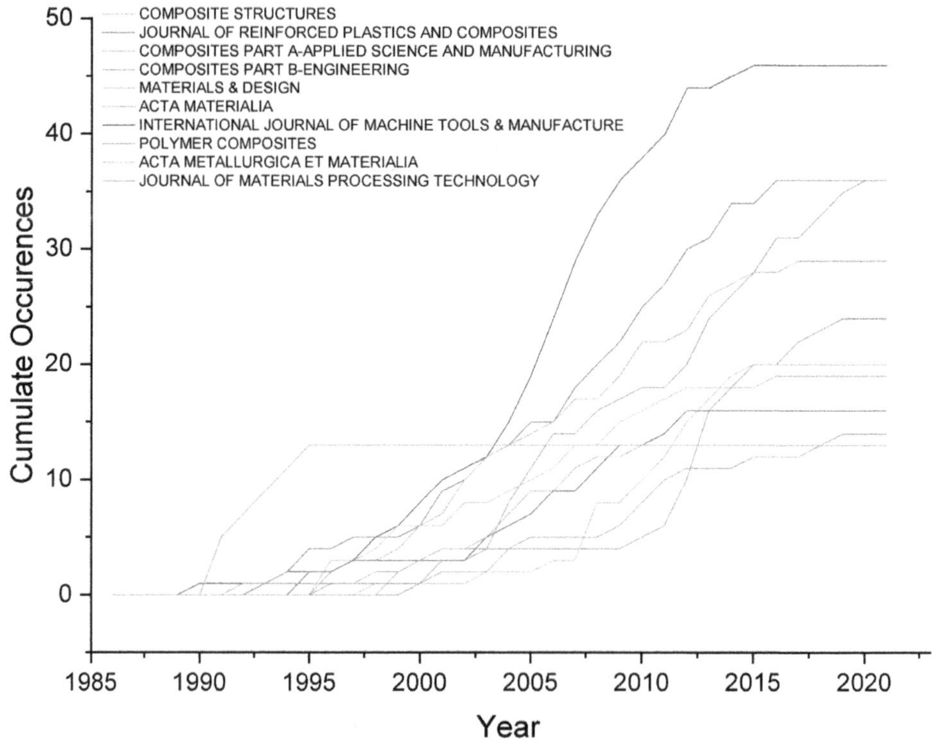

- **Emergence and Discontinuation:** Some journals, like "JOURNAL OF MATERIALS PROCESSING TECHNOLOGY," show a consistent number of publications over the years, possibly indicating a stable niche of research topics. On the other hand, some journals have shown an initial growth in publications but have stagnated or declined in recent years. This could be due to changes in the field's focus, competition from other journals, or shifts in research trends.

- **Consolidation of Research Areas:** The growth in the number of publications for specific journals might reflect a consolidation of research areas in the field of composites, materials, and related technologies. As the field matures, specific journals may become more established as go-to sources for certain types of research, leading to increased submissions and publication output.
- **Steady Increase in Impact Metrics:** The growth in publications over the years might contribute to an increase in the journals' impact metrics. More publications can lead to more citations, potentially boosting h-index, g-index, and other measures of influence.
- **Shifts in Research Focus:** The increase in publications in some journals and the decline in others might reflect shifts in research trends or the emergence of new subfields within the broader domain of composites and materials science. New areas of interest might lead researchers to publish in journals that align better with their specific research topics.
- **Research Collaboration and Multidisciplinarity:** Collaboration among researchers and interdisciplinary work could be contributing to the growth of certain journals. Journals that publish articles from multiple disciplines might attract a wider range of researchers, leading to increased submissions and publications.
- **Sustainability of Long-Running Journals:** Journals like "COMPOSITES SCIENCE AND TECHNOLOGY" have maintained a steady publication output over a long period. This indicates their sustained popularity and ability to adapt to changing research landscapes.
- **Impact of Open Access and Digital Publishing:** The advent of open access publishing and digital platforms might have influenced the ease of publishing and contributed to the overall growth in publications.

The provided list in figure 8 includes authors and the number of their publications in the field of composites and related areas. DAVIM JP (15 Publications): Having the highest number of publications indicates a substantial contribution to the field. This author's work might span various topics within composites science and technology, showcasing a diverse research portfolio. LI Q (10 Publications): With 10 publications, Li Q has also made a notable impact in the field.The focus of their research could encompass a range of areas, potentially indicating expertise in a specific subfield or technology. LLORCA J (9 Publications): Llorca J's 9 publications indicate a consistent presence in the field. This author might be known for research contributions in specific aspects of composites or materials science. MOHANTY AK (8 Publications): With 8 publications, Mohanty AK is a prominent contributor.

Their work might be significant in areas like composite materials manufacturing, characterization, or application. TSAO CC (8 Publications): Tsao CC's 8 publications suggest an active involvement in research. Their research topics could range from material development to structural behavior of composites. HOCHENG H (7 Publications): Hocheng H's 7 publications highlight consistent engagement in the field. They might be known for research in manufacturing processes, mechanics, or structural analysis of composites. MISRA M (7 Publications): With 7 publications, Misra M's work contributes to the field's knowledge base. They might have expertise in areas like polymer composites, nanocomposites, or sustainable materials.

Figure 8. Authors and the number of their publications

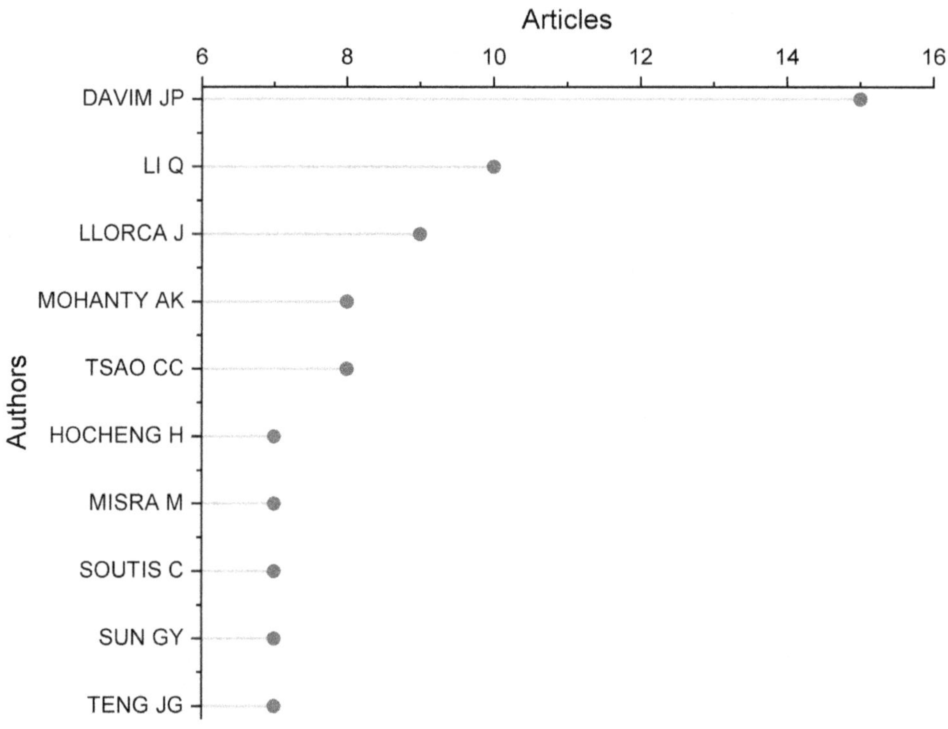

SOUTIS C (7 Publications): Soutis C's 7 publications indicate an active presence in the research community. Their work might span composite materials development, design, and applications. SUN GY (7 Publications): Sun GY's 7 publications highlight their contributions to the field. Their research might encompass areas such as mechanical behavior, durability, or multiscale modeling of composites. TENG JG (7 Publications): Teng JG's 7 publications indicate a significant research output. Their work might focus on composite structures, mechanics, or innovative applications. These authors have collectively contributed a considerable number of publications to the field of composites science and technology. Their impact can be attributed to their extensive research output, which likely covers a wide range of topics within the domain of composite materials, manufacturing, design, and applications.

It's important to note that the relevance of authors goes beyond just the number of publications. Their influence can also be assessed through factors such as citation counts, impact of their research on the field, collaborations with other researchers, leadership in projects, and contributions to advancing knowledge. Researchers with high publication counts often have diverse expertise, and their collective efforts contribute to the advancement of composites technology and its various applications.

Figure 9. Lists authors along with their local citations

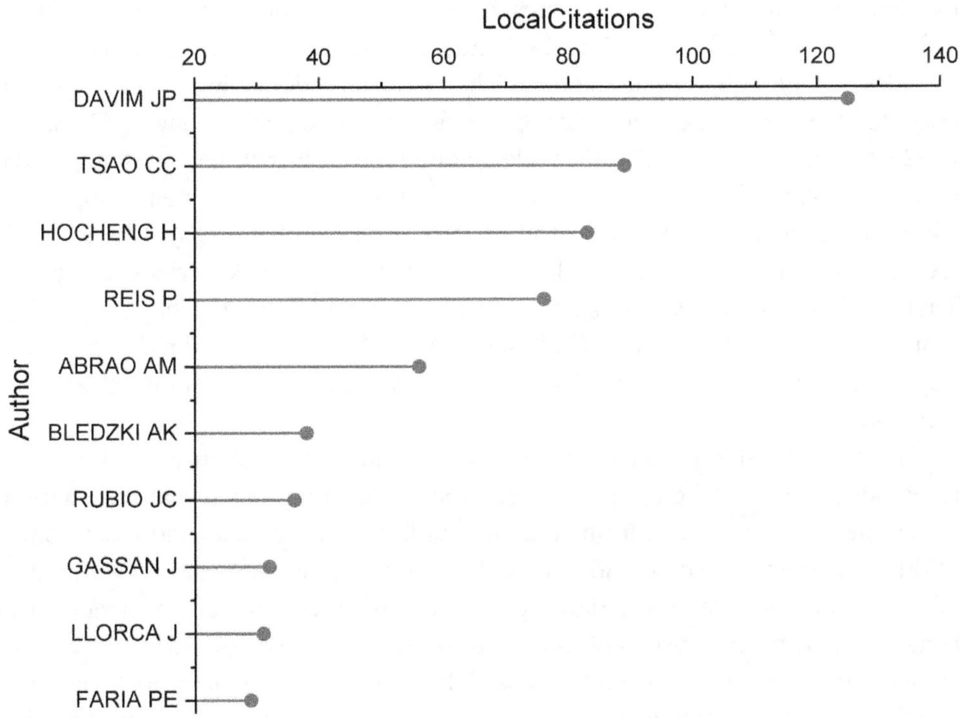

The provided data lists authors along with their local citations (figure 9). Local citations refer to the number of times an author's work has been cited within the same field or discipline. DAVIM JP (125 Local Citations): With the highest number of local citations, DAVIM JP is widely recognized within the field. Their work has significantly influenced and contributed to the research and knowledge base of composite materials and related areas. TSAO CC (89 Local Citations): TSAO CC's substantial local citations indicate their impact and recognition in the field. Their research might have focused on specific topics that resonate with fellow researchers in composites science. HOCHENG H (83 Local Citations): HOCHENG H's local citations reflect their valuable contributions within the field. Their work could span areas like composite manufacturing, mechanics, or structural analysis. REIS P (76 Local Citations): REIS P's work has garnered a noteworthy number of local citations. Their research might be particularly relevant to a specific aspect of composite materials or engineering. ABRAO AM (56 Local Citations): ABRAO AM's local citations indicate their impact on research conversations within the field. Their work might cover topics such as machining, manufacturing processes, or tribology of composites. BLEDZKI AK (38 Local Citations): BLEDZKI AK's local citations signify their contributions to the composites field. Their research might involve natural fiber composites, polymer matrix materials, or reinforcement technologies. RUBIO JC (36 Local Citations): RUBIO JC's local citations reflect the relevance of their work in the field. Their research might encompass areas such as composite materials characterization, testing, or mechanical behavior. GASSAN J (32 Local Citations): GASSAN J's local citations highlight their presence in the field's discourse. Their research might include topics like polymer composites, adhesion, or interfacial behavior. LLORCA J (31 Local Citations): LLORCA J's local citations suggest their research contributions have been acknowledged by peers. Their work might span various aspects of composite materials and their mechanical properties. FARIA PE (29 Local Citations): FARIA PE's local citations indicate their impact in the field. Their research might focus on specific applications, processing techniques, or characterization methods for composites.

The authors listed have earned recognition within the composites science and technology field based on their local citations. Local citations reveal the extent to which their work has been influential among fellow researchers and practitioners in the same domain. These authors' contributions might encompass a wide range of topics, including materials development, manufacturing processes, mechanical behavior, structural analysis, and application-specific studies. Beyond local citations, authors' impact can be further assessed by considering their global citations, collaborative efforts, leadership in research projects, and contributions to advancing the field's knowledge.

The provided data in figure 10 represents the author production over time, indicating the number of publications each author has contributed in different years. DAVIM JP: This author's production has shown variation over time. There is a peak in publications around 2008, with 4 publications. This could be due to a concentrated research effort during that period. The author's research productivity has been consistent, with contributions in multiple years. HOCHENG H's production also displays variation. The years 2003, 2005, and 2008 stand out with multiple publications. While the author's production is not consistent every year, their contributions are spread across several years. LI Q's publication frequency is relatively lower compared to some other authors. There is an increase in publications from 2018 to 2019, indicating a more active phase of research during that period. LLORCA J's production spans multiple years but is more sporadic. There is a cluster of publications in 2004, suggesting a concentrated research effort during that period. MISRA M's production is spread across various years. There is a concentration of publications in the early 2000s, with two publications in 2005. MOHANTY AK's publication pattern is consistent across years. There is a steady contribution of publications with multiple in 2004 and 2005. SOUTIS C's publications are distributed over different years. There is a concentration of publications in 2005 and 2013. SUN GY's publication frequency has increased over time. The years 2018 and 2019 show higher productivity, possibly indicating an active research phase. TENG JG's publication pattern is intermittent. There is a peak in publications in 2003, followed by a few contributions in other years. TSAO CC's production is relatively consistent across years.

Figure 10. The author production over time

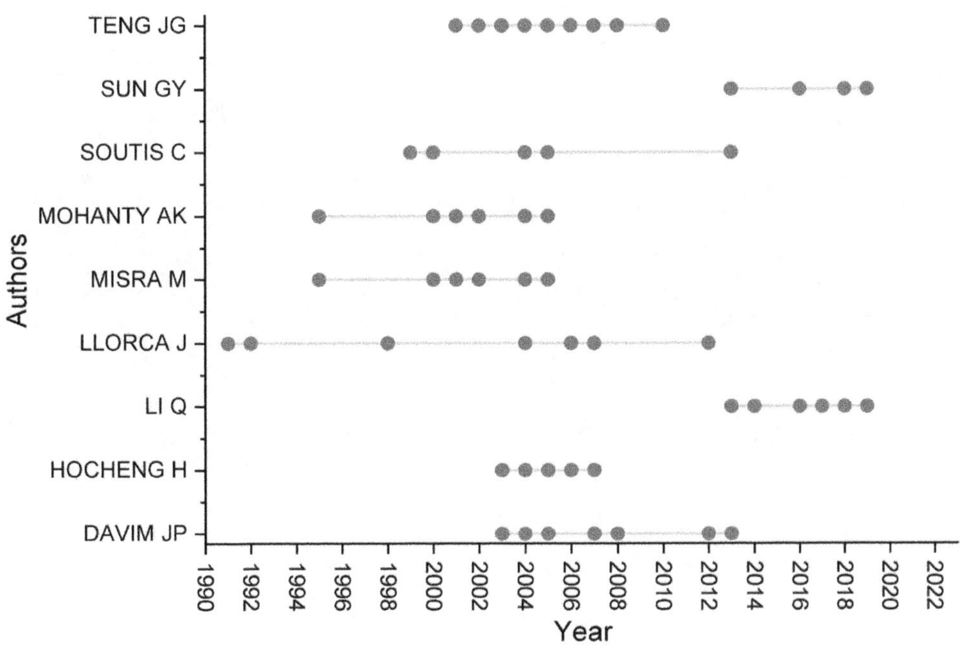

The years 2003, 2005, and 2008 have multiple publications. The analysis of author production over time reveals some common trends:

- Many authors have shown fluctuations in their publication output, with peak years indicating periods of heightened research activity.
- Some authors exhibit consistent contributions across multiple years, suggesting a steady engagement with their research area.
- Concentrations of publications in specific years might indicate focused research efforts or the emergence of key themes within the field.

While the dataset provides insights into individual author trends, it's important to consider factors beyond publication frequency, such as the impact of their work, collaborations, and the broader context of their research topics. This analysis can assist researchers, institutions, and publishers in understanding research dynamics and trends within the field of composites and related areas.

Lotka's Law, also known as the "inverse square law," is a principle in bibliometrics that describes the distribution of author productivity within a given field (Figure11). It suggests that the number of authors who have produced a certain number of documents decreases exponentially as the number of documents increases. Lotka's Law is commonly used to analyze and model author productivity patterns in scholarly publishing. Lotka's Law predicts that the number of authors producing a certain number of documents follows a power-law distribution.

Figure 11. Lotka's law

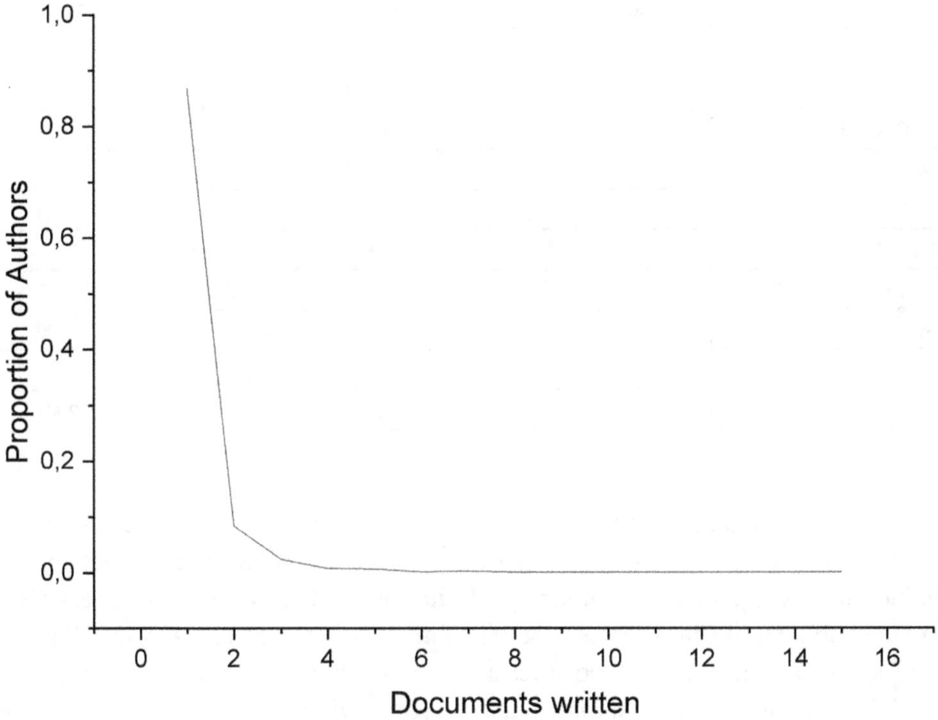

In this context, the number of authors who have written 'n' documents can be represented as 'a/n^b', where 'a' and 'b' are constants. The data shows a decreasing trend in the number of authors as the number of documents they have written increases. This aligns with the expectations of Lotka's Law. The data provides the proportion of authors for each document count. The numbers decrease rapidly as the number of documents increases, indicating a concentration of authors with lower productivity. The proportions highlight that a vast majority of authors (86.8%) have written only one document. This demonstrates the well-known phenomenon of a large number of authors contributing minimally to the overall literature. The provided data points could be used to fit a power-law curve that adheres to Lotka's Law. The relationship between the number of authors and the number of documents they've written might follow a specific power-law equation. In summary, the data provided aligns with the principles of Lotka's Law, highlighting the distribution of author productivity across different levels of document output.

Table 4. The authors and their impact metrics

Element	h_index	g_index	m_index	TC	NP	PY_start
DAVIM JP	15	15	0,714	3514	15	2003
LI Q	10	10	0,909	1252	10	2013
LLORCA J	9	9	0,273	1512	9	1991
MOHANTY AK	8	8	0,276	4819	8	1995
TSAO CC	8	8	0,381	1640	8	2003
HOCHENG H	7	7	0,333	1534	7	2003
MISRA M	7	7	0,241	4625	7	1995
SOUTIS C	7	7	0,28	1787	7	1999
SUN GY	7	7	0,636	887	7	2013
TENG JG	7	7	0,304	2203	7	2001

The provided data in Table 4 presents information about different elements (authors) and their impact metrics in the field of research. The impact metrics include h-index, g-index, m-index, total citations (TC), number of papers (NP), and the publication year start (PY_start). With an h-index of 15, DAVIM JP has 15 papers that have each been cited at least 15 times. Their total citation count (TC) is 3514, indicating a substantial impact in terms of citations. The g-index and m-index reflect the distribution of their citations. The fact that the h-index and g-index are equal suggests a balanced distribution of citations among their papers. LI Q has an h-index of 10, indicating 10 papers with at least 10 citations each. The high m-index value (0.909) suggests that a significant portion of their papers have a relatively high number of citations. Despite having a lower TC compared to some others, the m-index indicates consistently impactful work. LLORCA J has a relatively lower h-index (9), indicating fewer highly cited papers. The m-index is also lower (0.273), suggesting a variation in citation counts among their papers. Their TC (1512) aligns with their h-index. Similar analyses can be performed for MOHANTY AK, TSAO CC, HOCHENG H, MISRA M, SOUTIS C, SUN GY, and TENG JG. Each author's impact metrics reflect their individual contributions, influence, and the distribution of citations across their work. Impact metrics should be interpreted within the context of the field. Different disciplines have varying citation norms and standards. A "high" or "low" value can differ depending on the research area. Impact metrics do not capture the full scope of an author's contribution, such as collaboration, interdisciplinary work, or influence beyond citations.

The data presented in Figure 12 provides a comprehensive overview of the article production trends for various countries across different time periods. India exhibited a gradual increase in article production, experiencing accelerated growth from the late 1990s onwards. The United Kingdom demonstrated a steady rise in production after a slow start, with noticeable growth in recent years. France showed significant growth from the mid-2000s to the early 2010s, followed by a period of stable production. Japan experienced consistent growth after a slow start, with stabilization in recent years. The USA demonstrated remarkable growth and stability in production, with peak growth in the mid-2010s. Canada showcased steady growth over the years, while China's rapid and consistent growth was a notable trend. Germany's production surged in the early 2000s and stabilized in later years. Portugal's growth was prominent from the mid-2000s to the early 2010s, followed by stable production. Malaysia displayed growth from the mid-2000s to the early 2020s after a slow start. Overall, the data highlights the diverse trajectories of article production among these countries, reflecting their contributions to global research efforts and knowledge dissemination in the field.

The USA's impressive trajectory reflects its long-standing dominance in research, while Canada's steady growth signifies its dedication to maintaining a strong research presence. China's exponential growth reflects its emergence as a global research powerhouse, driven by significant investments in research and development.

Germany's spike in the early 2000s showcases its strong research ecosystem, while its subsequent stabilization highlights its sustained contributions. Portugal's rapid growth and stabilization indicate a period of heightened research activity followed by a focus on maintaining momentum. Malaysia's emergence as a research contributor is evident through its growth, signaling its efforts to establish itself on the global research map.

Figure 12. Number of articles produced by different countries

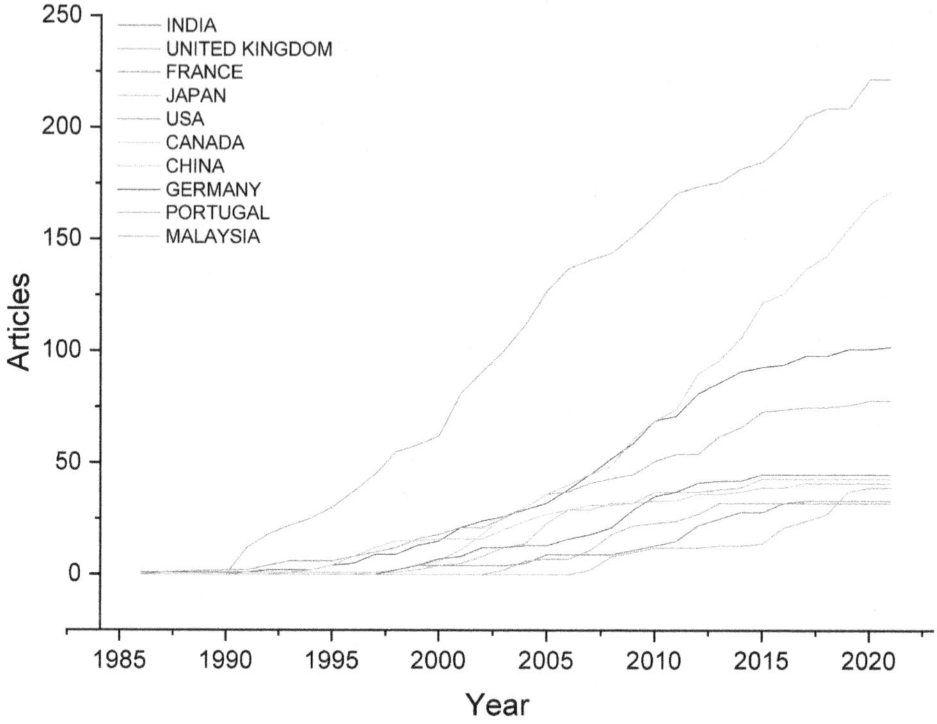

The data depicted in Figure 13 provides insight into the research output of various countries based on their article counts. The USA leads the list with the highest count, underscoring its prominent role in research and academic contributions across diverse fields. China's closely trailing count highlights its significant growth in research output, reflecting its strategic investments in education and research. India's growing count signifies its steady contribution to global knowledge and research efforts. The UK's count reflects its enduring presence in academia and research. Germany's count showcases its active involvement in research and innovation, a testament to its emphasis on education. Japan's significant count aligns with its substantial contributions to research and knowledge dissemination. Canada's steady count reflects its consistent role in advancing knowledge across various domains. Malaysia's growing count signals its increasing presence in the research landscape. France's count emphasizes its involvement in research and scholarly endeavors. Lastly, Portugal's count, while lower, still highlights its participation in knowledge creation and dissemination. Collectively, these article counts unveil the diverse and multifaceted contributions of different countries to the global research community.

Figure 13. List of countries along with their corresponding article

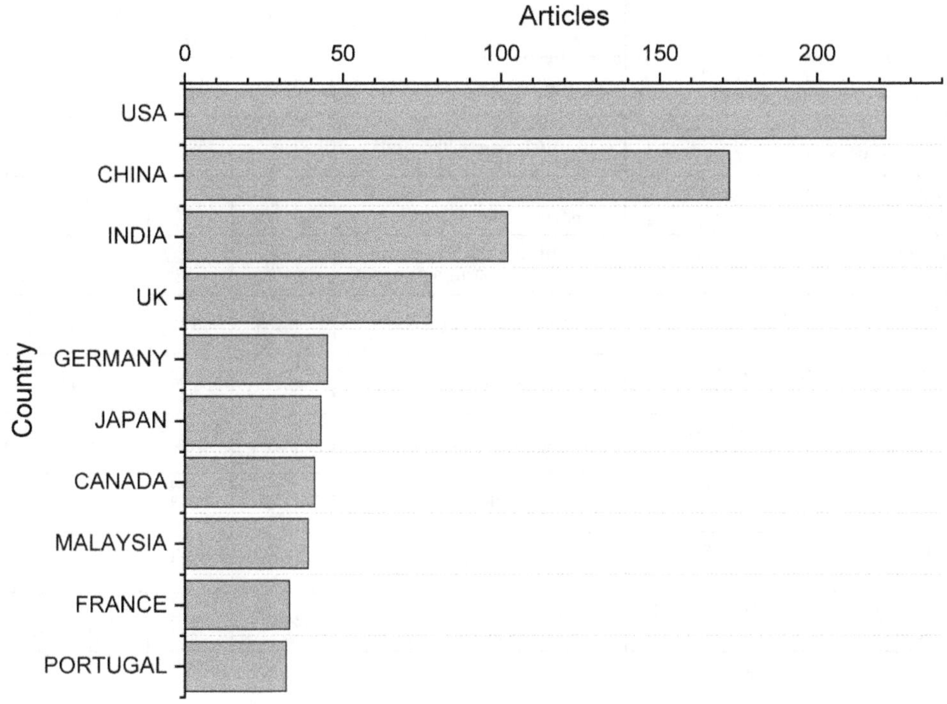

The provided data appears in table 5 to represent a list of research papers, along with their DOIs (Digital Object Identifiers) and corresponding citation-related metrics. The data seems to pertain to research papers in the field of materials science, polymers, composites, and related areas. The "Total Citations" metric indicates the number of times each paper has been cited by other researchers. "TC per Year" represents the average annual citation rate since the publication of the paper. "Normalized TC" provides a normalized value, possibly considering factors such as the journal's impact factor or the field's average citation rate.

Table 5. List of research papers corresponding citation-related metrics

Paper	DOI	Total Citations	TC per Year	Normalized TC
BLEDZKI AK, 1999, PROG POLYM SCI (Bledzki & Gassan, 1999)	10.1016/S0079-6700(98)00018-5	3285	131,40	7,87
SCHUH CA, 2007, ACTA MATER (Schuh et al., 2007)	10.1016/j.actamat.2007.01.052	2706	159,18	8,35
FARUK O, 2012, PROG POLYM SCI (Faruk et al., 2012)	10.1016/j.progpolymsci.2012.04.003	2349	195,75	10,23
MOHANTY AK, 2000, MACROMOL MATER ENG (Mohanty et al., 2000a)	10.1002/(SICI)1439-2054(20000301)276:13.0.CO;2-W	2270	94,58	7,30
WAMBUA P, 2003, COMPOS SCI TECHNOL (Wambua et al., 2003)	10.1016/S0266-3538(03)00096-4	1820	86,67	6,63
MOHANTY AK, 2002, J POLYM ENVIRON (Mohanty et al., 2002)	10.1023/A:1021013921916	1512	68,73	4,49
XIE YJ, 2010, COMPOS PART A-APPL S (Xie et al., 2010)	10.1016/j.compositesa.2010.03.005	1427	101,93	6,84
IBRAHIM IA, 1991, J MATER SCI (Ibrahim et al., 1991)	10.1007/BF00544448	1415	42,88	4,26
HAYS CC, 2000, PHYS REV LETT (Hays et al., 2000)	10.1103/PhysRevLett.84.2901	1251	52,13	4,02
BAKIS CE, 2002, J COMPOS CONSTR (Bakis et al., 2002)	10.1061/(ASCE)1090-0268(2002)6:2(73)	1183	53,77	3,51

The provided data in Table 6 seems to represent a list of cited references along with their corresponding citation counts. The data showcases the impact of specific research papers and authors within their respective fields. The number of citations reflects the level of influence these references have had on subsequent research and scholarly discussions. The references cover a range of topics including polymer science, composite materials, machining, metallurgy, and material design.

Table 6. List of the most cited references along with their corresponding citation counts

Cited References	Citations
CHEN WC, 1997, INT J MACH TOOL MANU, V37, P1097, DOI 10.1016/S0890-6955(96)00095-8 (Chen, 1997)	27
BLEDZKI AK, 1999, PROG POLYM SCI, V24, P221, DOI 10.1016/S0079-6700(98)00018-5 (Bledzki & Gassan, 1999)	24
DAVIM JP, 2003, COMPOS STRUCT, V59, P481, DOI 10.1016/S0263-8223(02)00257-X (Davim & Reis, 2003b)	22
KOPLEV A, 1983, COMPOSITES, V14, P371, DOI 10.1016/0010-4361(83)90157-X (Koplev et al., 1983)	20
ESHELBY JD, 1957, PROC R SOC LON SER-A, V241, P376, DOI 10.1098/RSPA.1957.0133 (Eshelby & Peierls, 1997)	19
HOCHENG H, 2003, J MATER PROCESS TECH, V140, P335, DOI 10.1016/S0924-0136(03)00749-0 (Hocheng & Tsao, 2003)	19
HOCHENG H, 2006, INT J MACH TOOL MANU, V46, P1403, DOI 10.1016/J. IJMACHTOOLS.2005.10.004 (Hocheng & Tsao, 2006)	19
MOHANTY AK, 2000, MACROMOL MATER ENG, V276, P1, DOI 10.1002/(SICI)1439-2054(20000301)276:13.0.CO (Mohanty et al., 2000b)	19
CHRISTMAN T, 1989, ACTA METALL MATER, V37, P3029, DOI 10.1016/0001-6160(89)90339-8 (Christman et al., 1989)	18
DAVIM JP, 2003, MATER DESIGN, V24, P315, DOI 10.1016/S0261-3069(03)00062-1 (Davim & Reis, 2003a)	18

In conclusion, the data presented in Figure 14 unveils the evolving landscape of research interests and trends within materials science and engineering over the span of three decades. Through the lens of various keywords and topics, the shifts and dynamics in research emphasis become discernible:

The consistent rise in occurrences of "Mechanical Properties" from 1991 to 2021 reflects a burgeoning interest in understanding materials' mechanical behaviors. This underscores a growing dedication to unraveling the mechanical intricacies of diverse materials. The unwavering ascent of "Composites" signals a sustained fascination with these materials. This trend echoes the progress in materials science and engineering and the expanding spectrum of composite applications across industries. The stability in occurrences of "Delamination" implies a steady research focus in this domain. It suggests that either the issues concerning delamination in composites have maintained a certain constancy or that research in this area has reached a degree of stability. The gradual increase in occurrences of "Drilling" mirrors continuous research growth in drilling-related techniques, tools, and technologies. This trend reflects steady advancements in this field. The increasing occurrences of "CFRP" in later years align with the rising popularity and applications of lightweight, high-strength CFRP materials in diverse industries. The consistent growth of "Polymer-Matrix Composites (PMCs)" underscores the abiding interest in polymer-based composite materials and their myriad applications. The sustained occurrences of "Confinement" point to an unwavering focus on investigating the effects of confinement on various materials, including composites, under diverse conditions. The emergence and subsequent rise of "Nanocomposites" reflect the ascending intrigue in the amalgamation of nanotechnology and composite materials, heralding a new frontier in material design.

Figure 14. The occurrences of various keywords over a span of three decades

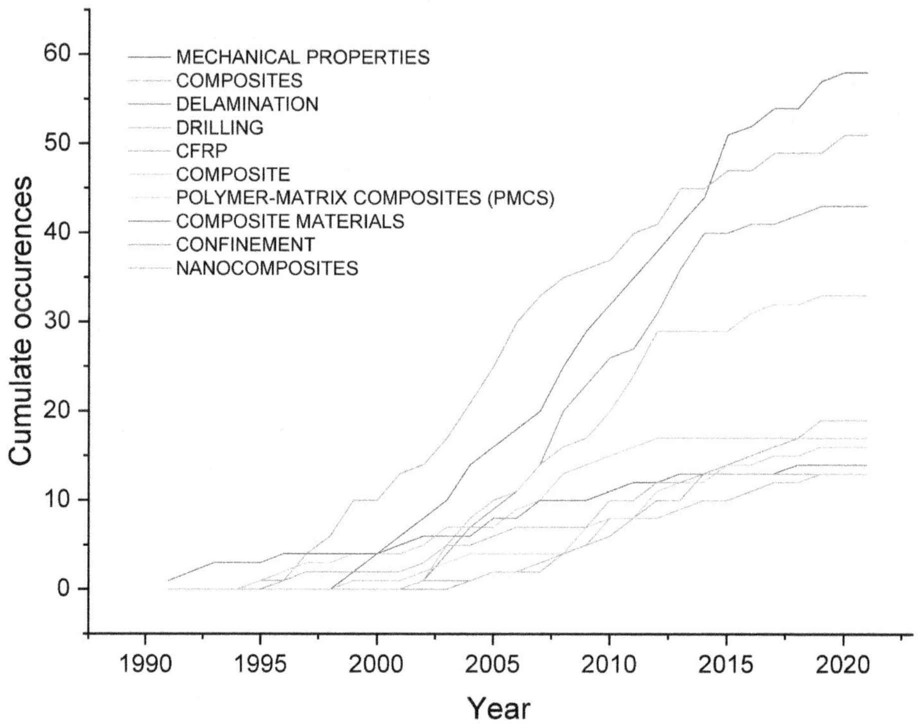

The data presented in Figure 15 outlines the temporal trends of specific keywords in materials science and engineering research literature. Analyzing these trends offers insights into evolving research interests.

- The keyword "ductility" trended in 1996, 1998, and 2005, indicating heightened interest in materials' plastic deformation ability.
- "Concrete" trends in 1997, 2003, and 2004, reflecting efforts to improve this crucial construction material's properties.

- The term "composite materials" is prominent from 1997 to 2009, signifying prolonged research into enhancing composite properties.
- "Finite element analysis" trends in 2007, 2008, and 2013, highlighting the growing use of computational methods in materials research.
- "Mechanical properties" remains a consistent trend from 2005 to 2014, underscoring the continuous focus on material behavior.
- "Fracture toughness" trended in 2004, 2008, and 2012, suggesting ongoing study of materials' fracture resistance.
- "Delamination" is observed in 2006, 2009, and 2013, reflecting sustained exploration of layer separation in composites.
- "Additive manufacturing" emerges from 2017 to 2020, showcasing the rise of 3D printing's applications.
- "Energy absorption" trends in 2018 and 2019, highlighting research into materials' capacity to dissipate energy.
- Both "natural fiber" and "natural fibres" trend from 2007 to 2016, indicating interest in sustainable composite materials.
- "CFRP" trends from 2010 to 2016, representing research into lightweight, strong materials like carbon fiber-reinforced polymers.
- "Non-destructive testing" is observed from 2009 to 2014, showcasing ongoing work on damage-free material assessment.

Figure 15. Trends in research literature

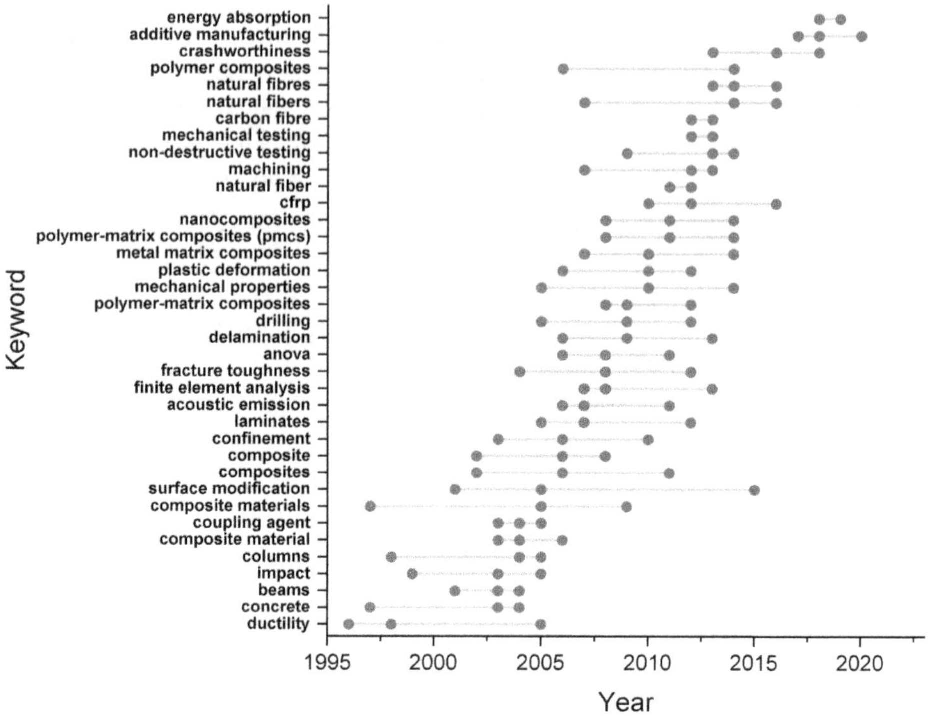

CONCLUSION

This dataset lays bare the intricate dance of research publication within the spheres of machining composites and materials science. It not only paints the canvas of journal evolution and the shifting focus of research endeavors but also offers guidance to researchers grappling with decisions. The presence of authors with diverse productivity levels and the noteworthy concentration of highly prolific contributors mirrors the broader patterns seen across academic domains. Across the globe, the initial gradual climb in article production followed by the explosive growth in the late 1990s and early 2000s resonates universally, albeit with China's distinct consistency standing out against the backdrop of steady progress in countries such as the USA, Canada, and India. Moreover, the data's resonance extends to the impact and visibility of research papers, testifying to the recognition and influence they wield within their respective fields. These findings reflect the shared recognition and significance of research, spanning across the spectrum of scholarly communities. Ultimately, this narrative culminates in a striking testament to the power of influential research papers that have indelibly shaped various domains. Their high citation counts echo their lasting impact and enduring recognition within the tapestry of research discourse.

REFERENCES

Ahmadifar, M., Benfriha, K., Shirinbayan, M., Fitoussi, J., & Tcharkhtchi, A. (2022). Mechanical behavior of polymer-based composites using fused filament fabrication under monotonic and fatigue loadings. *Polymers & Polymer Composites*, *30*. doi:10.1177/09673911221082480

Bakis, C. E., Bank, L. C., Brown, V. L., Cosenza, E., Davalos, J. F., Lesko, J. J., Machida, A., Rizkalla, S. H., & Triantafillou, T. C. (2002). Fiber-Reinforced Polymer Composites for Construction—State-of-the-Art Review. *Journal of Composites for Construction*, *6*(2), 73–87. doi:10.1061/(ASCE)1090-0268(2002)6:2(73)

Bledzki, A. K., & Gassan, J. (1999). Composites reinforced with cellulose based fibres. *Progress in Polymer Science*, *24*(2), 221–274. doi:10.1016/S0079-6700(98)00018-5

Boey, J. Y., Lee, C. K., & Tay, G. S. (2022). Factors Affecting Mechanical Properties of Reinforced Bioplastics : A Review. *Polymers*, *14*(18), 18. Advance online publication. doi:10.3390/polym14183737 PMID:36145883

Cabrera, F. M., Hanafi, I., Khamlichi, A., Jabbouri, A., & Bezzazi, M. (2010). Sur l'usinabilité des composites à matrices polymères renforcée par des fibres. *Mechanics & Industry*, *11*(2), 2. Advance online publication. doi:10.1051/meca/2010021

Chen, W.-C. (1997). Some experimental investigations in the drilling of carbon fiber-reinforced plastic (CFRP) composite laminates. *International Journal of Machine Tools & Manufacture*, *37*(8), 1097–1108. doi:10.1016/S0890-6955(96)00095-8

Chohan, J. S., Boparai, K. S., Singh, R., & Hashmi, M. S. J. (2022). Manufacturing techniques and applications of polymer matrix composites : A brief review. *Advances in Materials and Processing Technologies*, *8*(1), 884–894. doi:10.1080/2374068X.2020.1835012

Christman, T., Needleman, A., & Suresh, S. (1989). An experimental and numerical study of deformation in metal-ceramic composites. *Acta Metallurgica*, *37*(11), 3029–3050. doi:10.1016/0001-6160(89)90339-8

Davim, J. P., & Reis, P. (2003a). Drilling carbon fiber reinforced plastics manufactured by autoclave—Experimental and statistical study. *Materials & Design*, *24*(5), 315–324. doi:10.1016/S0261-3069(03)00062-1

Davim, J. P., & Reis, P. (2003b). Study of delamination in drilling carbon fiber reinforced plastics (CFRP) using design experiments. *Composite Structures*, *59*(4), 481–487. doi:10.1016/S0263-8223(02)00257-X

Eshelby, J. D., & Peierls, R. E. (1997). The determination of the elastic field of an ellipsoidal inclusion, and related problems. *Proceedings of the Royal Society of London. Series A, Mathematical and Physical Sciences*, *241*(1226), 376–396. doi:10.1098/rspa.1957.0133

Faruk, O., Bledzki, A. K., Fink, H.-P., & Sain, M. (2012). Biocomposites reinforced with natural fibers : 2000–2010. *Progress in Polymer Science*, *37*(11), 1552–1596. doi:10.1016/j.progpolymsci.2012.04.003

Frketic, J., Dickens, T., & Ramakrishnan, S. (2017). Automated manufacturing and processing of fiber-reinforced polymer (FRP) composites : An additive review of contemporary and modern techniques for advanced materials manufacturing. *Additive Manufacturing*, *14*, 69–86. doi:10.1016/j.addma.2017.01.003

Fuentes Molina, N., Fragozo Brito, Y., & Polo Benavides, J. M. (2021). Recycling of Residual Polymers Reinforced with Natural Fibers as a Sustainable Alternative : A Review. *Polymers*, *13*(21), 21. Advance online publication. doi:10.3390/polym13213612 PMID:34771169

Hanafi, I., Cabrera, F. M., Bahaoui, J. E., Rubio, J. C. C., & Leite, W. de O. (2023). 3D-Printed Polymer Composite : A Bibliometric Approach – What, Where, and Who Is Doing Research on This Topic. In Development, Properties, and Industrial Applications of 3D Printed Polymer Composites (pp. 9-30). IGI Global. doi:10.4018/978-1-6684-6009-2.ch002

Hanafi, I., Khamlichi, A., Cabrera, F. M., Almansa, E., & Jabbouri, A. (2012). Optimization of cutting conditions for sustainable machining of PEEK-CF30 using TiN tools. *Journal of Cleaner Production*, *33*, 1–9. doi:10.1016/j.jclepro.2012.05.005

Hays, C. C., Kim, C. P., & Johnson, W. L. (2000). Microstructure Controlled Shear Band Pattern Formation and Enhanced Plasticity of Bulk Metallic Glasses Containing in situ Formed Ductile Phase Dendrite Dispersions. *Physical Review Letters*, *84*(13), 2901–2904. doi:10.1103/PhysRevLett.84.2901 PMID:11018971

Hocheng, H., & Tsao, C. C. (2003). Comprehensive analysis of delamination in drilling of composite materials with various drill bits. *Journal of Materials Processing Technology*, *140*(1), 335–339. doi:10.1016/S0924-0136(03)00749-0

Hocheng, H., & Tsao, C. C. (2006). Effects of special drill bits on drilling-induced delamination of composite materials. *International Journal of Machine Tools & Manufacture*, *46*(12), 1403–1416. doi:10.1016/j.ijmachtools.2005.10.004

Ibrahim, I. A., Mohamed, F. A., & Lavernia, E. J. (1991). Particulate reinforced metal matrix composites—A review. *Journal of Materials Science*, *26*(5), 1137–1156. doi:10.1007/BF00544448

Jha, K., Kataria, R., Verma, J., Pradhan, S., Jha, K., Kataria, R., Verma, J., & Pradhan, S. (2019). Potential biodegradable matrices and fiber treatment for green composites : A review. *AIMS Materials Science*, *6*(1), 119–138. doi:10.3934/matersci.2019.1.119

Koplev, A., Lystrup, A., & Vorm, T. (1983). The cutting process, chips, and cutting forces in machining CFRP. *Composites*, *14*(4), 371–376. doi:10.1016/0010-4361(83)90157-X

Kumar, A., Sharma, K., & Dixit, A. R. (2020). Carbon nanotube- and graphene-reinforced multiphase polymeric composites : Review on their properties and applications. *Journal of Materials Science*, *55*(7), 2682–2724. doi:10.1007/s10853-019-04196-y

Mohanty, A. K., Misra, M., & Drzal, L. T. (2002). Sustainable Bio-Composites from Renewable Resources : Opportunities and Challenges in the Green Materials World. *Journal of Polymers and the Environment*, *10*(1), 19–26. doi:10.1023/A:1021013921916

Mohanty, A. K., Misra, M., & Hinrichsen, G. (2000). Biofibres, biodegradable polymers and biocomposites : An overview. *Macromolecular Materials and Engineering*, *276-277*(1), 1–24. doi:10.1002/(SICI)1439-2054(20000301)276:1<1::AID-MAME1>3.0.CO;2-W

Nurazzi, N. M., Harussani, M. M., Aisyah, H. A., Ilyas, R. A., Norrrahim, M. N. F., Khalina, A., & Abdullah, N. (2021). Treatments of natural fiber as reinforcement in polymer composites—A short review. *Functional Composites and Structures*, *3*(2), 024002. doi:10.1088/2631-6331/abff36

Olonisakin, K., fan, M., Xin-Xiang, Z., Ran, L., Lin, W. S., Zhang, W., & Wenbin, Y. (2022). Key Improvements in Interfacial Adhesion and Dispersion of Fibers/Fillers in Polymer Matrix Composites; Focus on PLA Matrix Composites. *Composite Interfaces*, *29*(10), 1071–1120. doi:10.1080/09276440.2021.1878441

Rajak, D. K., Pagar, D. D., Kumar, R., & Pruncu, C. I. (2019). Recent progress of reinforcement materials : A comprehensive overview of composite materials. *Journal of Materials Research and Technology*, *8*(6), 6354–6374. doi:10.1016/j.jmrt.2019.09.068

Rajak, D. K., Pagar, D. D., Menezes, P. L., & Linul, E. (2019). Fiber-Reinforced Polymer Composites : Manufacturing, Properties, and Applications. *Polymers*, *11*(10), 10. Advance online publication. doi:10.3390/polym11101667 PMID:31614875

Raspall, F., Velu, R., & Vaheed, N. M. (2019). Fabrication of complex 3D composites by fusing automated fiber placement (AFP) and additive manufacturing (AM) technologies. *Advanced Manufacturing: Polymer & Composites Science*, *5*(1), 6–16. doi:10.1080/20550340.2018.1557397

Ribeiro, P. de O., Krahl, P. A., Carrazedo, R., & Bernardo, L. F. A. (2023). Modeling the Tensile Behavior of Fiber-Reinforced Strain-Hardening Cement-Based Composites : A Review. *Materials (Basel)*, *16*(9), 9. Advance online publication. doi:10.3390/ma16093365 PMID:37176256

Schuh, C. A., Hufnagel, T. C., & Ramamurty, U. (2007). Mechanical behavior of amorphous alloys. *Acta Materialia*, *55*(12), 4067–4109. doi:10.1016/j.actamat.2007.01.052

Seydibeyoğlu, M. Ö., Dogru, A., Wang, J., Rencheck, M., Han, Y., Wang, L., Seydibeyoğlu, E. A., Zhao, X., Ong, K., Shatkin, J. A., Shams Es-haghi, S., Bhandari, S., Ozcan, S., & Gardner, D. J. (2023). Review on Hybrid Reinforced Polymer Matrix Composites with Nanocellulose, Nanomaterials, and Other Fibers. *Polymers*, *15*(4), 4. Advance online publication. doi:10.3390/polym15040984 PMID:36850267

Wambua, P., Ivens, J., & Verpoest, I. (2003). Natural fibres : Can they replace glass in fibre reinforced plastics? *Composites Science and Technology*, *63*(9), 1259–1264. doi:10.1016/S0266-3538(03)00096-4

Xie, Y., Hill, C. A. S., Xiao, Z., Militz, H., & Mai, C. (2010). Silane coupling agents used for natural fiber/polymer composites : A review. *Composites. Part A, Applied Science and Manufacturing*, *41*(7), 806–819. doi:10.1016/j.compositesa.2010.03.005

Zhang, X., Zhao, N., & He, C. (2020). The superior mechanical and physical properties of nanocarbon reinforced bulk composites achieved by architecture design – A review. *Progress in Materials Science*, *113*, 100672. doi:10.1016/j.pmatsci.2020.100672

Chapter 2
Effect of Fiber Orientation and Matrix Type on Machining Behavior and Structural Integrity of Glass and Basalt Fiber-Reinforced Polymer Composites

Jalal El Bahaoui
Faculty of Sciences, Abdelmalek Essaadi University, Tetouan, Morocco

Mohamed Chairi
https://orcid.org/0000-0002-9474-6633
Faculty of Sciences, Abdelmalek Essaadi University, Tetouan, Morocco

Guido Di Bella
https://orcid.org/0000-0002-3663-9303
Department of Engineering, University of Messina, Italy

Chiara Borsellino
https://orcid.org/0000-0003-1940-8607
Department of Engineering, University of Messina, Italy

Federica Favaloro
Department of Engineering, University of Messina, Italy

ABSTRACT

The marine industry is adopting composite materials with a polymer matrix for shipbuilding due to new technologies and high-performance expectations. However, selecting the right material is challenging, especially considering the machining behavior required for manufacturing cylindrical composite structures. This study

DOI: 10.4018/978-1-6684-9927-6.ch002

focused on evaluating the turning operation and mechanical properties of composites made of glass and basalt fibers with a polymer matrix specifically for maritime construction. The composites consisted of two fabric types, six layers, and two fiber orientation lay-ups, using vinyl ester resin and polyester as the matrix. Test specimens were machined and analyzed, revealing that the hardness and mechanical properties had an impact on the machining process. Glass fiber reinforced polymer (GFRP) samples were successfully machined, resulting in well-formed and undamaged FRP bars. In contrast, basalt fiber reinforced polymer (BFRP) specimens faced difficulties during machining, with matrix splitting, interfacial delamination, fiber ripping.

1. INTRODUCTION

Machining is any of various operations that include cutting a piece of beginning material into the required final form and dimension through a regulated material removal technique (Cabrera et al., 2010). Machining is utilized in the production of many different metal items, but it can also be used composites structures with matrix polymer. A polymer matrix composite material is being used more frequently in the marine industry as a result of new technologies and high-performance standards. Machining is a procedure that uses a cutting tool to remove small amounts of material from a part. The operation necessitates rotation between the instrument being used and the work item. In most machining operations, this relative movement is produced through a primary movement cutting speed and a secondary movement feed. tool's shape and penetration into the work surface, in conjunction with these movements, produce the desired shape of the final work surface. However, their non-homogeneous, anisotropic, and brittle character makes ensuring turning quality (F. Mata et al., 2012).

Sharp tools work well on plastics. With a modest lead, try to get as deep a pass as feasible. A modest rounding of the cut's edge profile prevents the formation of sharp-edged grooves on the room's surface. Cooling is frequently unnecessary (Hanafi et al., 2012). It is sufficient to blow compressed air on the cutting tip if necessary. The majority of requirements are met by ordinary high-speed steels. The use of sintered hard metal inserts is not only advantageous in certain circumstances, such as high quantities, tight tolerances, or plastics incorporating mineral fillers or glass fibers.

It should be mentioned that machining composite materials, particularly CFRP, is different than machining metallic materials (Mata et al., 2012). This is due mostly to the abrasive character of the reinforcements as well as the variability of composite structures. Indeed, throughout the machining process, damage to the machined item occurs, such as die breakage, torn fibers, and fiber/matrix delamination. Machining procedures are frequently modified to composites in order to avoid or reduce this damage. Drilling and milling for contouring or clipping are two of the most typical machining techniques. Turning is a method of moving material against a cutting tool by rotating the workpiece. The lathe is the most common machine tool used in turning (Hanafi et al., 2012). This is one of the simplest machining procedures. In other words, while the workpiece rotates, a cutting tool with a single tip is rotating parallel to the axis of revolution.

Variables Affecting Fiber-Reinforced Polymeric Polymers' Machinability and Characteristics (Hanafi et al., 2013). This study takes into consideration the fundamental research on fiber reenforced polymer composites machining, the characteristics associated with quality that affect FRP composites machinability, considering the existing body of information concerning the ability to be machined of fiber made from natural materials reinforced mixed composites made of polymer and polymer composites strengthened by artificial fibers. The discussion as well highlights the unusual manufacturing method used for FRP composites machining (Alarifi, 2023).

(Rajasekaran et al., 2012a) the aim of this research is to look at the impact of the machining factor combinations on obtaining a high-quality surface finish while turning composites made of carbon polymers with fiber reinforcement (CFRP) using cubic nitride of boron (CBN) cutting tool, as well as to use fuzzy modeling to forecast surface roughness values. The conclusions suggest that the approach for modeling with fuzzy logic may to estimate surface roughness be employed in CFRP composite machining successfully.

Çelik & Türkan (2020) investigated the mechanical characteristics of chopped glass fiber GFRP composites and the use of Taguchi techniques in turning operations. By mixing polyester matrix substance with different lengths of glass fibers (GF), GFRP composites were produced. Taguchi L27 standard orthogonal cutting different speeds, feed rates, and cut depths also used in turning studies. Machining forces and surface roughness were assessed in relation to fiber length, ratio, and cutting parameters.

This work primarily concerns the creation of models of cutting forces for cutting composites with fiber reinforcement made of polymer matrix (PMCs) (Wan et al., 2019). Research developments in four typical processes are examined and introduced in depth: orthogonal cutting, milling, drilling, and turning. This reviews study offers a thorough evaluation of the existing research on machining studies on fiber-matrix materials. The improvement of parameter settings in composite materials with variable machining variables is a major focus of the review research. This study takes into account fiber-reinforced material composites as well as traditional methods of machining including turning, drilling, and milling.

Rajasekaran et al. (2012b) conducted an experiment using a ceramic cutting tool to investigate the machining process for machining fiber reinforced polymeric composites. The objective of this experiment was to see which machining parameter had the greatest impact on surface roughness. Find the machining parameters that, when combined, produce the desired surface abrasion. The surface roughness of it was revealed that a ceramically machining tool existed to be appropriate.

The impact of the procedure settings on the abrasion of the glass fiber-reinforced surface composite polymer pipelines fabricated employing the method of filament winding process was investigated using a high-speed rotating CNC machine. The technique variables investigated hey adjusted the cutting velocity, and feed, and cutting level (Mansour et al., 2021).

The methods and experimental findings reported by (Hutyrova et al., 2016) show the suitability of abrasive water-jetting rough machining procedures for WPCs and associated composite materials. Additionally, this method efficiently solves the problem of polymer matrix melting and following attachment to the operational components of a machining tool formed by traditional turning.

The radial-mode abrasion waterjet (AWJ) machining technology was shown and investigated (W. Li et al., 2013). This cutting mode is found to be more favorable for high material removal rate (MRR) when compared to ordinary offset-mode turning due to the efficient utilization of jet energy for the removal of material, and the exploitation of the proper jet angle of impact and high workpiece surface speed.

Equbal et al. (2020) focuses regarding manufacturing GFRP composites to find the ideal cutting parameters for the best results. A complete multifactorial design for an experiment is used to examine into the impact of three cutting variables, spindle rotation velocity, rate of feed, and depth cut, as well as their relationships (Palanikumar et al., 2006a). The purpose of this research was to ascertain the consequences of parameter settings on the machining of GFRP composites. Experimentation was carried out utilizing the concept of design of experiment. All geared lathes were subjected to machining tests making use of coated cermet tool components and two levels of factors. Cutting speed, fiber placement angles the cutting piece, the level of cut, and the feed rate were all taken into consideration.

Aside from its low weight, strong mechanical properties, and optical openness, composites of polymers reinforced with the cellulose nanofiber (CeNF) provide a diverse range of possibilities applications in the manufacture of optical and mechanical elements (Kamada & Yan, 2021). The study contributes to our understanding of the mechanisms that drive ultraprecision machining of CeNF-reinforced composites and makes recommendations for increasing machined surface quality.

Palanikumar et al. (2006b) the intention of the research is to delve into machining's a parametric influence. settings on surface finish attained when glass fiber reinforced manufacturing of composites made of polymers. According to the findings, the rate at which feed is given has the biggest effect on roughness of the surface. For turning carbon fiber-reinforced (epoxy) composites, establish the best process parameter setting. the study of (Abhishek et al., 2017) combines nonlinear regression modeling, the JAYA optimization technique, and the fuzzy inference network (FIS). This demonstrates the utility of a multi-response optimization strategy.

The current experimental work investigates the machining of composite materials with matrix polymer reinforcement reinforced with fibers of glass and basalt fibers that have high strength and are sustainable in nature. These materials are used in marine building. The usage of basalt fiber in maritime engineering is extremely debatable. Basalt fiber, according to specific research, is a natural fiber obtained from a mineral source because it does not require any additive material during the manufacturing process (Dhand et al., 2015). Composite laminate with matrix polymer cylindrical tube is used in a variety of engineering applications, including desalination plants, power transmission systems, paper mills, and maritime industries. Microscopic analyses were performed in order to comprehend the interactions between different failure causes in the fiber, matrix, and fiber/matrix interface.

2. MATERIAL PREPARATION AND METHODS

2.1 Laminate Composite Panels Preparation

Five composite laminates, labeled as GFRP1, GFRP2, BFRP1, BFRP2, and BFRP3, were employed in this study and they are shown in Figure 1. These laminates were manufactured by Intermarine Spa and consist of six layers of bidirectional woven fabrics combined with thermoplastic plastics. Two types of reinforcement fibers, namely glass and basalt, were used, along with two different resins, polyester and vinylester.

The fabrication techniques employed for these laminates varied depending on the type. GFRP1, BFRP1, and BFRP3 were manufactured using the manual hand lay-up lamination technique, while GFRP2 and BFRP2 were fabricated using the transfer vacuum resin technique. The facesheets were made of two unique fabric kinds (1100 g/m2 glass fabric and basalt layers), and the post-manufacturing material has six layers with two fiber orientations lay-up [0/90°]6 and [+45°/-45°/0°/+45°/-45°/90°]. As the matrix for the FRP facesheets, Atlac® 580 AC 300 Vinyl ester resin with NOROX® and Polyester was used.

Figure 1. Glass and basalt laminates made with different techniques (manual/ infusion) and orientation

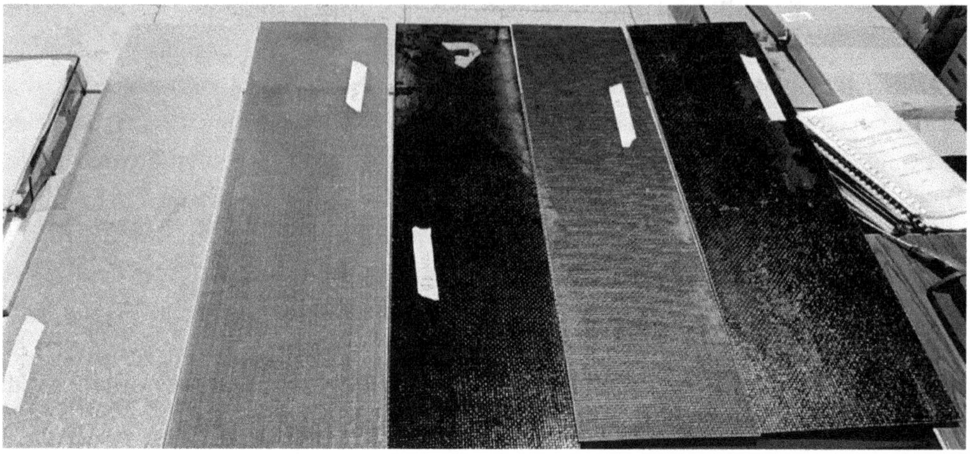

The variation in production and material constituents resulted in differences in the thickness of the laminate composites across the specimens. A detailed description of the laminate composites, including their thicknesses, is provided in Table 1.

Table 1. Description of the laminate composites

Specimen Designation	Laminate Materials		Laminate Sequence	Laminate Thickness	Production Techniques
	Fibers	Resin			
GFRP1	Glass	Polyester	$[0/90]_6$	10 mm	Hand lay-up
GFRP2	Glass	Vinylester	$[0/90]_6$	8 mm	Vacuum resin transfer
BFRP1	Basalt	Polyester	$[0/90]_6$	10 mm	Hand lay-up
BFRP2	Basalt	Vinylester	$[0/90]_6$	8 mm	Vacuum resin transfer
BFRP3	Basalt	Polyester	[+45/-45/0/+45/-45/90]	8 mm	Hand lay-up

2.2 Cutting Planned of Rectangular Specimens

The specimens are sliced into rectangular bars of constant section. The composites laminates specimens were cut with a manually operated machine for cutting bands that was 250 mm long and 25 mm wide as illustrated in the Figure 2. The cutting orientation was parallel to the longitudinal axis of the 0/90° fabric on the face sheet, as well as the through thickness direction of the end-graine fiber orientation. The saw may produce simple cuts without an operator using the weight of the head under the control of a hydraulic brake in addition to cutting in manual mode. The bow must first be manually elevated till the bar moves forward after the cut has been made.

Figure 2. The cutting machine

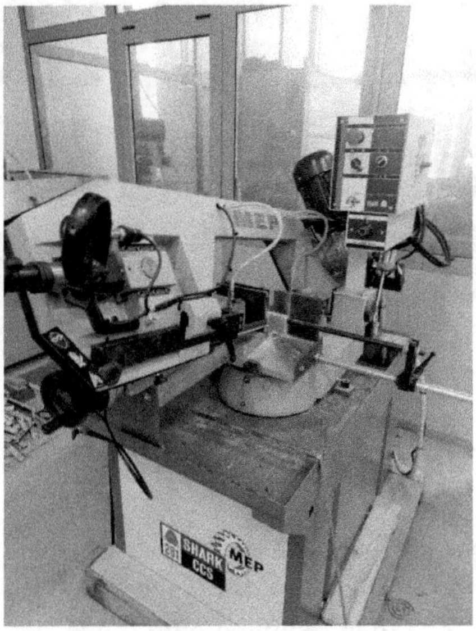

Figure 3. Specimens dimension of GFRP1, GFRP2, GFRP3, GFRP4, and GFRP5

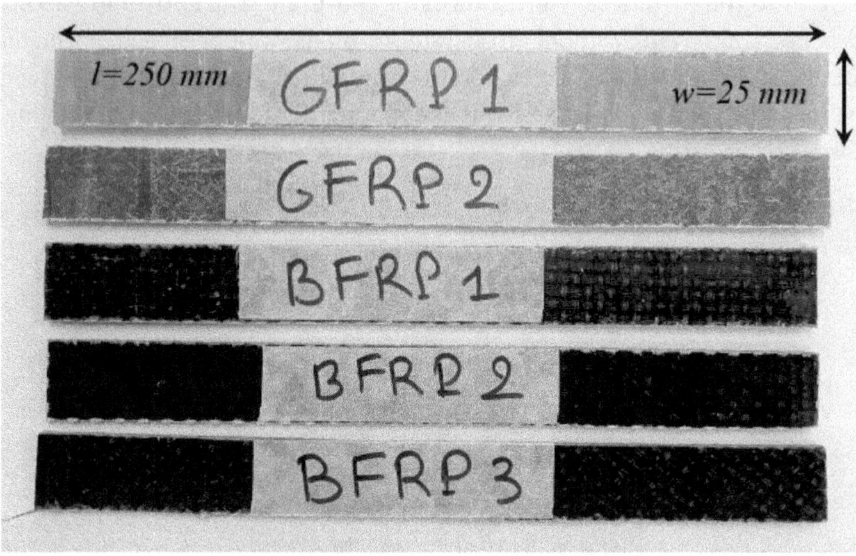

3. EXPERIMENT TURNING MACHINING

To prepare the specimens for the machining process, small pieces with dimensions of 80 mm in length and 10 mm in width (Figure 3) GFRP and BFRP were cut using a cutting machine and then mounted onto the H. Ernault Somua Cholet 435 conventional turning machine. This machine was used for the machining of the composite specimens. Figure 4 depicts the difference between the conventional turning process of a cylindrical specimens (Figure 2.a) and the adaptive rectangular specimen (Figure 2.b), which was used in this study. The machining parameters employed in this study are provided in Table 2.

Table 2. Turning parameters

Feed Speed	**800 to1600 tr/min**
Cutting speed	900 to 1200 tr/min
Depth of feed	Varying depend on the thickness of the laminate
Cutting temperature	Ambient temperature

Figure 4. Turning machining configuration: (a) typical cylinder specimen, (b) adaptative rectangular specimen

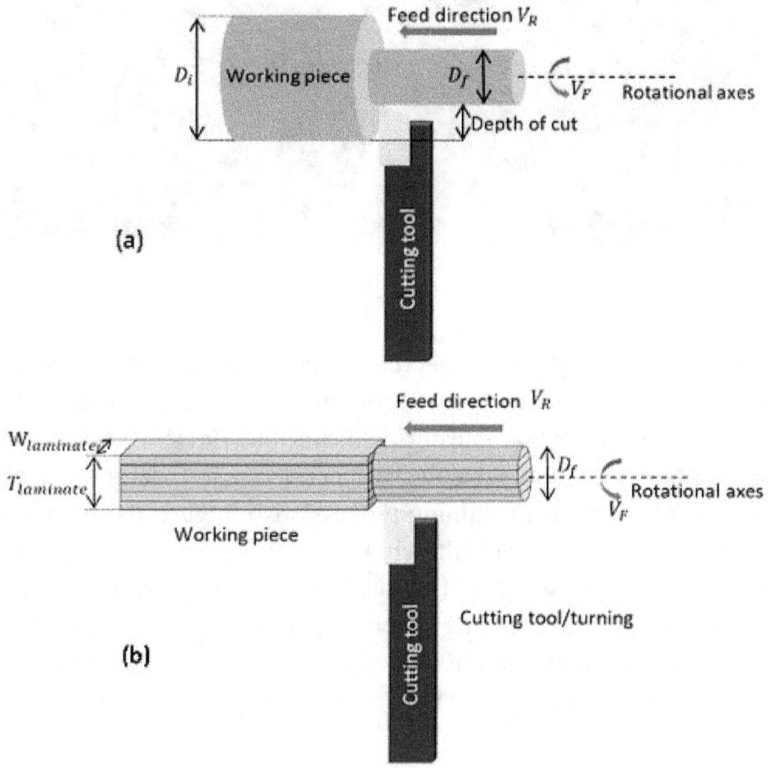

Cutting tool used in this process is speed steel, hi is the cheapest to buy but they have a good lifespan, especially with glass fiber reinforced materials (Figure 5). Their cutting quality is good to average. High speed steel is an alloy steel comprising carbon.

Figure 5. Typical geometry of a turning tool

The machining of glass and basalt fiber reinforced laminates were investigated in this work, which used to turn machining for producing FRP bars. Figure 6 showed visual representation of BFRP2 sample during the turning process, providing both a front view (Figure 6.a) and a side view of the specimen as it undergoes the turning process, giving insight into the machining process. also, Figure 7 showed a visual representation of a GFRP1 sample following the turning process. The results of the research focused on investigating the machinability of glass and basalt fiber reinforced laminates using the turning machining method to produce FRP bars. Figure 3 provided visual representation of the BFRP2 and GFRP1 specimen during the turning process, allowing for better understanding of the machining procedure.

Figure 6. BFRP2 during turning machining: (a) the frontal view, (b) the lateral view

Figure 7. GFRP1 during turning machining: (a) the frontal view, (b) the lateral view

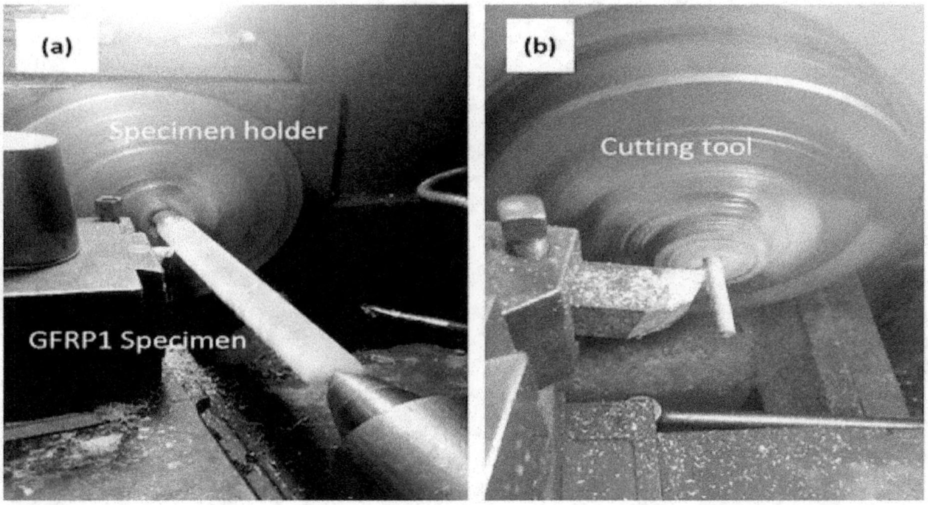

To assess the results of the machined samples, The figure 8 illustrates the obtained FRP bars from the different laminate materials. Important observations can be made:

- Glass fiber reinforced polymer (GFRP) samples were successfully machined, as evidenced the well-formed and intact FRP bars derived from the turning process.
- In contrast, basalt fiber reinforced polymer (BFRP) specimens encountered difficulties during the machining process. The three types of BFRP samples experienced fracturing, leading to the inability to produce intact FRP bars, more particularly BFRP1.

Figure 8. The turning of FRP cylindrical bars

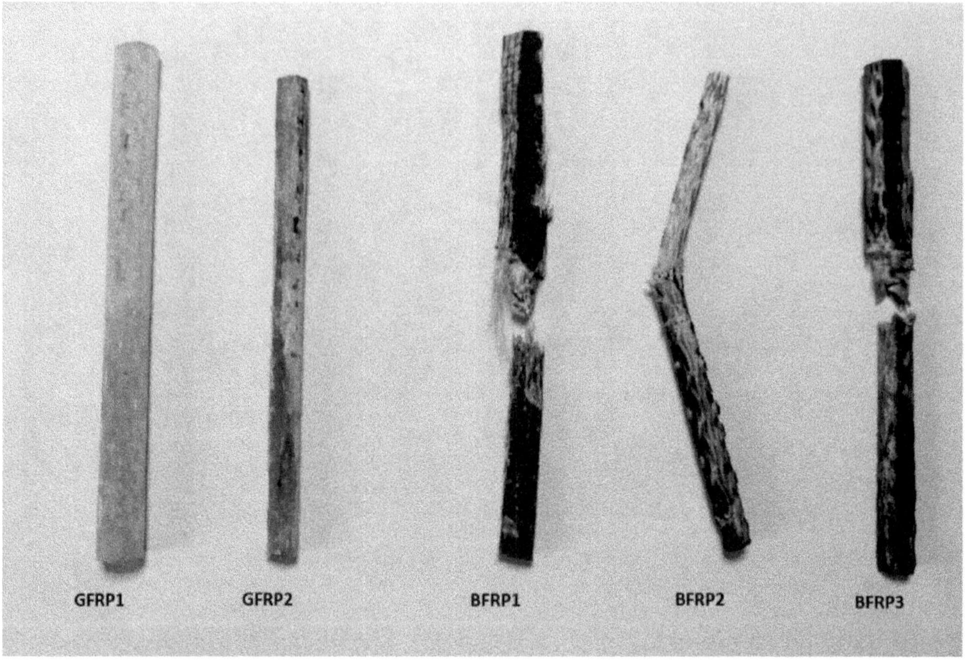

The assessment of the machined samples was done by examining the obtained FRP bars, as depicted in Figure 4. Important observations can be made based on these results. Firstly, the glass fiber reinforced polymer (GFRP) samples exhibited successful machining, evident from the well-formed GFRP bars produced through the turning process. However, in contrast, the basalt fiber reinforced polymer (BFRP) specimens encountered difficulties during machining. All three types of BFRP samples experienced fracturing, resulting in the inability to produce FRP cylindrical bars.

To gain further insights, Figure 9 presented an analysis of the fractured location within the BFRP1 specimen, pinpointing the specific areas where fractures occurred. Moreover, upon closer examination, it was observed that the fracture was mainly caused by the deflection of the specimen during turning. According to (X. Li & Du, 2002a) (X. Li & Du, 2002b) and (Schindler et al., 2013) the deflection of the workpiece during the turning is highly dependent on the machining operation especially the force of cutting.

Figure 9. Fracture location in BFRP1

Several factors also contribute to the difficulties encountered in machining BFRP specimens. High forces for cutting and wrong feed rate, spindle speed, and other parameters related to machining, and depth of cut can cause vibration, errors, and flaws. In addition, inadequate clamping of the specimen can also cause shifting and affect the desired cuts (Teti, 2002).

Also, factors connected to the composite characteristics, structure, and material composition influence FRP turning. The surface quality is influenced by fiber orientation in relation to the cutting direction and cutting methods, which vary based on the texture of the composite. (GIROT et al., 2009). The mechanical behavior between the composite and the cutter, leading to bending deformation, cracks, and failure modes (Islam Shyha, 2021).(Shyha & Huo, 2021)

Figure 10 illustrates the fracture of BFRP1 with fibers oriented at 0° and 90°, highlighted fiber-pull-out and breakage. Additionally, the fracture of BFRP3 with fibers oriented at 45°, 0°, and 90° demonstrates resin-fiber debonding. These fractures patterns align with specific failure modes observed in composites based on fiber orientation and material characteristics during turning machining process (Islam Shyha, 2021).

Figure 10. Damage failure mechanisms of BFRP1 and BFRP3

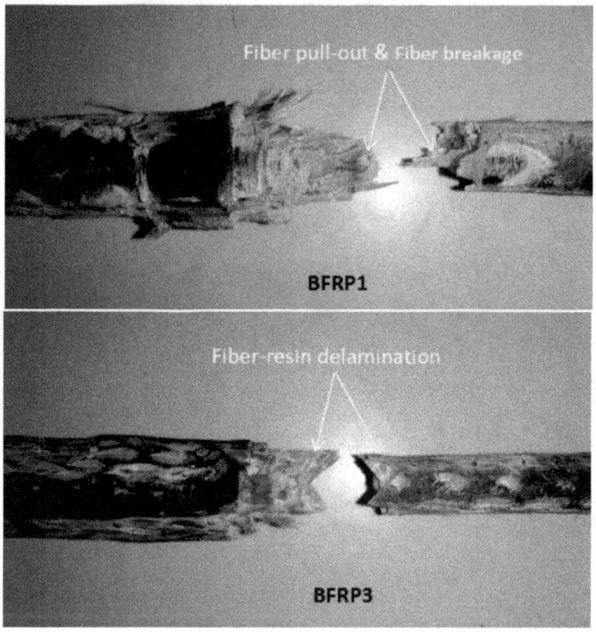

We employed a scanning electron microscope to analyze distinct surface states in order to completely understand the microscopic mechanism of damage to polymer matrix composite materials used in this work. Figures 11 and Figure 12 illustrate the failure sequence of GFRP1 and GFRP2 composite specimens after turning procedures-based failure specimens were examined using a electron microscope (SEM).

Figure 11. Scanning electron micrographs reveal various fiber/matrix contacts: (a) GFRP1 hand lay-up production techniques, (b) GFRP1 vacuum resin production techniques

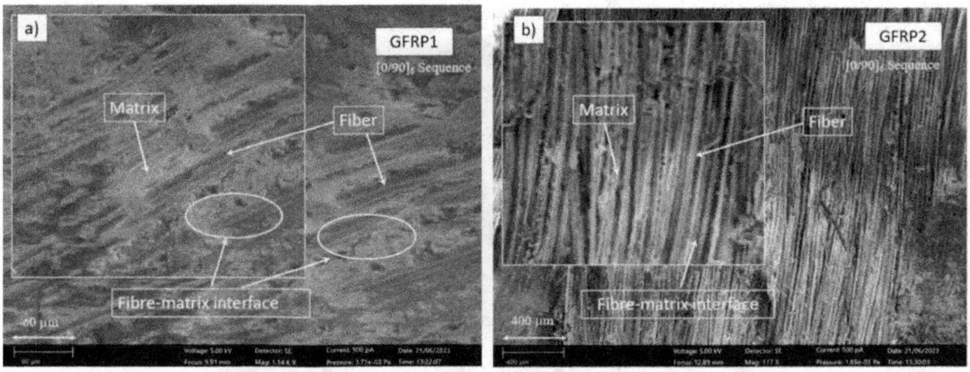

Figure 12. SEM machined basalt fiber composite laminate specimens: (a) fiber breakage and interface matrix delamination of BFRP1 specimen, (b) fiber debonding and matrix cracking of BFRP2 specimen, (c) bending fiber deformation of BFRP3 specimen

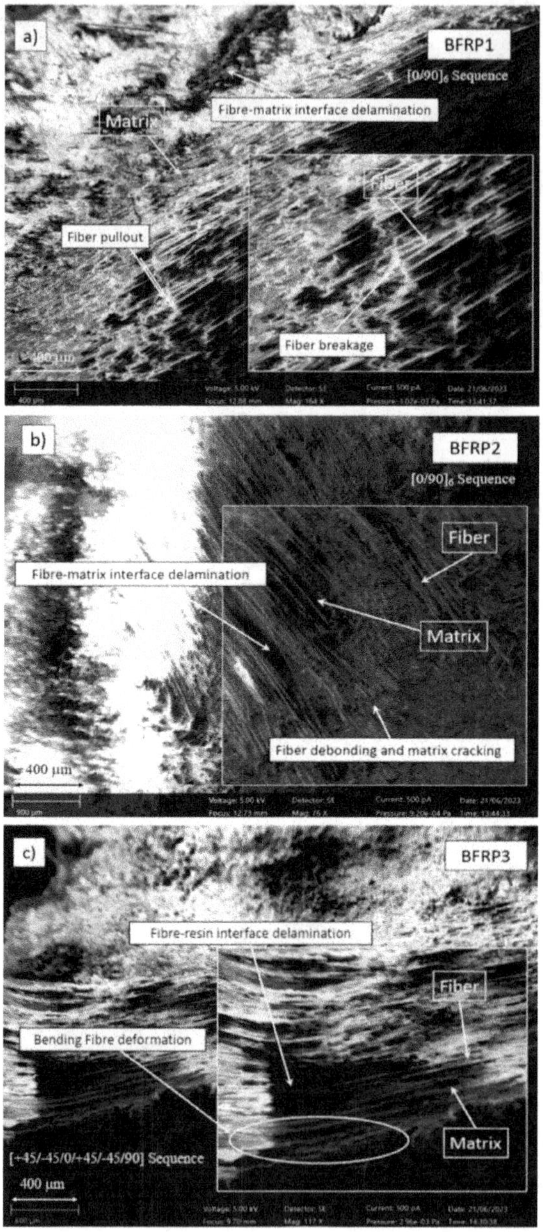

The geometry and topography of microscopic state GFRP are given in figure 11. After the turning.test, the samples' surfaces were examined under SEM to determine the dominant failure mechanisms. Micro structural assessments can that the surface state of this type of composite materials does not undergo any change, the materix keeps its structural shape and does not suffer from any mode of failure.

For Basalt fibre BFRP1 and BFRP2 composites laminates, As demonstrated in Figures 12 (a) and (b), there has been matrix cracking at the fiber and matrix interface., resulting in fibre debonding and pullout from the matrix. Fibres failed catastrophically at the upper levels of turning process.

Also for BFRP2 composites specimen, fiber extraction and matrix delamination as observed, whereas the fbending of ineterface fiber matrix was observed from the SEM for the BFRP3 figure 12.c. BFRP are insulating materials, The heat generated during the milling process concentrates on the machined area, causing significant thermal distortion and degradation.

In the instance of directed fiber composites, the structure may be anisotropic and exhibit varied behavior depending on the desired orientation, complicating machining. The longitudinal stiffness of the fibers will be greater than the vertical stiffness., causing undesired deformations during machining (Al-Turaihi et al., s. d.).

Tools speed steel causes typical failure of Basalt fibers. Carbide and even diamond tools can used to replace the tools used in our machining. Furthermore, the turning process may result in damage to the machined area's edges of BFRP, such as fiber ripping, apparent delamination, internal delamination, thermal degradation, and a rough surface.

As a result of this, an investigation to optimize the machining parameters is necessary. Based on the results, we are considering redoing the machining using a different process, such as laser, water jet, or ultrasonic machining.

4. EXPERIMENTAL MECHANICAL MATERIAL CHARACTERIZATION

4.1 Flexural Behavior of GFRP and BFRP Composites Laminates

The ASTM D7264/D7264M reference standard was used for conducting the three-point bending tests. In particular, the three-point bending tests with central loading on a supported beam were conducted in accordance with the Type A technique. Utilizing a 2 mm/min constant loading rate, the experiments were run. The standard three-point bending test was performed using an "MTS" tensile testing instrument. Flexural tests were carried out in accordance with the experimental strategy. With three repetitions being conducted for each specimen at the t span lengths of 180 mm. The experimental results obtained are as follows in figure 13.

Figure 13. Three-point mechanical properties of the laminate's beams: flexural modulus, flexural strength, and stress failure

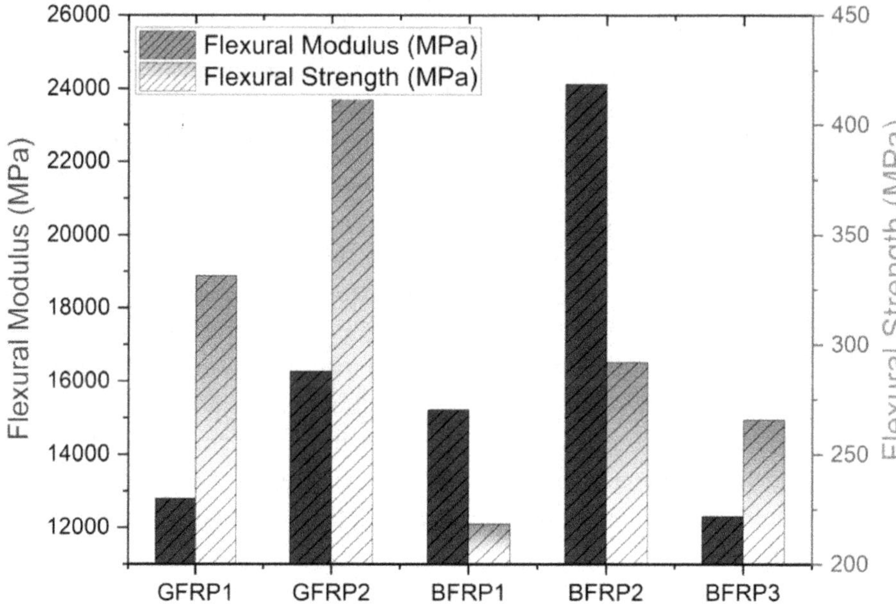

The flexural strength behavior of glass fiber and basalt fiber with matrix polymer reinforced composites was described in this experiment. The results of three-point bending strength test with the variation in the resin polymer content of the fabricated composites and Laminate sequence orientation of fiber, production techniques used are presented.

It was found from this result, the bending strength increased from331 MPa to 450 MPa with glass fiber and 218 MPa to 292 MPa for Glass fiber, the maximum value of bending strength reached to 450 MPa at the glass fiber and Basalt fiber with Vacuum resin transfer technique production. In accordance with the data provided below, polyester is brittle and has lower flexural strength when compared to vinylester polymer in various configuration and tests. But after replacement of the production methodology by replacing the manual method with that of Vacuum resin the flexural strength will be aided for the glass and basalt fiber, because the elevated modulus of elasticity of these fibers will aid to carry a huge quantity of loads and improve this strength. Results reveal that type of reinforcements show good stiffness and bending strength.

The Basalt fiber laminates, specifically samples BFRP2 based on the vacuum resin transfer production technique, demonstrate a significantly higher flexural modulus in comparison to the glass fiber laminate samples, for BFRP2, the minimal value is observed based on land lay un technique production and [+45/-45/0/+45/-45/90] sequence orientation.

Considering the polymer matrix composite materials reinforced by basalt have a lower bending strength than fiberglass, the matrices and glass fibers are exposed to bending deformation during the machining operation. SEM photographs show fiber debonding and matrix cracking in the BFRP2 specimen and bending fiber deformation in the BFRP3 specimen figure 12-b and 12-c.

4.2 Tensile Behavior of GFRP and BFRP Composites Laminates

Tensile characteristics of GFRP and BFRP composites were measured utilizing g displacement-controlled tensile tests on an MTS universal testing machine outfitted in accordance with ASTM D3239, with data collection equipment at crosshead speed of 2 mm/min.

Figure 14. Tensile mechanical properties of the laminate's beams: tensile modulus, tensile strength

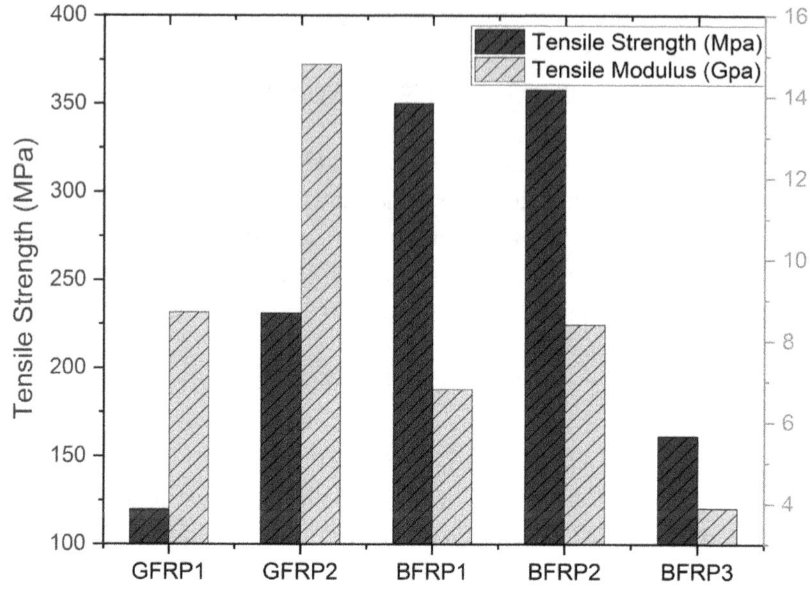

The reinforcement materials have an impact on the composites' characteristics. Figure 14 demonstrates the tensile mechanical properties and considering the modulus of elasticity of glass basalt fiber laminates composites. The results of tensile strength show the maximal composite's tensile strength with matrix polymer reached to 360 MPa, this value is for BFRP2. the minimal one is observed in the case GFRP1 with 89 MPa. The behavior of the material, the method and technology of production, the sequence of fiber orientation of the layers, preparation, as well as the kind of reinforcement, all have significant effects on tensile strength properties of laminate composites with matrix polymer. It was found from Figure 14, the tensile strength increased from 98 to 230 MPa with the maximum for the composite with glass fiber using the vacuum technique production. The tensile strength of manufactured composite laminates is heavily influenced by the interfacial between matrix reinforcement and the polymer used during the fabrication processes. We show that the vacuum technique with Vinylester, the tensile modulus will be significantly enhanced, where the fibers will sustain the majority of stresses, hence increasing the strength of composite laminates for the two situations of glass and basalt fiber reinforcement. In overall, the tensile behavior of glass fiber specimens is distinct from that of basalt fiber. the tensile and flexural test of composites laminate helps to understand the behavior of this composites when they are in machining operation. Also, materials with more strain are likely to fail safe in turning.

4.3 Hardness Behavior of GFRP and BFRP Composites Laminates

Using The MT 3037 mechanical testing machine, Brinell hardness test based on a ball indenter are investigated for GFRP and BFRP. The tests consist of driving an indenter into the material. The load is constant for a given time and the size of the footprint is measured. The impression is all the larger as the metal is soft. The Brinell Hardness (HB) is expressed by the ratio of the force on the surface of the impression. Hardness test is used to characterize materials composite with matrix polymer and determine whether they are suitable for their machining use. The figure 15 shows the hardness variation values of reinforcement with glass fiber and basalt fiber laminate composites

Figure 15. The Brinell hardness values of glass and basalt fiber reinforcing laminate composites

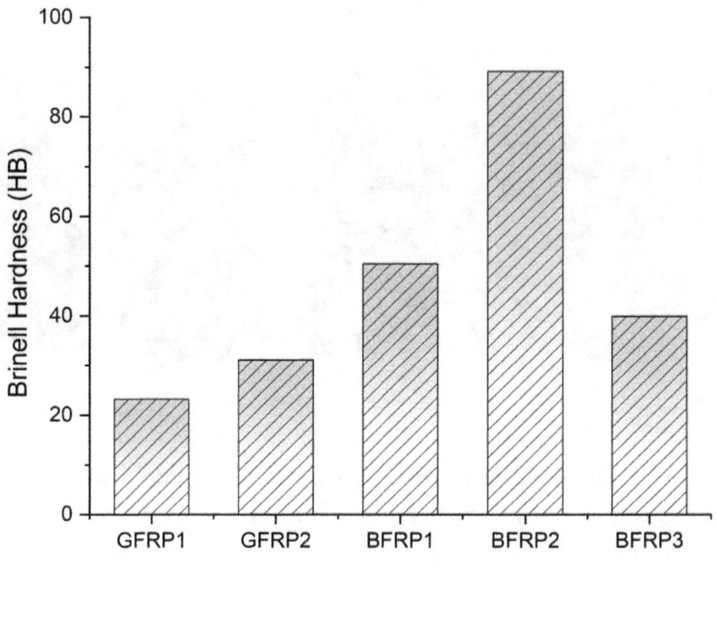

This figure showed the hardness of the composites with glass fiber increases from 23 BH to 31 BH with of glass fiber when changing the production technique as well as the resin. The maximum Brinell hardness obtained at 89 HB for BFRP2. We notice that hardness change with the technique of production and the resin used, Glass fiber reenforced polymer composites have low hardness by comparing with basalt fiber reenforced polymer composite, the lowest value for Polyester resin and hand lay-up technique was 23 BH for GFRP1. However, if we replace the fiberglass with basalt fiber while using the same manufacturing procedure, the hardness value will skyrocket. The distribution of the test load on glass reduces the penetration of the test ball to the surface of the manufactured composite material while increasing its hardness.

The hardness failure mode with different specimens shows in Figure 16. The hardness test on the five configurations shows that the basalt fiber has a longer duration than the fiberglass. The materials' rupture mode is catastrophic, with matrix deformation and fiber delaminate. However, the penetration of the ball does not result in the same condition on the basalt fiber; the structure stays stable and there is no actual change at the interface matrix reinforcement level. A specific situation for BFRP3 is observed, resulting in fiber orientation. The high hardness values of basalt fiber have an impact on the machining operation of this material.

Figure 16. The Brinell hardness test specimens: (a) before and (b) after test

Based on our mechanical experiments, we can conclude that the mechanical behavior of composite materials with a polymer matrix is considered as a parameter to be highlighted in the machining operation of this type of material.

CONCLUSION

In this experimental work, the turning of composite laminates strengthened with glass and basalt fibers has been investigated. used turning machining method to produces FRP bars. Glass fiber and basalt fiber composite mechanical characteristics with various resins, sequence fiber orientation and production technique of composites was investigated. The vacuum resin transfer technique of basalt and glass fibers jointly boosted the tensile and flexural characteristics. Glass fiber reinforced polymer (GFRP) samples were effectively machined, as shown by the well-formed and undamaged FRP bars obtained from the turning process. In contrast, basalt fiber reinforced polymer (BFRP) specimens encountered difficulty throughout the machining process. The three types of BFRP samples fractured, making it impossible to make unbroken FRP bars, notably BFRP1. Based on our mechanical examinations, we can conclude that the mechanical behavior of composite materials with a polymer matrix is a parameter to be highlighted in this type of material's machining operation. In future work. In order to ensure the turning operation for c basalt fiber reinforced composite laminates, we recommend optimizing the speed of cutting, the feeding depth, and cutting tool are turning parameters.

REFERENCES

Abhishek, K., Kumar, V. R., Datta, S., & Mahapatra, S. S. (2017). Application of JAYA algorithm for the optimization of machining performance characteristics during the turning of CFRP (epoxy) composites : Comparison with TLBO, GA, and ICA. *Engineering with Computers, 33*(3), 457–475. doi:10.1007/s00366-016-0484-8

Al-Turaihi, A. S., Hunain, M. B., Hamzah, A. F., & Fadhel, E. Z. (n.d.). *Experimental and Numerical Investigation of Fatigue Behavior of Chopped GFRP Composite Rod under Rotating Bending Load*. Academic Press.

Alarifi, I. M. (2023). A Review on Factors Affecting Machinability and Properties of Fiber-Reinforced Polymer Composites. *Journal of Natural Fibers, 20*(1), 2154304. doi:10.1080/15440478.2022.2154304

Cabrera, F. M., Hanafi, I., Khamlichi, A., Jabbouri, A., & Bezzazi, M. (2010). On machinability of fiber reinforced polymeric composites. *Mecanique et Industries, 11*(2), 93-103. Advance online publication. doi:10.1051/meca/2010021

Çelik, Y. H., & Türkan, C. (2020). Investigation of mechanical characteristics of GFRP composites produced from chopped glass fiber and application of taguchi methods to turning operations. *SN Applied Sciences*, *2*(5), 849. doi:10.1007/s42452-020-2684-5

Dhand, V., Mittal, G., Rhee, K. Y., Park, S.-J., & Hui, D. (2015). A short review on basalt fiber reinforced polymer composites. *Composites. Part B, Engineering*, *73*, 166–180. doi:10.1016/j.compositesb.2014.12.011

Equbal, A., Shamim, M., Badruddin, I. A., Equbal, I., Sood, A. K., Ghazali, N. N. N., & Khan, Z. A. (2020). Application of the Combined ANN and GA for Multi-Response Optimization of Cutting Parameters for the Turning of Glass Fiber-Reinforced Polymer Composites. *Mathematics*, *8*(6), 947. doi:10.3390/math8060947

Girot, F., López de Lacalle, L. N., Lamikiz, A., Iliescu, D., & Gutiérrez, M. E. (2009). Machinabilty aspects of polymer matrix composites. In J. P. Davim (Éd.), Machining Composites Materials. ISTE Publishing Knowledge and John Wiley & Sons Inc.

Hanafi, I., Abdellatif, K., Cabrera, F. M., & Manzanares, J. T. (2013). Application of Desirability Function Based on Neural Network for Optimizing the Process Parameters in Turning of PEEK CF30 Composites. *Materials Science Forum*, *766*, 1–20. . doi:10.4028/www.scientific.net/MSF.766.1

Hanafi, I., Khamlichi, A., Cabrera, F. M., Almansa, E., & Jabbouri, A. (2012). Optimization of cutting conditions for sustainable machining of PEEK-CF30 using TiN tools. *Journal of Cleaner Production*, *33*, 1–9. doi:10.1016/j.jclepro.2012.05.005

Hanafi, I., Khamlichi, A., Cabrera, F. M., Nuñez López, P. J., & Jabbouri, A. (2012). Fuzzy rule based predictive model for cutting force in turning of reinforced PEEK composite. *Measurement*, *45*(6), 1424–1435. doi:10.1016/j.measurement.2012.03.022

Hutyrova, Z., Scucka, J., Hloch, S., Hlavacek, P., & Zelenak, M. (2016). Turning of wood plastic composites by water jet and abrasive water jet. *International Journal of Advanced Manufacturing Technology*, *84*(5-8), 1615–1623. doi:10.1007/s00170-015-7831-6

Islam Shyha, D. H. (2021). Advances in Machining of Composite Materials. In I. Shyha & D. Huo (Eds.), *Springer International Publishing*. Springer International Publishing. doi:10.1007/978-3-030-71438-3

Kamada, Y., & Yan, J. (2021). Machinability Investigation for Cellulose Nanofiber-Reinforced Polymer Composite by Ultraprecision Diamond Turning. *International Journal of Automotive Technology*, *15*(4), 475–482. doi:10.20965/ijat.2021.p0475

Li, W., Zhu, H., Wang, J., Ali, Y. M., & Huang, C. (2013). An investigation into the radial-mode abrasive waterjet turning process on high tensile steels. *International Journal of Mechanical Sciences*, *77*, 365–376. doi:10.1016/j.ijmecsci.2013.05.005

Li, X., & Du, R. (2002). Analysis and compensation of workpiece errors in turning. *International Journal of Production Research*, *40*(7), 1647–1667. doi:10.1080/00207540110100685

Mansour, G., Kyratsis, P., Korlos, A., & Tzetzis, D. (2021). Investigation into the Effect of Cutting Conditions in Turning on the Surface Properties of Filament Winding GFRP Pipe Rings. *Machines*, *9*(1), 16. doi:10.3390/machines9010016

Mata, F., Hanafi, I., Beamud, E., Khamlichi, A., & Jabbouri, A. (2012). Modelling of machining force components during turning of PEEK CF30 by TiN coated cutting tools using artificial intelligence. *International Journal of Machining and Machinability of Materials*, *11*(3), 263. doi:10.1504/IJMMM.2012.046885

Palanikumar, K., Karunamoorthy, L., & Karthikeyan, R. (2006a). Assessment of factors influencing surface roughness on the machining of glass fiber-reinforced polymer composites. *Materials & Design*, *27*(10), 862–871. doi:10.1016/j.matdes.2005.03.011

Palanikumar, K., Karunamoorthy, L., & Karthikeyan, R. (2006b). Parametric optimization to minimise the surface roughness on the machining of GFRP composites. *Journal of Materials Science and Technology*, *22*(1), 66–72.

Rajasekaran, T., Palanikumar, K., & Vinayagam, B. (2012a). Experimental investigation and analysis in turning of CFRP composites. *Journal of Composite Materials*, *46*(7), 809–821. doi:10.1177/0021998311410500

Rajasekaran, T., Palanikumar, K., & Vinayagam, B. K. (2012b). Turning CFRP Composites with Ceramic tool for Surface Roughness Analysis. In R. Rajesh, K. Ganesh, & S. C. L. Koh (Éds.), *International Conference on Modelling Optimization and Computing* (Vol. 38, p. 2922-2929). Elsevier Science Bv. 10.1016/j.proeng.2012.06.341

Schindler, S., Zimmermann, M., Aurich, J. C., & Steinmann, P. (2013). Modeling deformations of the workpiece and removal of material when turning. *Procedia CIRP*, *8*, 39–44. doi:10.1016/j.procir.2013.06.062

Shyha, I., & Huo, D. (Eds.). (2021). *Advances in Machining of Composite Materials : Conventional and Non-conventional Processes*. Springer International Publishing. doi:10.1007/978-3-030-71438-3

Teti, R. (2002). Machining of composite materials. *CIRP Annals - Manufacturing Technology, 51*(2), 611-634. doi:10.1016/S0007-8506(07)61703-X

Wan, M., Li, S.-E., Yuan, H., & Zhang, W.-H. (2019). Cutting force modelling in machining of fiber-reinforced polymer matrix composites (PMCs): A review. *Composites. Part A, Applied Science and Manufacturing, 117*, 34–55. doi:10.1016/j.compositesa.2018.11.003

Chapter 3
Accurate Estimation of Mechanical Properties in PMC Milling:
Factors, Limitations, and Solutions

Guido Di Bella
https://orcid.org/0000-0002-3663-9303
Department of Engineering, University of Messina, Italy

Jalal El Bahaoui
Faculty of Sciences, Abdelmalek Essaadi University, Tetouan, Morocco

Mohamed Chairi
https://orcid.org/0000-0002-9474-6633
Faculty of Sciences, Abdelmalek Essaadi University, Tetouan, Morocco

ABSTRACT

In contrast to traditional materials, polymer matrix-based composites (PMCs) have become increasingly popular due to their versatility in a wide range of industrial applications. These composites possess the necessary properties and offer endless possibilities for construction and composite levels. This chapter aims to provide a comprehensive overview of the machining process for composite materials with a polymer matrix. It will cover the machining procedures involved and shed light on the challenges faced during this process. These challenges encompass the varying properties of the composites, their complex structural composition, and the difficulty in achieving high-quality surface finishes. Furthermore, the chapter delves into various research studies that have addressed different aspects of polymer matrix-based composites. These studies include investigations into material handling and properties, encompassing physical, mechanical, electrical, thermal, and tribological characteristics.

DOI: 10.4018/978-1-6684-9927-6.ch003

1. INTRODUCTION

Machining is a comprehensive process that involves various methods to transform raw materials into the desired shape and size by carefully removing material (Cabrera et al., 2010). Unlike additive manufacturing, which involves adding materials, machining falls under subtractive manufacturing, where materials are precisely eliminated. This versatile manufacturing technique is widely used in the production of a diverse range of metal products. However, its application is not limited to metals alone; it can also be utilized for materials like wood, plastics, ceramics, and composites. In modern machining, computer numerical control (CNC) technology plays a vital role in governing the movement and operation of cutting equipment such as mills, lathes, and other machinery. CNC technology enables precise and automated control, significantly enhancing efficiency and accuracy throughout the machining process(Hanafi et al., 2012) (Hanafi et al., 2013) (V. John, 1992) (Patel G. C et al., 2020) (Helmi & El-Hofy, 2020).

Machining is a procedure that entails the application of a cutting implement to selectively extract small quantities of material from a component, commonly referred to as a "workpiece". The undertaking requires the creation of relative motion between the implement and the workpiece. Typically, this relative motion is accomplished through two fundamental movements: the first being the cutting speed, and the second being the feed speed. By combining the specific shape and depth of penetration of the tool into the work surface with these movements, the desired final shape of the work surface is attained.

When dealing with composite materials, the machining process can present various challenges due to several factors that can influence the manufacturing process. These factors include the composition of the materials, their inherent properties, and the composition of the material parameters. These considerations are crucial to ensure successful machining of composites. Given the unique characteristics and composition of composites, careful attention must be given to these factors to optimize the machining process and achieve the desired results (Helmi & El-Hofy, 2020).

Composite materials consist of two primary constituents: reinforcements and a matrix. The reinforcements can be taking various forms and exhibit diverse characteristics. They can exist as a particles, short fibers, or long fibers, with the later being oriented in a specific direction, such as unidirectional or bidirectional (as found in woven fabric). Alternatively, the fibers can have a random orientation within the matrix. On the other hand, the matrix component of composites can be classified into three main types: polymer matrix, ceramic matrix, and metallic matrix. These matrices provide the structural integrity and bonding for the reinforcement materials within the composite (Chairi et al., 2023) (Rajak et al., 2019).

Regarding polymer-based reinforced fiber composites, the non-homogeneous nature, anisotropic properties, and the abrasive nature of the reinforcements presents a unique challenge. The presence of different materials with varying hardness, such as the soft polymer matrix and the harder fibers, requires the cutting tool to alternate between these two phases during machining. This non-uniformity in material composition can make it difficult to achieve consistent and precise machining results. Additionally, the abrasive nature of the fibers can lead to increased tool wear and reduced tool life (Teti, 2002) (Mohamed et al., 2019). Special attention must be given to selecting appropriate cutting tools, optimizing machining parameters, and selecting appropriate cutting tools, optimizing machining parameters, and employing effective cooling and chip evolution strategies to overcome these challenges when machining polymer-based reinforced fiber composites (H. Wang et al., 2018).

The machining of composite materials often relies on conventional machining practices such as turning (Davim & Mata, 2005) (Chang, 2006), drilling (Caggiano et al., 2019) (Voss et al., 2016) (Abrão et al., 2007), grinding (Gao et al., 2022) (H. Wang, Ning, et al., 2016) milling (Çelik & Alp, 2022; Sun et al., 2023), and trimming (Elfarhani et al., 2022; Haddad et al., 2014). These methodologies were adopted due to the availability of suitable equipment and extensive experience in traditional fabrication techniques. Meanwhile, non-traditional machining techniques have also been explored for composite materials. These include abrasive water machining (AWJM) (Vigneshwaran et al., 2018) (Patel & Shaikh, 2013), laser beam machining (LBM) (El-Hofy & El-Hofy, 2019) (Jäschke et al., 2017) (Z. Wang et al., 2023), and electrical discharge machining (EDM) (Abdallah et al., 2018) (Dutta et al., 2020), ultrasonic-assisted machining (UAM) (Uhlmann et al., 2017) (Uhlmann & Protz, 2019), and elliptic vibration-assisted cutting (EVAC) (W. Xu et al., 2014) are also being explored for their potential in machining fiber-reinforced composites. The non-traditional techniques offer alternative approaches to machining composites and have been utilized to address specific challenges associated with the material's unique properties. he selection of fabrication methodologies is contingent upon several factors including the specific composite material, its composition, desired tolerances, surface finish requirements, and the availability of equipment and expertise. Each fabrication technique possesses unique advantages and limitations, and the decision regarding which method to employ should be based on the specific requirements of the composite material and the desired machining outcome.

2. POLYMER MATRIX REINFORCED WITH FIBERS

Polymer matrix composites (PMCs) have been extensively researched and are widely utilized materials in the scientific field. These composites are classified based on their matrix type and the form of reinforcements employed. PMCs can have either thermosetting or thermoplastic resin matrices, and they can be reinforced with fibers, whiskers, or particles. The distinct properties and advantages offered by each type of reinforcement result in a diverse range of material characteristics. PMCs with thermosetting resin matrices are commonly utilized in the production of high-performance structural composites, while PMCs with thermoplastic resin matrices are employed in the manufacturing of insulation boards and similar products. Continuous and discontinuous fiber-reinforced composites are the two types of PMCs that can be further classified based on the type of fiber used such as glass, carbon, aramid, and natural fibers. Each fiber has its own unique properties and applications. Notably, natural fibers derived from plants, animals, or minerals possess advantageous qualities, including low density, high toughness, and environmental friendliness. (Jialiang et al., 2019) (Chawla, 2012) (Kessler, 2012). Fiber-reinforced composites utilize fibers or strands, which can either be oriented or arranged randomly, in order to augment their mechanical properties. Structural composites are produced by bonding multiple layers together, thereby generating laminate structures. The properties of these composites are influenced by various factors, including the type of fibers and matrix used, the fiber content, the adhesion between the fiber and matrix, as well as the orientation and arrangement of the fibers in conjunction with the specific manufacturing process employed.

The adhesion of fibers to a matrix is a crucial aspect that exerts a direct influence on the characteristics of composites, ultimately impacting their mechanical behavior. This adhesion encompasses the bonding between the fibers and matrix and governs the transfer of loads within the composite, thereby affecting its mechanical properties. The adhesion mechanisms that enable such bonding can vary depending on the specific fiber/matrix system and are influenced by factors such as the fiber and matrix's chemical makeup, the fiber's surface properties, and the processing conditions employed during composite production. The manufacturing techniques utilized in the fabrication of composites play a critical role in determining their properties. Certain challenges can arise, such as poor wettability, inadequate bonding, degradation at the fiber/matrix interface, and fiber damage during manufacturing, all of which can significantly impact the strength of the resulting composites. The direction and distribution of fibers can also influence properties such as strength, stiffness, and anisotropy. By orienting fibers in specific directions, the composite's response to applied load can be tailored accordingly. (Erden & Ho, 2017) (G. Yang et al., 2019).

Fiber-reinforced polymer composites are prone to various forms of damage and failure modes, encompassing delamination, fiber breakage, matrix cracking, and fiber-matrix interfacial debonding. Typically, the damage progression in these composites transpires in multiple phases, which entail the onset of initial damage, its accumulation, and eventually leading in catastrophic damage and rupture (L. Yang et al., 2012) (Arumugam et al., 2014)

3. MILLING OF FIBER REINFORCED POLYMER COMPOSITES

A. Milling Process

In the most basic explanation of the milling process, milling machining employs rotating cutter tools in order to efficiently eliminate material that may produce a diverse range of features on a part by precisely cutting away undesired material. Milling allows for the creation of parts with varied features and may function independently as well as a secondary process for precision manufacturing. The milling process necessitates three main components, namely a milling machine, a workpiece, and a cutter tool. By introducing the workpiece into the rotating cutter, material is excised from the workpiece in the form of small chips, ultimately yielding the desired shape (Leite et al., 2015) (De Oliveira-Leite et al., 2014).

There are two primary types of milling machine: horizontal milling machine, which entails the rotation of the cutter tool around a horizontal axis and the extraction of material from the workpiece through its sides, and the vertical milling machine, which employs a cutter tool oriented vertically and extracts material by using both the bottom and sides of the cutter. These milling machines have two primary structures: manual milling machines refers as "conventional machine" and computer numerical control milling machines refers as "CNC machine". One the other hand, CNC milling machine is a computer-operated, with the workpiece and tool executing precise movements based on programmed directives guided by G-codes or NC-codes, which provides unparalleled precision and reproducibility in their overall operations.

During the process cycles of milling, various operations are performed to shape the workpiece according to desired specifications. The end milling technique necessitates the use an end mill to execute peripheral or slot cuts through the entirety of the workpiece, facilitating the machining of specific features like profiles, slots, pockets, or intricate surface forms. The extent of these features may be attained through a single stroke or through multiple strokes with reduced axial cutting depths. Conversely, chamfer milling, utilizes a chamfer end mill to create angled surfaces along edges or features, producing chamfers. This can be done on the exterior or interior of a part, following a path that is either linear or curves. Face milling, on the other hand, employs a face mill to contour level surfaces, achieving a smooth finish.

Milling cutters are available in a variety of materials, each with distinct properties that make them suitable for specific workpiece materials. The choice of cutter material is influenced by factors such as the hardness, toughness, and wear resistance required. Some commonly used cutter materials include:

- High-speed steel (HSS): HSS cutters exhibit versatility and can be employed for a diverse range of materials. They offer commendable toughness and are relatively cost-effective. (Example of applications for FRP milling: (Suh et al., 2002) (Geier, 2020) (Kumar & Prakash, 2021))
- Carbide: Carbide cutters are renowned for their exceptional hardness and wear resistance. (Example of applications for FRP milling: (Bayraktar & Turgut, 2016) (Azmi et al., 2012) (Muhamad Khairussaleh et al., 2016) (Davim et al., 2004) (Knápek et al., 2022) (Kim et al., 2010))
- Carbon steel: Carbon steel cutters are economical and ideal for softer materials. They provide moderate hardness and toughness. (Example of applications for FRP milling: (Nieminen et al., 1996) (De Lacalle López & Lamikiz, 2010))
- Diamond and polycrystalline diamond (PCD): Diamond and polycrystalline diamond (PCD) cutting tools are renowned for their outstanding hardness, resistance to abrasion and unparalleled precision. They are frequently employed in applications where high-quality surface finishes are required. (Example of applications for FRP milling: (D. Liu et al., 2011) (Kitajima et al., 2014) (Sorrentino et al., 2017) (Sa, 2020) (Inoue et al., 2009)).

B. Milling in the Context of Fiber Reinforced Polymers

As defined in the last section, milling of fiber reinforced polymers is an operation used to removes material using a rotary cutting tool. The composite workpiece can be fed into the cutting tool using two approaches: conventional milling, also known as up milling, and climb milling, referred to as down milling. The conventional method is characterized by the rotation of the cutter in a direction that opposes the feed, causing the cutter and the workpiece to move against each other. In contrast, when performing climb milling, the cutter rotates in direction of the feed, leading to movement in the same direction for both the cutter and the workpiece (Mohamed et al., 2019). The two milling processes can also refer to up milling for the conventional machining and down milling for climb machining. Side milling is a machining process used for tasks such as machining side walls.

Millings tools of fiber reinforced plastics have different designs to increase processing efficiency and reduce tool wear. One design consists of a shaft portion and a cutting portion, with multiple lands separated by chip flutes, forming helical segments at different pitch angles (Bílek et al., 2016). Another design includes a drilling part, a remaining and forming part, an abrasive machining part, a tool holder, and a tool matrix (J. D. Liu et al., 2017). There is also a milling tool with a twist angle that decreases the cut from the front side end to the shaft side end (Y. D. Zhang et al., 2013). Additionally, a milling cutter design includes main cutting edges with levels and relief angles situated in front of the main free surface (Bílek et al., 2016). Finally, a method for milling long fiber reinforced composite plastics is described, involving a rotary milling tool and an advancing movement parallel to the work piece cutting face (Hintze et al., 2013).

Several studies have centered their attention on the process of milling fiber-reinforced polymer composites (FRPCs), whereby the selection of milling tools plays a crucial role in the process. A range of milling tools have been subjected to an evaluation based on their cutting force, surface quality, as well as tool wear. Back to 1983, Richard W. Paige (Paige, 1983) invented a specific milling cutter consists of a body with multiple cutting edges arranged in a specific pattern. The milling cutter is intended to address the challenges associated with machining composite materials used in aerospace applications, such as delamination and fraying. H. Hocheng et al. (Hocheng et al., 1993) provided an initial investigation on milling of continuous carbon fiber-reinforced epoxy. This study highlighted the complexity aspect of milling of composite materials, which is attributable to the anisotropic nature of the material and the presence of two constituents with varying hardness and strength.. H. Y. Puw and H. Hocheng (Puw & Hocheng, 1993) presented a fundamental study on milling of epoxy matrix composites reinforced with continuous carbon fibers, which can help in achieving tolerance assurance of components made of composite materials. An attention has been paid for examining the impact of fiber orientation on the quality of the cut and the magnitude of the cutting force. J. Paulo Davim et al. (Davim et al., 2004) conducted a study concerning the milling process of two GFRP composite materials manufactured by hand-lay up technique. In (Davim & Reis, 2005) a plan of experiments based on Taguchi's method and an analysis of variance (ANOVA) were used to investigate the cutting characteristics of CFRP composites including cutting velocity ad feed rate. The study focused on varying cutting velocity and feed rate while conducting experiments on a laminate plate composed of 16 alternating layers of fibers, possessing a thickness of 4mm

Jamal Sheikh-Ahmed and Rahul Yadav (Sheikh-Ahmad & Yadav, 2005) presented a predictive model for determine cutting forces in the milling process of both uni- and multi-directional carbon fiber reinforced polymers. The model is leveraging the cutting energy principle and encompasses the broad range of fiber orientations and chip thickness.. N. Lopez de Lacalle et al. (López De Lacalle et al., 2009) in their work, examined the advancement of a novel variety of router milling implements for the purpose of high-performance milling of plastic composite materials reinforced with carbon fibers. Additionally, the impact of each characteristic of the milling tools on their efficacy was also investigated. Devi Kalla et al (Kalla et al., 2010) proposed a methodology for predicting cutting forces during helical milling procedure of CFRP using mechanistic modeling techniques. This modeling approach involves the use of physical properties and mathematical models to design a cutting tool and cutting parameters for simulating the machining process of carbon fiber reinforced polymers using a helical end mill. F. Cheddani et al. (Chegdani et al., 2015) conducted an analysis on the impact of distinct verities of organic fibers, specifically bamboo, sisal, and miscanthus fibers on tribological characteristics of natural fiber reinforced plastic materials during the profile milling machining. The study identifies the effect of natural fibers on the quality of machined surface, including all scales, spanning from roughness to waviness.

In (Jia et al., 2018) an empirical analysis was performed regarding the relationship between cutting area temperature and surface quality, and integrity during end milling of muti-directional CFRP composites. Fu-Ji Wang et al. (F. J. Wang et al., 2020) proposed a novel milling too for the aerospace industry that operates along left-right axis. The tool was specifically designed to mitigate surface damage when milling carbon fiber polymer composite. Through its innovative design, the tool minimizes cutting force and reduces surface damage during the milling process. Jie Xu et al . (J. Xu et al., 2021) dedicated their attention in improving the precision of the machining process for composites made of aramid fiber reinforced plastic. They utilized longitudinal-torsional ultrasonic vibration milling to solve issues related to subsurface damage and significant burr. Congle Liu et al. (C. Liu et al., 2023) investigated the model for the evolution of fracture mechanism in unidirectional and multi-directional carbon fiber reinforced polymer throughout the milling machining. By analyzing the machining force, surface morphology and roughness after the slot milling of UD-CFRP and two types of MD-CFRP, the researchers were able to obtain significant findings can be used to design and process CFRP with different laying orders, which can help to reduce the defects in the machining process. Zhefei Sun et al. (Sun et al., 2023) proposed a strategy for achieving effective performance milling of carbon fiber reinforced plastic by integrating ultrasonic vibration assistance with optimized tool architectures. This approach led to a productive and minimally detrimental milling of CFRP.

4. VARIABLES INFLUENCING MILLING OF FIBER REINFORCED POLYMERS

A. The Effects of Milling Parameters and Milling Conditions

In the milling operations, the fundamental machining parameters include the cutting speed, feed rate, and axial depth. These parameters, along with cutting tool characteristics such as tool geometry, have been shown to significantly influence the milling process of fiber-reinforced polymer (FRP) composites. As shown a study by (Ghidossi et al., 2006), the effects of altering cutting parameters on the degree of damage and mechanical behavior of off-axis unidirectional glass fibers reinforced epoxy coupons machined by side milling are noteworthy. The side-milling operation involved a single pass on either side of rectangular roughs to attain the final specimen for testing purposes.. In a study conducted by Rafal Rusinek (Rusinek, 2010), the effect of two governing parameters, specifically feed rate and rotational speed, on the cutting forces during the milling action of an CFRP composite material was experimentally investigated. The milling operation was performed on a CNC machine, with the feed rates between 200 to 720 mm per minute and rotational speed fluctuating from 2000 to 8000 revolutions per minute. Higher feed rates correspond to increased cutting resistance. Furthermore, the rotational speed was found to have an impact on the stability of the cutting process, with regions of both stable and unstable cutting observed.

Jia Ge et al (Ge et al., 2022), conducted a comparative analysis of conventional drilling and helical milling techniques for hole making, focusing on the effects of distinctive feed rates on various parameters. The study investigated various factors such as the resultant thrust force, machining temperature, delamination damage, hole microstructure, and chip morphology and crystallinity.. However, the helical milling process resulted in significantly elevated machining temperatures, which Exhibited a more prominent influence on the production of delamination damage. Cococcetta et al. (Cococcetta et al., 2020) conducting an investigating onto the milling of Composites made of CFRP that have been produced using 3D printing methods, with a focus on manipulating machining process parameters and the conditions of lubrification. The researchers performed a two consecutive slot milling, one with constant feed rate and another with increased feed rate, to evaluate every material specimen machinability. The findings of this investigation indicated that the finished surface of 3D composites exhibited fewer imperfections and protrusions. Notably, the degree of burr formation was reduced through the use of MQL techniques in the machining of CFRP composites. Burr formation is a common challenge encountered in machining processes and can be influenced by various factors, including material properties, machining process parameters, and tool wear.

During the machining of hard composite materials, specifically those that are reinforced with fibers, tool wear represents a significant challenge. In their work, Dervis Ozkan et al. (Ozkan et al., 2019) discussed the milling behavior analysis of CFRP composites, which are extensively utilized in a variety of industries duo to their exceptional properties. The study primary focuses on the surface quality of the machined components, understanding the mechanisms behind tool wear. A number of primary tool wear mechanisms have been identified in literature, including abrasive wear, adhesive wear, and delamination wear. Abrasive wear arises because of the abrasive nature of the CFRP composite material, which leads to an accelerated tool wear rate. The milling process can cause the layers to separate and peel off, leading to undesirable delamination effects. Bartek Stawiszynski et al. (Stawiszynski et al., 2021) investigated the wear resistance of four distinct kinds of milling tools from silicon nitride ceramics, which were intended for use as end mills for milling composite materials reinforced with glass and carbon. Additionally, the paper also addressed the propagation of crack formulation in diamond-coated silicon nitrides during the milling these materials. The researchers measured and evaluated the wear profile and constant cutting forces of the diamond-coated substrates to assess their suitability for milling the composite materials.. The application of mechanical stress during the cutting process led to the development of progressive cracks at these imperfections, which were triggered by different mechanisms of wear, such as adhesion, abrasion, and surface fatigue.

Tomas Knapek et al (Knápek et al., 2022) aimed to examine the impact of different types of cutters on the wear caused by milling process of CFRP laminates with twill weave fabric. The researchers explored the practicability of utilizing conventional coated tools, which are not specifically engineered for CFRP milling, versus a specified DLC coated cutting tool. The study revealed that inappropriate tool selection resulted on occurrence of delamination between laminate plies within the cut and irreversible damage in the cutting tool.

Heat generation is a crucial factor to consider during the milling of fiber-reinforced polymer (FRP) composites, as it can significantly impact both the workpiece and cutting tool. Excessive heat during milling can adversely affect the quality of the machined part, leading to issues such as thermal damage, material degradation, and reduced tool lifespan. Makram Elfarhani et al. (Elfarhani et al., 2022) conducted a research to investigate how the laminating arrangement in glass fiber-reinforced polymer (GFRP) composites and the presence of silica sand particles within the polymer affect the generation of heat that occurs during the dry edge trimming process. For the purpose, temperature was measured for various GFRP specimens using evenly distributed type-K thermocouples along the trimming paths. According to the results, milling perpendicular to the fibers resulted in a significant increase in temperature, with maximum recorded value of 119°C. In contrast, milling parallel to the fibers yielded a maximum temperature of 41°C, similar to that observed in pure epoxy tests. The addition of silica sand particles into the GFRP polymer resulted in decreased temperatures for both tool and specimen, with reduction reached up to 14% for 0° plies and 67% in the case of 90° plies compared to composites that did not contain silica sand. Zhenyuan Jia et al. (Jia et al., 2018) made z noteworthy discovery that regulating the temperature of the cutting area temperature within an appropriate range is crucial for enhancing machining outcomes in end milling of multi-directional carbon fiber-reinforced polymers. During the end milling process, the temperature of the cutting area has a significant role influencing several factors such as cutting force, surface integrity, and subsurface quality. A significant reduction in tool life may result from extremely low temperatures leading to a dramatic increase in cutting force. Conversely, machining at excessively high temperatures can compromise overall machining quality. Milling at lower temperatures is generally preferred when working with FRP composites as it helps minimize thermal damage, reduce material degradation, and improve dimensional accuracy and the refinement of surface finish of the milled workpiece. In (H. Wang, Sun, et al., 2016), the utilization of response surface methodology (RSM) was employed to investigate the correlation between cutting force and cutting temperature in milling carbon fiber-reinforced polymer composites. The primary objective was to analyze their influence on surface quality of the composite material. The specific composite analyzed in the research was a 16-ply ([0/90]2S) laminate with a workpiece material thickness of 3.1 mm. The findings of the research indicated that the spindle speed played the key parameter in influencing the cutting temperature during the milling process. conversely, it was observed that surpassing the glass-transition temperature during cutting resulted in the composite matrix being unable to offer support to the fibers, which led to inappropriate machining quality. These observations suggest the significant of keeping the cutting temperature below the Tg to ensure satisfactory surface quality when milling CFRP materials.

B. The Effects of Composite Material Constituents

The anisotropic nature of composite materials poses a notable concern due to the combination of the different phases and the orientation of fibers, which leads to varying cutting angles between the fibers and the cutting surface of the tool. In their research, Changying Wang et al. (C. Wang et al., 2018) addressed the challenges associated with machining high-strength CFRPs, with particular emphasis on the critical factor of fiber direction. The aim of the study was to the impacts of the fiber cutting angle in milling process of an unidirectional CFRP laminates with high-strength. It was observed that the fiber cutting angle substantially impact on both cutting forces and surface roughness during milling. Specifically, the researchers identified an optimal fiber cutting angle of approximately 30 degrees, which resulted induced cutting forces and improved surface roughness. Furthermore, the fiber cutting angle affects the quality of the machined surface, with the optimal angle resulting in lower surface roughness. Norbert Geier (Geier, 2020) emphasized the significance of taking into account the orientation of the fibers in the course of machining CFRP composites, particularly in the approaches of up milling and down milling. These considerations can lead to improved results and reduce the risk of material damage. The milling experiments were executed using a VF 22 vertical spindle milling machine. Uncoated high-speed steel (HSS) end mills having a diameter of 90 mm and specific cutting-edge characteristics (clearance angle of 10°, rake angle of 25°, and helix angle of 40°) were employed for the milling trials. The investigation exposed that the angle of fiber orientation holds a significant sway on the cutting force. Nevertheless, it was observed that the variations in the fiber orientation angle does not have a notable impact on the passive force component, which refers to the cutting force that acts perpendicular to the cutting edge of the tool. However, the radial force component was found to be more responsive to the change in fiber orientation angle during up milling compared to down milling.

Abburi Lakshman Kumar and M Prakash (Kumar & Prakash, 2021), conducted a study to examine the impact of fiber orientation of the milling of GFRP composite materials throughout end milling process. The experiments focused on GFRP laminates with different orientations (0°/90° and ±45°) and a thickness 10 mm. The milling operations were performed by employing a four-fluted high-speed steel end mill. The findings indicated that the fiber orientation had a momentous impact on the cutting forces encountered during the end milling process. However, the machining parameters, such as cutting speed, transverse feed rate, and axial depth of cut, were found to have minimal impact on the cutting forces. Changying Wang et al. (C. Wang et al., 2017) revealed that surface cavities occur specially in the area of fiber orientation is perpendicular to the cutting direction ($0° < \Theta < 90°$) during circular orthogonal milling of composite laminate disks reinforced with unidirectional carbon fibers. In this particular area, the fibers are aligned perpendicular to the cutting direction, making them more vulnerable to damage. The maximum depth of these surface cavities was found to increase with higher fiber cutting angles, which is directly related to the change in fiber orientation. Surface cavities are considered a type of machining defect that commonly arises during the milling of FRP laminates. They represent a significant manifestation of subsurface damages, which ultimately lead to a decline in the overall quality of the machined surface. Wolfgang Hintze et al (Hintze et al., 2015) conducted a study regarding the association between undulation of fibers induced by weaving and delamination while milling CFRP with woven fabric. The researchers had a particular interest on the effect of the woven patten on the edge quality of CFRP when milling full-groove cuts into the composite material using plain weave fabric. The milling trajectory was positioned at a slight inclination of 1° in relation to the fill yarn, and the relative orientation of the trimmed edge to the undulation of the weave was changed by slightly tilting the milling path slightly in relation to the orientation of fibers. According to the results, the fluctuation of the fiber rovings in the woven fabric led to diverse cutting behavior at the cutting edge. In consideration of the relative position of the milled edge with respect to the undulation, it either encouraged or prevented delamination in the edge.

Majid Haddad et al (Haddad et al., 2014) observed two primary categories of imperfections: fiber pull-out and resin degradation, which occurred during trimming machining process with using a burr tool. These flaws were primarily observed in the layers with fiber orientation of -45° and 90° relative to the carbon fiber reinforced epoxy matrix. These orientations are particularly susceptible to defect generation during the trimming of the composite material. The shape of these defects, such as fibers pull-out and resin degradation, can create stress concentrations within the composite material. This, in turn, can result in a degradation of the mechanical properties of the composite. Nicholas M. Cococcetta et al. (Cococcetta et al., 2020) explore the impact of matrix arrangement on the machining response of composites materials. It was achieved different matrix layouts by utilizing a nylon matrix with chopped carbon fibers as a matrix, along with continuous carbon fibers as the reinforcement material. During the study, it was determined that the maximum cutting force was experienced when the continuous carbon fibers were oriented at an angle of 135° in the sample.

The impacts of fiber kind and fiber proportion on milling of fiber reinforced plastics has been also studied. Raveen John et al. (R. John et al., 2023) explored the effect of kenaf and glass fibers on the machining of polypropylene-based composites. The findings revealed that even though glass fibers are more rigid and abrasive as compared to kenaf fibers, cutting glass fiber reinforced polypropylene composites led to relatively lesser tool wear. However, the machined surface exhibited adverse tearing-dragging mechanisms along with unclear textures. This suggests that the interaction between the glass fibers and the cutting process may lead to surface irregularities and reduced surface quality. The study suggested also that the superior mechanical properties and the adhesion between fibers and matrix adhesion of GF/PP affected by fiber dimensions and their higher aspect ratios, played a crucial role in limiting tool wear. Chegdani et al. (Chegdani et al., 2015) conducted an investigation to find out the impact of divers form of natural fiber (such as bamboo, sisal, and miscanthus) on the tribological behavior in profile milling of polypropylene composites. It was found that bamboo fibers, which possess high contact stiffness, contributed to achieving a smoother surface finish after milling process. This indicates that the choice of natural fiber types can significantly influence the surface quality of machined plastic composites. Dave Kim et al. (Kim et al., 2010) examined the influenced of varying fiber volume fractions from 25% to 40% on cutting forces and the quality of machined surfaces during orthogonal edge trimming process. The research utilized carbide inserts for the experimental investigations. The outcomes showed that an increase in fiber content in the random short GFRP composites led to a decrease in matrix damage and fiber protrusions at the machined edge. Furthermore, the levels of cutting forces and surface textures were found to be influenced by the varying fiber contents.

5. CHALLENGES, DEVELOPMENTS AND STRATEGIES FOR OPTIMIZING MILLING OF FIBER REINFORCED POLYMERS

In (Ishida et al., 2014), the authors introduced an innovative milling procedure that hybrid the ultrasonic vibration and cryogenic tool cooling for machining CFRP. The utilization of ultrasonic vibration in the machining process contributes to a reduction in thrust force, while the cryogenic tool cooling method effectively mitigates delamination at the machined surface. Thrust force refers to the force that acts on the cutting tool along the cutting axis during the milling operation. Experimental results from the cutting tests clearly demonstrated the effectiveness of the recommended cutting method, showcasing significant reductions in thrust force and successful suppression of delamination occurred on the machined surface. Additionally, in another research effort by Xiao Bo Wang et al. (X. B. Wang & Zhao, 2014) a novel technique for machining carbon fiber composites was developed. The approach entailed the utilization of ultrasonic vibration and high-speed milling, with the aim to ensure superb quality for the surface, while simultaneously limiting cutting forces, tool wear formulation and temperatures elevation.

In the study referenced as (Stawiszynski et al., 2021), the ultrasonic assisted milling technique was employed to enhance the production quality of milling glass fiber and carbon fiber reinforced plastics. By incorporating ultrasonic vibrations during the milling process, this technique effectively reduces cutting forces and the manufacturing quality. In addition, the suitability of utilizing ultrasonic machining as a method for slot milling of CFRP was thoroughly examined by Muhammad amin et al. (Amin et al., 2017). A comparative analysis between ultrasonic machining and rotary ultrasonic milling (RUM) was conducted, marking the first time such an analysis has been performed. Ultrasonic assisted milling is characterized by the application of high-frequency oscillations to the cutting tool, resulting in a decrease in cutting forces and a superior quality of the machined surface. Conversely, RUM is a hybrid machining process that combines the material removal mechanisms of diamond grinding and ultrasonic machining. The feasibility of ultrasonic machining or rotary ultrasonic machining is primarily determined by the elastic modulus of the material, with ultrasonic machining being the preferred choice for CFRP materials with low elastic modulus values. Jian Liu et al. (J. Liu et al., 2020), carried out a comparative investigation of various forms of ultrasonic vibration helical milling processes employed for removal of fibers from CFRP composites. The evaluated processes included helical milling (without ultrasonic vibration), axial ultrasonic vibration helical milling (A-UVHM), and longitudinal and torsional ultrasonic vibration helical milling (LT-UVHM). The results demonstrated that the utilization of ultrasonic vibration in helical milling processes led to several significant improvements compared to traditional helical milling. These improvements included reduced axial forces, minimized burrs, and enhanced fiber removal. Moreover, the ultrasonic vibration processes yielded a refined surface finish characterized by regular fractured fiber morphology and the generation of minute fragments. Overall, the findings underscored the effectiveness of ultrasonic vibration helical milling techniques in achieving superior machining outcomes and enhancing the quality of CFRP composite components.

Richard Zemann et al. (Zemann et al., 2014) explored the application of vibration-assisted machining as a means to mitigate fraying issues encountered during process of the milling of CFRP. Fraying is one of primary types of damage faced when machining fiber reinforced polymers (FRP). Vibration-assisted machining involves the utilization of high-frequency vibrations to aid the cutting process. The vibration system effectively breaks up the chips and diminishes the forces exerted on the cutting tool, consequently reducing the occurrence of fraying. In (Szallies et al., 2017), an aim of investigating the feasibility of oscillated milling for chipping unidirectional CFRP with a thickness of 1.5 mm while evaluating the influence of tool characteristics and the orientation of fibers on factors such as delamination, chip formation, and finished surface. Oscillated milling is a machining technique that utilizes a low-frequency tool movement to minimize delamination and enhance the surface quality of CFRP materials. In this process, a conventional 5-axis milling machine is employed, with the z-axis responsible for the oscillatory motion. The tool oscillates at a frequency below 30 Hz, while the workpiece is moved using a linear actuator. The findings of the study indicate that oscillated milling demonstrates the capability to achieve high quality machining of plastic composites, thereby showcasing its potential as an effective approach for improving the machining quality of CFRP components.

Grisol de Melo et al. (De Melo et al., 2020) highlighted the challenges encountered when machining carbon fiber-reinforced polymers arising from their inherent anisotropy and inhomogeneity. These material characteristics result in various issues such as wear on the cutting tool, spalling, delamination, fuzzy surface, fiber pull-out, matrix cracking, and thermal degradation. To address these challenges, the authors explored the potential of utilizing an industrial robot as a viable option for pocket milling in polymer composites. By employing an industrial robot for machining operations, the aim was to mitigate the detrimental effects associated with traditional milling processes and improve the overall machining quality of carbon fiber-reinforced polymers. In their study, Qiang Wang et al. (Q. Wang et al., 2018) introduced a new technique known as tilt helical milling (THM). This method was developed specifically for the precise and high-quality fabrication of holes in carbon fiber reinforced polymers. Through experimental investigation, the researchers demonstrated that THM offers several advantages over conventional helical milling (CHM), including reduced thrust force, lower delamination factor, and improved hole surface finish. This method replaces the conventional revolving motion of the tool in helical milling is replaced with a conical pendulum motion. This achieved by tiling the tool axis towards the hole axis at a specific angle. The tool attitude is dynamically adapting the tool position, it maintains a consistent off-center tilt angle against the hole axis, producing a conical pendulum motion. By implementing the THM technique, the tool rotates around its own axis while exhibiting a conical pendulum motion around the hole axis, enabling exceptional hole quality in CFRP materials.

Moreover, Deyuan Zhang et al. (D. Zhang et al., 2021) proposed a new technique known as wave-motion milling (WMM) for the production of holes in carbon fiber reinforced polymeric matrix materials, with the aim of enhancing hole quality. The WMM method utilizes a wave-like motion path model, derived from the kinematics of the process. The tool is moved in back-and-forth motion, creating a wave-like pattern on the surface of the drilled hole. The unique motion pattern offers several advantages, including the reduction of thrust force and tool wear. Furthermore, this motion contributes to improved hole surface quality and mitigates damages at the hole exist.

S. A. Sundi et al. (Sundi et al., 2021) focused on the recent advancements in tool geometrical design, specifically tailored for milling and trimming operations involving CFRPs. The study emphasizes the significance of cutting tool geometry in achieving high-quality machining outcomes for CFRP materials. Notably, recent progress in tool geometrical design has led to the development of innovative features, including variation in helix angle, rake angle, and edge radius. The new tool geometries aim to optimize the cutting process and enhance the performance and efficiency of machining CFRP materials. In their research, K. Kishore et al. (Kishore et al., 2020) discovered that the incorporation of graphene into hybrid natural composites comprising jute and basalt offered significant benefits in terms of lubrification and surface roughness after milling. The study investigated the effects of adding varying weight percentages (0.2 wt.%, 0.4 wt%, and 0.6 wt%) of graphene to natural composites.

CONCLUSION

In conclusion, this chapter has presented a thorough and all-encompassing overview of the milling machining process for fiber-reinforced plastics (FRPs). It has shed light on the various challenges and considerations that arise when working with composite materials, emphasizing the need for a deep understanding of the interplay between milling parameters and the constituents of fiber-reinforced composites. Throughout the chapter, we have explored numerous factors that influence the milling process, including heat generation, thermal damage, material degradation, and tool life. Additionally, the effects of fiber orientation, cutting angle, weave structure, undulation, fiber types, and volume content have been thoroughly investigated, highlighting their impact on surface quality, cutting forces, and overall machining performance.

By delving into these factors, this chapter has provided valuable insights into the complexities of machining FRPs, offering guidance for optimizing the milling process. Understanding the unique characteristics of fiber-reinforced composites and their interactions with milling parameters allows for improved surface finishes, reduced tool wear, and enhanced mechanical properties of the machined components. Moving forward, further research and advancements in machining techniques for FRPs are crucial. Continued exploration of new fiber types, matrix materials, cutting tools, and machining strategies will contribute to the ongoing improvement and development of milling processes for fiber-reinforced plastics.

REFERENCES

Abdallah, R., Soo, S. L., & Hood, R. (2018). A feasibility study on wire electrical discharge machining of carbon fibre reinforced plastic composites. *Procedia CIRP*, *77*, 195–198. doi:10.1016/j.procir.2018.08.284

Abrão, A. M., Faria, P. E., Rubio, J. C. C., Reis, P., & Davim, J. P. (2007). Drilling of fiber reinforced plastics : A review. In Journal of Materials Processing Technology (Vol. 186, Numéros 1-3, p. 1-7). Elsevier. doi:10.1016/j.jmatprotec.2006.11.146

Amin, M., Yuan, S., Khan, M. Z., Zhang, C., & Israr, A. (2017). Parametric analysis for slot milling of carbon fiber reinforced polymers based on ultrasonic machining. *Proceedings of 2017 14th International Bhurban Conference on Applied Sciences and Technology, IBCAST 2017*, 91-99. 10.1109/IBCAST.2017.7868040

Arumugam, V., Sidharth, A. A. P., & Santulli, C. (2014). Failure modes characterization of impacted carbon fibre reinforced plastics laminates under compression loading using acoustic emission. *Journal of Composite Materials*, *48*(28), 3457–3468. doi:10.1177/0021998313509504

Azmi, A. I., Lin, R. J. T., & Bhattacharyya, D. (2012). Experimental study of machinability of GFRP composites by end milling. *Materials and Manufacturing Processes*, *27*(10), 1045–1050. doi:10.1080/10426914.2012.677917

Bayraktar, S., & Turgut, Y. (2016). Investigation of the cutting forces and surface roughness in milling carbon-fiber-reinforced polymer composite material. *Materiali in Tehnologije*, *50*(4), 591–600. doi:10.17222/mit.2015.199

Bílek, O., Žaludek, M., & Čop, J. (2016). Cutting tool performance in end milling of glass fiber-reinforced polymer composites. *Manufacturing Technology*, *16*(1), 12–16. doi:10.21062/ujep/x.2016/a/1213-2489/MT/16/1/12

Cabrera, F. M., Hanafi, I., Khamlichi, A., Jabbouri, A., & Bezzazi, M. (2010). On machinability of fiber reinforced polymeric composites. *Mecanique et Industries*, *11*(2), 93-103. *Scopus*. Advance online publication. doi:10.1051/meca/2010021

Caggiano, A., Nele, L., & Teti, R. (2019). Drilling of Fiber-Reinforced Composite Materials for Aeronautical Assembly Processes. In *Characterizations of Some Composite Materials*. IntechOpen., doi:10.5772/intechopen.80466

Chairi, M., El Bahaoui, J., Hanafi, I., Mata, F., & Di Bella, G. (2023). Composite Materials : A Review of Polymer and Metal Matrix Composites, Their Mechanical Characterization, and Mechanical Properties. In Next Generation Fiber-Reinforced Composites—New Insights. IntechOpen. doi:10.5772/intechopen.106624

Chang, C. S. (2006). Turning of glass-fiber reinforced plastics materials with chamfered main cutting edge carbide tools. *Journal of Materials Processing Technology*, *180*(1-3), 117–129. doi:10.1016/j.jmatprotec.2006.05.011

Chawla, K. K. (2012). Composite materials : Science and engineering, third edition. In Composite Materials: Science and Engineering, Third Edition (Numéro 1994). doi:10.1007/978-0-387-74365-3

Chegdani, F., Mezghani, S., El Mansori, M., & Mkaddem, A. (2015). Fiber type effect on tribological behavior when cutting natural fiber reinforced plastics. *Wear*, *332-333*, 772–779. doi:10.1016/j.wear.2014.12.039

Cococcetta, N. M., Pearl, D., Jahan, M. P., & Ma, J. (2020). Investigating surface finish, burr formation, and tool wear during machining of 3D printed carbon fiber reinforced polymer composite. *Journal of Manufacturing Processes*, *56*, 1304–1316. doi:10.1016/j.jmapro.2020.04.025

Davim, J. P., & Mata, F. (2005). A new machinability index in turning fiber reinforced plastics. *Journal of Materials Processing Technology*, *170*(1-2), 436–440. doi:10.1016/j.jmatprotec.2005.05.047

Davim, J. P., & Reis, P. (2005). Damage and dimensional precision on milling carbon fiber-reinforced plastics using design experiments. *Journal of Materials Processing Technology*, *160*(2), 160–167. doi:10.1016/j.jmatprotec.2004.06.003

Davim, J. P., Reis, P., & António, C. C. (2004). A study on milling of glass fiber reinforced plastics manufactured by hand-lay up using statistical analysis (ANOVA). *Composite Structures*, *64*(3-4), 493–500. doi:10.1016/j.compstruct.2003.09.054

De Lacalle López, L. N., & Lamikiz, A. (2010). Milling of carbon fiber reinforced plastics. *Advanced Materials Research*, *83-86*, 49–55. . doi:10.4028/www.scientific.net/AMR.83-86.49

De Melo, E. G., Klein, T. B., Reinkober, S., Gomes, J. D. O., & Uhlmann, E. (2020). Pocket milling of composite fibre-reinforced polymer using industrial robot. *Procedia CIRP*, *85*, 180–185. doi:10.1016/j.procir.2019.09.006

De Oliveira-Leite, W., Campos-Rubio, J. C., Mata-Cabrera, F., Tejero-Manzanares, J., & Hanafi, I. (2014). Utilization of artificial neural networks to predict the influence of milling type on the quality product. *Dyna (Spain), 89*(4), 457-466. *Scopus*. Advance online publication. doi:10.6036/6981

Dutta, H., Debnath, K., & Sarma, D. K. (2020). A Study of Wire Electrical Discharge Machining of Carbon Fibre Reinforced Plastic. In M. S. Shunmugam & M. Kanthababu (Eds.), *Advances in Unconventional Machining and Composites* (pp. 451–460). Springer Singapore. doi:10.1007/978-981-32-9471-4_36

El-Hofy, M. H., & El-Hofy, H. (2019). Laser beam machining of carbon fiber reinforced composites : A review. *International Journal of Advanced Manufacturing Technology*, *101*(9-12), 2965–2975. doi:10.1007/s00170-018-2978-6

Elfarhani, M., Guesmi, F., Mkaddem, A., Ghazali, S., Rubaiee, S., & Jarraya, A. (2022). Thermal Aspects in Edge Trimming of Bio-Filled GFRP: Influence of Fiber Orientation and Silica Sand Filler in Heat Generation. *Materials (Basel)*, *15*(14), 4792. Advance online publication. doi:10.3390/ma15144792 PMID:35888259

Erden, S., & Ho, K. (2017). Fiber reinforced composites. In *Fiber Technology for Fiber-Reinforced Composites* (pp. 51–79). Woodhead Publishing. doi:10.1016/B978-0-08-101871-2.00003-5

Gao, T., Zhang, Y., Li, C., Wang, Y., Chen, Y., An, Q., Zhang, S., Li, H. N., Cao, H., Ali, H. M., Zhou, Z., & Sharma, S. (2022). Fiber-reinforced composites in milling and grinding : Machining bottlenecks and advanced strategies. In Frontiers of Mechanical Engineering (Vol. 17, Numéro 2, p. 1-35). Springer. doi:10.1007/s11465-022-0680-8

Ge, J., Catalanotti, G., Falzon, B. G., McClelland, J., Higgins, C., Jin, Y., & Sun, D. (2022). Towards understanding the hole making performance and chip formation mechanism of thermoplastic carbon fibre/polyetherketoneketone composite. *Composites. Part B, Engineering*, *234*, 109752. doi:10.1016/j.compositesb.2022.109752

Geier, N. (2020). Influence of fibre orientation on cutting force in up and down milling of UD-CFRP composites. *International Journal of Advanced Manufacturing Technology*, *111*(3-4), 881–893. doi:10.1007/s00170-020-06163-3

Ghidossi, P., El Mansori, M., & Pierron, F. (2006). Influence of specimen preparation by machining on the failure of polymer matrix off-axis tensile coupons. *Composites Science and Technology*, *66*(11-12), 1857–1872. doi:10.1016/j.compscitech.2005.10.009

Haddad, M., Zitoune, R., Bougherara, H., Eyma, F., & Castanié, B. (2014). Study of trimming damages of CFRP structures in function of the machining processes and their impact on the mechanical behavior. *Composites. Part B, Engineering*, *57*, 136–143. doi:10.1016/j.compositesb.2013.09.051

Hanafi, I., Cabrera, F. M., Khamlichi, A., Garrido, I., & Manzanares, J. T. (2013). Artificial neural networks back propagation algorithm for cutting force components predictions. *Mechanics & Industry*, *14*(6), 6. Advance online publication. doi:10.1051/meca/2013084

Hanafi, I., Khamlichi, A., Cabrera, F. M., Nuñez López, P. J., & Jabbouri, A. (2012). Fuzzy rule based predictive model for cutting force in turning of reinforced PEEK composite. *Measurement*, *45*(6), 1424–1435. doi:10.1016/j.measurement.2012.03.022

Helmi, Y., & El-Hofy, H. (2020). Design for Machining. In Non-Traditional and Advanced Machining Technologies (2nd Editio, p. 62). CRC Press. doi:10.21062/ujep/x.2009/a/1213-2489/MT/9/1/81

Hintze, W., Cordes, M., & Koerkel, G. (2015). Influence of weave structure on delamination when milling CFRP. *Journal of Materials Processing Technology*, *216*, 199–205. doi:10.1016/j.jmatprotec.2014.09.004

Hintze, W., Hartmann, D., & Schütte, C. (2013). *Method for Milling Long Fiber Reinforced Composite Plastic* (Patent N° US20130020735A1).

Hocheng, H., Puw, H. Y., & Huang, Y. (1993). Preliminary study on milling of unidirectional carbon fibre-reinforced plastics. *Composites Manufacturing*, *4*(2), 103–108. doi:10.1016/0956-7143(93)90077-L

Inoue, T., Hagino, M., Matsui, M., & Gu, L. (2009). Cutting characteristics of CFRP materials with end milling. *Key Engineering Materials*, *407-408*, 710–713. . doi:10.4028/www.scientific.net/KEM.407-408.710

Ishida, T., Noma, K., Kakinuma, Y., Aoyama, T., Hamada, S., Ogawa, H., & Higaino, T. (2014). New production technologies in aerospace industry—5th machining innovations conference (MIC 2014) helical milling of carbon fiber reinforced plastics using ultrasonic vibration and liquid nitrogen. *Procedia CIRP*, *24*(C), 13–18. doi:10.1016/j.procir.2014.07.139

Jäschke, P., Wippo, V., Bluemel, S., Staehr, R., & Dittmar, H. (2017). Laser Machining of Carbon Fiber-Reinforced Plastic Composites. In Advances in Laser Materials Processing : Technology, Research and Applications (p. 121-152). Woodhead Publishing. doi:10.1016/B978-0-08-101252-9.00006-6

Jia, Z., Fu, R., Wang, F., Qian, B., & He, C. (2018). Temperature effects in end milling carbon fiber reinforced polymer composites. *Polymer Composites*, *39*(2), 437–447. doi:10.1002/pc.23954

Jialiang, Y., Zhigang, Z., & Hongzhuan, Z. (2019). Polymer Matrix Composites. In Highway Engineering Composite Material and Its Application (Vol. 70, Numéro 7, p. 1282-1283). Springer Singapore. doi:10.1007/s11837-018-2917-x

John, R., Lin, R., Jayaraman, K., & Bhattacharyya, D. (2023). Investigation on microstructure characteristics of tool wear and machined surface mechanisms while milling : Kenaf vs glass fiber-reinforced composites. *Journal of Materials Research and Technology*, *23*, 4716–4733. doi:10.1016/j.jmrt.2023.02.115

John, V. (1992). Material Removal Processes. In *Introduction to Engineering Materials* (pp. 360–380). Palgrave Macmillan. doi:10.1007/978-1-349-21976-6_25

Kalla, D., Sheikh-Ahmad, J., & Twomey, J. (2010). Prediction of cutting forces in helical end milling fiber reinforced polymers. *International Journal of Machine Tools & Manufacture*, *50*(10), 882–891. doi:10.1016/j.ijmachtools.2010.06.005

Kessler, M. R. (2012). Polymer matrix composites : A perspective for a special issue of polymer reviews. In Polymer Reviews (Vol. 52, Numéros 3-4, p. 229-233). Taylor & Francis Group. doi:10.1080/15583724.2012.708004

Kim, D., Kim, Y. H., Gururaja, S., & Ramulu, M. (2010). Processing and fiber content effects on the machinability of compression moulded random direction short GFRP composites. *International Journal of Automotive Technology*, *11*(6), 849–855. doi:10.1007/s12239-010-0101-6

Kishore, M., Amrita, M., & Kamesh, B. (2020). Experimental investigation of milling on basalt-jute hybrid composites with graphene as nanofiller. *Materials Today: Proceedings*, *43*, 726–730. doi:10.1016/j.matpr.2020.12.847

Kitajima, T., Kusuyama, J., Yui, A., Fujii, K., & Ito, Y. (2014). Development of PCD milling tool for carbon-fiber-reinforced plastics. *Advanced Materials Research*, *1017*, 411–414. . doi:10.4028/www.scientific.net/AMR.1017.411

Knápek, T., Dvořáčková, Š., & Knap, A. (2022). Wear Study of Coated Mills during Circumferential Milling of Carbon Fiber-Reinforced Composites and Their Influence on the Sustainable Quality of the Machined Surface. *Coatings*, *12*(10), 1379. doi:10.3390/coatings12101379

Kumar, A. L., & Prakash, M. (2021). The effect of fiber orientation on mechanical properties and machinability of GFRP composites by end milling using cutting force analysis. *Polymers & Polymer Composites*, *29*(9), S178–S187. doi:10.1177/0967391121991289

Leite, W. O., Rubio, J. C. C., Tejero, J., Mata, F., & Hanafi, I. (2015). A comparison of milling cutting path strategies for aluminium 6262-T6 alloys fabrication. In Manufacturing Engineering : New Research (p. 77-90). Scopus.

Liu, C., Ren, J., Shi, K., & Zhang, Y. (2023). Investigation of fracture mechanism evolution model for UD-CFRP and MD-CFRP during the milling process. *Composite Structures*, *306*, 116585. doi:10.1016/j.compstruct.2022.116585

Liu, D., Wang, H., Xu, H., & Zhang, C. (2011). Study of milling force in high speed milling Carbon Fibre Reinforced Plastics(CFRP) using diamond tool. *Applied Mechanics and Materials*, *50-51*, 368–371. . doi:10.4028/www.scientific.net/AMM.50-51.368 PMID:31267820

Liu, J., Chen, G., Ren, C., Qin, X., Zou, Y., & Ge, J. (2020). Effects of axial and longitudinal-torsional vibration on fiber removal in ultrasonic vibration helical milling of CFRP composites. *Journal of Manufacturing Processes*, *58*, 868–883. doi:10.1016/j.jmapro.2020.08.071

Liu, J. D., Yang, X. F., Xu, Z. L., Li, Y. S., Yan, G. H., & Huang, L. K. (2017). Experimental study on the tool wear in milling carbon fiber reinforced plastics. *Materials Science Forum, 893 MSF*, 57-61. https://doi.org/ doi:10.4028/www.scientific.net/MSF.893.57

López De Lacalle, N., Lamikiz, A., Campa, F. J., Valdivielso, A. F., & Etxeberria, I. (2009). Design and test of a multitooth tool for CFRP milling. *Journal of Composite Materials*, *43*(26), 3275–3290. doi:10.1177/0021998309345354

Mohamed, S. B., Rashid, R. A., Muhamad, M., & Ismail, J. (2019). Composite Materials and Types of Machining. In *SpringerBriefs in Applied Sciences and Technology* (pp. 1–14). Springer Verlag. doi:10.1007/978-981-13-1804-7_1

Muhamad Khairussaleh, N. K., Che Haron, C. H., & Ghani, J. A. (2016). Study on wear mechanism of solid carbide cutting tool in milling CFRP. *Journal of Materials Research, 31*(13), 1893–1899. doi:10.1557/jmr.2016.21

Nieminen, I., Paro, J., & Kauppinen, V. (1996). High-speed milling of advanced materials. *Journal of Materials Processing Technology, 56*(1-4), 24–36. doi:10.1016/0924-0136(95)01817-4

Ozkan, D., Sabri Gok, M., Oge, M., & Cahit Karaoglanli, A. (2019). Milling Behavior Analysis of Carbon Fiber-Reinforced Polymer (CFRP) Composites. *Materials Today: Proceedings, 11*, 526–533. doi:10.1016/j.matpr.2019.01.024

Paige, R. W. (1983). *US Patent for Milling cutter for composite laminates Patent (Patent # 4,810,136 issued March 7, 1989)—Justia Patents Search* (Patent N° 4810136).

Patel, G. C. M., Chate, G. R., Parappagoudar, M. B., & Gupta, K. (2020). Introduction to Hard Materials and Machining Methods. In *SpringerBriefs in Applied Sciences and Technology* (pp. 1–24). Springer. doi:10.1007/978-3-030-40102-3_1

Patel, S. R., & Shaikh, A. A. (2013). A review on machining of fiber reinforced plastic using abrasive waterjet. *International Journal of Innovative Technology & Adaptive Management, 1*(3).

Puw, H. Y., & Hocheng, H. (1993). Machinability Test Of Carbon Fiber-Reinforced Plastics In Milling. *Materials and Manufacturing Processes, 8*(6), 717–729. doi:10.1080/10426919308934875

Rajak, D. K., Pagar, D. D., Kumar, R., & Pruncu, C. I. (2019). Recent progress of reinforcement materials : A comprehensive overview of composite materials. *Journal of Materials Research and Technology, 8*(6), 6354–6374. doi:10.1016/j.jmrt.2019.09.068

Rusinek, R. (2010). Cutting process of composite materials : An experimental study. *International Journal of Non-linear Mechanics, 45*(4), 458–462. doi:10.1016/j.ijnonlinmec.2010.01.004

Sa, M.-W. (2020). Comparison Study on Side Milling of CFRP with AlCrN-based, Diamond-Like-Carbon(DLC), and Diamond-Coated End Mill. *The Korean Society of Manufacturing Process Engineers, 19*(4), 9–15. doi:10.14775/ksmpe.2020.19.04.009

Sheikh-Ahmad, J., & Yadav, R. (2005). Force prediction in milling of carbon fiber reinforced polymers. *American Society of Mechanical Engineers, Materials Division (Publication). MD Medical Newsmagazine*, *100*(MD), 429–436. doi:10.1115/IMECE2005-81909

Sorrentino, L., Turchetta, S., & Bellini, C. (2017). Analysis of carbon fibre reinforced polymers milling by diamond electroplated tool. *Diamond and Related Materials*, *76*, 184–190. doi:10.1016/j.diamond.2017.05.005

Stawiszynski, B., Protz, F., & Uhlmann, E. (2021). Selection of diamond coated silicon nitrides for use as ceramic end mills when machining glass and carbon fiber reinforced plastics. *International Journal of Refractory & Hard Metals*, *96*, 105487. doi:10.1016/j.ijrmhm.2021.105487

Suh, J. D., Lee, D. G., & Kegg, R. (2002). Composite machine tool structures for high speed milling machines. *CIRP Annals - Manufacturing Technology, 51*(1), 285-288. doi:10.1016/S0007-8506(07)61518-2

Sun, Z., Geng, D., Zheng, W., Liu, Y., Liu, L., Ying, E., Jiang, X., & Zhang, D. (2023). An innovative study on high-performance milling of carbon fiber reinforced plastic by combining ultrasonic vibration assistance and optimized tool structures. *Journal of Materials Research and Technology*, *22*, 2131–2146. doi:10.1016/j.jmrt.2022.12.054

Sundi, S. A., Izamshah, R., Kasim, M. S., Jaafar, M. F., & Hassan, M. H. (2021). Milling/Trimming of Carbon Fiber Reinforced Polymers (CFRP): Recent Advances in Tool Geometrical Design. In Machining and Machinability of Fiber Reinforced Polymer Composites (p. 101-128). doi:10.1007/978-981-33-4153-1_4

Szallies, K., Siebert, N., & Bergmann, J. P. (2017). Low Frequency Oscillated Milling of Carbon Fiber-reinforced Plastics. *Procedia CIRP*, *66*, 153–158. doi:10.1016/j.procir.2017.03.298

Teti, R. (2002). Machining of composite materials. *CIRP Annals - Manufacturing Technology, 51*(2), 611-634. doi:10.1016/S0007-8506(07)61703-X

Uhlmann, E., & Protz, F. (2019). Ultrasonic assisted peripheral milling of fiber reinforced plastics in consideration of clamping and cutting edge geometry. *Procedia Manufacturing*, *40*, 56–63. doi:10.1016/j.promfg.2020.02.011

Uhlmann, E., Protz, F., Stawiszynski, B., & Heidler, S. (2017). Ultrasonic Assisted Milling of Reinforced Plastics. *Procedia CIRP*, *66*, 164–168. doi:10.1016/j.procir.2017.03.278

Vigneshwaran, S., Uthayakumar, M., & Arumugaprabu, V. (2018). Abrasive water jet machining of fiber-reinforced composite materials. *Journal of Reinforced Plastics and Composites*, *37*(4), 230–237. doi:10.1177/0731684417740771

Voss, R., Henerichs, M., & Kuster, F. (2016). Comparison of conventional drilling and orbital drilling in machining carbon fibre reinforced plastics (CFRP). *CIRP Annals - Manufacturing Technology*, *65*(1), 137-140. doi:10.1016/j.cirp.2016.04.001

Wang, C., Liu, G., An, Q., & Chen, M. (2017). Occurrence and formation mechanism of surface cavity defects during orthogonal milling of CFRP laminates. *Composites. Part B, Engineering*, *109*, 10–22. doi:10.1016/j.compositesb.2016.10.015

Wang, C., Wen, L., Ming, W., An, Q., & Chen, M. (2018). Experimental study on effects of fiber cutting angle in milling of high-strength unidirectional carbon fiber–reinforced polymer laminates. *Proceedings of the Institution of Mechanical Engineers. Part B, Journal of Engineering Manufacture*, *232*(10), 1813–1824. doi:10.1177/0954405416679868

Wang, F. J., Yan, J. B., Zhao, M., Wang, D., Wang, X. N., & Hao, J. X. (2020). Surface damage reduction of dry milling carbon fiber reinforced plastic/polymer using left–right edge milling tool. *Journal of Reinforced Plastics and Composites*, *39*(11-12), 409–421. doi:10.1177/0731684420912018

Wang, H., Ning, F., Hu, Y., Fernando, P. K. S. C., Pei, Z. J., & Cong, W. (2016). Surface grinding of carbon fiber-reinforced plastic composites using rotary ultrasonic machining : Effects of tool variables. *Advances in Mechanical Engineering*, *8*(9), 1–14. doi:10.1177/1687814016670284

Wang, H., Qin, X., Wu, D., & Song, A. (2018). Optimization of Cutting Parameters in Helical Milling of Carbon Fiber Reinforced Polymer. *Transactions of Tianjin University*, *24*(1), 91–100. doi:10.1007/s12209-017-0079-5

Wang, H., Sun, J., Li, J., Lu, L., & Li, N. (2016). Evaluation of cutting force and cutting temperature in milling carbon fiber-reinforced polymer composites. *International Journal of Advanced Manufacturing Technology*, *82*(9-12), 1517–1525. doi:10.1007/s00170-015-7479-2

Wang, Q., Wu, Y., Lu, D., Bitoh, T., & Feng, M. (2018). Experimental investigation on the tilt helical milling of carbon fiber reinforced plastics (CFRP). *Key Engineering Materials*, *792*, 173–178. . doi:10.4028/www.scientific.net/KEM.792.173

Wang, X. B., & Zhao, C. Y. (2014). Design of special ultrasonic device in composite materials processing. *Advanced Materials Research*, *1027*, 331–335. . doi:10.4028/www.scientific.net/AMR.1027.331

Wang, Z., Ma, Y., Yuan, B., Wu, C., Li, C., & Sun, S. (2023). Development of Laser Processing Carbon-Fiber-Reinforced Plastic. In Sensors (Basel, Switzerland) (Vol. 23, Numéro 7). Sensors (Basel). doi:10.3390/s23073659

Xu, J., Feng, P., Feng, F., Zha, H., & Liang, G. (2021). Subsurface damage and burr improvements of aramid fiber reinforced plastics by using longitudinal–torsional ultrasonic vibration milling. *Journal of Materials Processing Technology*, *297*, 117265. doi:10.1016/j.jmatprotec.2021.117265

Xu, W., Zhang, L. C., & Wu, Y. (2014). Elliptic vibration-assisted cutting of fibre-reinforced polymer composites : Understanding the material removal mechanisms. *Composites Science and Technology*, *92*, 103–111. doi:10.1016/j.compscitech.2013.12.011

Yang, G., Park, M., & Park, S. J. (2019). Recent progresses of fabrication and characterization of fibers-reinforced composites : A review. In *Composites Communications* (Vol. 14, pp. 34–42). Elsevier. doi:10.1016/j.coco.2019.05.004

Yang, L., Yan, Y., Liu, Y., & Ran, Z. (2012). Microscopic failure mechanisms of fiber-reinforced polymer composites under transverse tension and compression. *Composites Science and Technology*, *72*(15), 1818–1825. doi:10.1016/j.compscitech.2012.08.001

Zemann, R., Kain, L., & Bleicher, F. (2014). Vibration assisted machining of carbon fibre reinforced polymers. *Procedia Engineering*, *69*, 536–543. doi:10.1016/j.proeng.2014.03.023

Zhang, D., Shao, Z., Geng, D., Jiang, X., Liu, Y., Zhou, Z., & Li, S. (2021). Feasibility study of wave-motion milling of carbon fiber reinforced plastic holes. *International Journal of Extreme Manufacturing*, *3*(1), 010401. doi:10.1088/2631-7990/abc675

Zhang, Y. D., Xiao, H. H., & Teng, Y. (2013). Study of surface roughness in milling carbon fiber reinforced plastics using PCD tool. *Key Engineering Materials*, *531-532*, 63–66. . doi:10.4028/www.scientific.net/KEM.531-532.63

Chapter 4
Fiber-Reinforced Polymer Machining:
A Comprehensive Scientific Exegesis

Issam Hanafi
https://orcid.org/0000-0001-8765-2631
ENSAH, Université Abdelmalek Essaâdi, Morocco

Wanderson de Oliveira Leite
https://orcid.org/0000-0002-5723-8525
Instituto Federal de Minas Gerais, Brazil

Francisco Mata Cabrera
Universidad de Castilla-La Mancha, Spain

Juan Carlos Campos Rubio
https://orcid.org/0000-0001-6187-6442
Universidade Federal de Minas Gerais, Brazil

ABSTRACT

Fiber-reinforced polymer (FRP) matrices have gained widespread traction across diverse industries for their exceptional attributes and lightweight nature, outperforming traditional materials and alternative composites. Consequently, the study of machining these composite matrices has gained significant attention. This review aims to establish robust physical and statistical models, opening avenues for optimizing the machining of polymer matrix composites. Encompassing various machining aspects, including crucial variables and criteria, the synthesis notably focuses on refining surface finish and cutting force dynamics. Curated from acclaimed scientific literature, this review provides an exhaustive overview of the prevailing landscape and intricate nuances inherent to FRP machining. The gleaned insights are poised to steer future empirical endeavors and refine precision cutting models.

DOI: 10.4018/978-1-6684-9927-6.ch004

INTRODUCTION

Machining is a production method wherein a cutting tool is utilized to eliminate surplus material to obtain the specified shape and dimensions of a mechanical part (Jawahir & Balaji, 2018). Turning is a highly prevalent machining process extensively applied in the industrial sector (Sousa & Silva, 2020) (Fernando et al., 2022). To produce a wide variety of mechanical components that meet diverse needs in the manufacturing field. Frequently, surfaces that function under mechanical contact conditions for various tribological applications are obtained using this process (G. Petropoulos & Pandazaras, 2003). Currently, the turning process is no longer limited to metallic materials and is also used to work with organic materials (Chaubey & Gupta, 2022) (Breidenstein et al., 2022). The interest in machining techniques for organic materials has significantly increased. The majority of research in this field focuses primarily on characterizing the turning process to control its application on an industrial manufacturing scale (Rance et al., 2019) (Singh et al., 2021). The objective is to obtain mechanical components with precisely controlled dimensional characteristics at the lowest possible cost. Among the organic materials processed by turning, we find Carbon and Glass Fiber Reinforced (CFRPs/GFRPs). The special conditions of using these composites require particular precautions during their manufacturing process. It should be noted that turning composite materials, consisting of a polymeric matrix containing reinforcements, is very different from turning metals and alloys (Bhatnagar et al., 1995) (Caggiano, 2018a) (Henry Ononiwu et al., 2021).

The theoretical and experimental knowledge acquired in the field of metals does not directly apply to these types of composite materials because they are generally heterogeneous and anisotropic. This is due, for example, to the fact that they are prepared in stratified form or extruded before considering machining (Somireddy & Czekanski, 2020). Composite materials consist of two phases with very dissimilar thermal and mechanical properties. Complex interactions occur between the reinforcement and the matrix during the turning manufacturing process, which significantly affects the machining characteristics of these materials compared to those composed of a single phase, such as metals (Voss & Friedrich, 1987) (Cabrera et al., 2010) (Rahman, Ramakrishna, Prakash, et al., 1999) (Rahman, Ramakrishna, & Thoo, 1999) (Sabet et al., 2018) (Mohamed et al., 2019).

The intricate interplay of composite material machinability inherently hinges upon the nuanced interplay of constituent properties. The disposition of fibers and their prescribed orientation, the proportional volumetric contribution of reinforcement, and the distinctive attributes intrinsic to the matrix material collectively conspire to orchestrate this behavior. In the crucible of the machining endeavor, the cutting implement navigates the complex terrain encompassing both fibers and matrix constituents, each distinct in their receptive retort to the imposed machining stimuli.

The concomitant genesis of chip formation emerges as a convoluted tapestry interwoven with shear forces, fractural dynamics, or the harmonious interlacing of these modalities, contingent upon the intricate contours of the tool geometry and the prevailing orientation of the fibers. Of marked significance, the presence of fibers, intrinsically endowed with abrasive proclivities, bears the potential to prematurely exacerbate tool attrition, therein underscoring the imperativeness of pivotally pivoting toward the deployment of polycrystalline diamond (PCD) tools (Uhlmann et al., 2016) (Xu & El Mansori, 2017).

Due to the matrix's lower thermal conductivity compared to inorganic materials and other metals, heating of the workpiece can occur during machining. This phenomenon can lead to dimensional instability and promote the occurrence of harmful vibrations. Additionally, it is essential to take into account the presence of a lubricant as it can be absorbed by the matrix, causing a local variation in properties, which could lead to instabilities (Mittal et al., 2017). For this reason, it is often recommended not to use any lubricant during the turning of organic materials.

In the realm of contemporary manufacturing, particularly aligned with the prevalent thrust toward miniaturization, the precision machining of diminutive components possessing cylindrical geometries assumes paramount significance, particularly in the context of machine and robotic componentry fabrication. Lu and Yoneyama have etched a significant imprint within this landscape, delineating a micro-turning framework that resides as an exemplar of meticulous methodology. A cornerstone of this endeavor lay in the critical calibration of cutting tools, a strategic endeavor essential to orchestrate the seamless execution of the micro-turning process (Lu & Yoneyama, 1999) (Shastri et al., 2021) (Selvakumar et al., 2021). Picard et al. proffer a pioneering paradigm, introducing an innovative regime for fabricating diminutive cutting implements through the ingenious medium of "sputtering" technology. Their astute methodology culminated in the realization of cutting tools spanning high-speed steel, tungsten carbide, and diamond crystalline substrates. These intricate tools, distinguished by diverse configurations and geometries, underwent rigorous testing, subsequently unveiling auspicious outcomes within the microcosm of the micro-cylindrical turning process (Picard et al., 2003).

Polymeric matrix composite materials exhibit a manifold spectrum of failure modes, encompasing phenomena such as fiber fracture, matrix plastic deformation, and delamination, among others (Spencer, 1972). Research has shown that surface texture and manufacturing defects influence static and dynamic service loads and affect fatigue strength (Arola & Ramulu, 1997). Hence, meticulous attention necessitates allocation towards the meticulous curation of machined surfaces and their stringent regulation, poised as an avenue to amplify the efficacy of components forged through the lathe-turning process. Within the pantheon of scholarly endeavors orbiting this realm, a notable nod pertains to inquiries that strive to prognosticate cutting forces by dint of scrutinizing the prevailing failure modalities encountered amidst the intricate matrix of chip formation dynamics (Caprino & Tagliaferri, 1995) (Mathew et al., 1999a) (Caprino et al., 1998) (Mathew et al., 1999b).

2. TURNING OF COMPOSITES

The characteristics of materials have a considerable influence on the conditions encountered during the machining operation. Many factors that affect material machinability, particularly those related to material properties, parameters, and cutting tools. Machinability is a term used to describe how easily a material can be shaped, cut, or otherwise worked during the machining process. It considers various factors, including cutting forces, the quality of the finished surface, and tool wear, to assess how efficiently the material can be machined (Groover, 2012) (Ullen et al., 2020). The relative ability of a material to be machined is usually expressed through a coefficient known as the machining index. This index serves as a scale to control the machining process. It is established by comparing different types of materials to identify the degree of ease that describes the machining conditions of a given material.

While conventional materials have a universally accepted machining index, there is no universally recognized criterion for reinforced plastic materials. As this research work suggests, it is necessary to delve into the analysis of turning mechanisms specific to composite materials in order to establish rules and criteria that will contribute to creating such widely anticipated references in the field.

Various research studies have addressed the machining of thermoset plastic materials when reinforced with fibers. An et al. investigated the cutting process characteristics (An et al., 1997) (Chegdani & Mansori, 2018) (Choi et al., 2021), paying particular attention to the influence of material and tool geometry. Ferreira et al. analyzed different aspects related to the machining of composites reinforced with carbon fibers (Ferreira et al., 2001). Hocheng and Tsao conducted series of tests on polymer matrix composite materials reinforced with carbon fibers. By observing chip characteristics, they assessed the machinability of these materials based on fiber direction and cutting parameters. They found, in particular, the cutting force component parallel to the fibers was found to be smaller than the force component perpendicular to them. (Hocheng & Tsao, 2005) (Hocheng & Tsao, 2006), and these various results were established using the Koplev model (Koplev et al., 1983). Chang conducted a study on machining fiberglass-reinforced plastic materials using cemented carbide tools and developed models to predict cutting forces (Chang, 2006).

3. COMPOSITES CUTTING MODELS

The inaugural foray into theoretical inquiry pertaining to the machination of FRP materials was undertaken by Everstine and Rogers (Everstine & Rogers, 1971). They established a prognostic model aimed at delineating the minimal threshold of cutting force, specifically when fibers are meticulously aligned in a parallel configuration, denoted by a precise 0° angular orientation.

Nearly 10 years later, Koplev conducted tests with a cutting configuration ranging from the parallel case to the orthogonal case with respect to the direction of the reinforcing fibers (Koplev, 1980). In his scientific undertaking, he delved into the realm of CFRP subject to the ministrations of single-edge machining implements. Via a comprehensive dissection of the intricate interplay underpinning chip generation and the post-machining topographical disposition, he unlocked a revelatory vista. This discernment illuminated an unequivocal correlation between the resultant surface texture, as sculpted by the machining procedure, and the inherent orientation of the embedded fibers. Notably, an exquisite synergy was discerned—wherein a parallel cutting trajectory synchronized with the fiber alignment elicited conspicuously smoother surface topographies. Conversely, as the trajectory of machining converged towards the orthogonal orientation vis-à-vis the fiber arrangement, a progressively augmented surface roughness tangibly ensued.

Koplev, in his scholarly discourse, further accentuated the salient verity that the frictional forces arising from the interplay between the tool and fiber entities stood as the catalytic agent governing the tensile rupture of the fibers. Subsequently, in a subsequent scholarly exploration, Koplev and his collaborators embarked upon a rigorous expedition, unraveling the intricate web interlinking tool geometry, the nuanced mechanics of chip generation, and the consequential forces orchestrating the cutting endeavor (Koplev et al., 1983). Their empirical scrutiny unveiled a discernible correspondence, where the principal cutting force exhibited a direct correlation with the depth of incision while demonstrating a marginal attenuation with an augmented rake angle. In the spectrum of rake angles spanning from 3 to 15 degrees, a conspicuous diminishment in the feed force was noted, a phenomenon which they explicatively rationalized through a framework tethered to force dynamics contingent upon the juxtaposed surface of contact between the tool and workpiece.

Furthermore, their meticulous inquiry substantiated the invariance of the cutting force in the face of tool wear, unveiling an intriguing facet—any surges in cutting force could be attributed to the amplification of frictional forces emergent amidst the interplay of workpiece and tool interfaces. This intricate interaction was, in turn, incited by perturbations in the feed force, thus threading a complex tapestry interlinking force dynamics, contact mechanics, and material interactions.

Others researchers conducted unidirectional turning tests on GFRP tubes to study the effects of tool wear on cutting forces and glass fiber orientation. They observed chip formation progressions comparable to those observed by Koplev. They also report that for minor angles defining the fiber orientation, the fibers were first bent under the tool's progress before breaking due to tension. For large angles defining the fiber orientation, the fibers were separated from the cutting surface and fractured by shear. Sakuma and Seto explained the variation in cutting forces with fiber orientation by the fact that the tangential strength of glass fibers is smaller than the normal strength (Sakuma & Seto, 1983).

Takeyama and Iijima conducted research on the chip formation process during turning of fiberglass-reinforced plastic materials at different fiber orientation angles. They put forward the idea that a predictive model for cutting forces according to Merchant's minimum energy theory. Their findings revealed that at an orientation angle of 0 degrees, fracture occurred in the fiber direction, whereas at an angle of 90 degrees, chip fracture was observed (Takeyama & Iijima, 1988). Conversely, Arola and Ramulu embarked upon a pioneering trajectory, engineering a two-dimensional finite element analysis paradigm meticulously calibrated to emulate the intricate dynamics underpinning chip formation within the realm of unidirectional composite materials (Arola & Ramulu, 1997). They employed a serious traction condition based on the strength of the constituents of the composite material to simulate chip separation. The model's accuracy was confirmed through experimental cutting force measurements, which showed good agreement with the model's predicted cutting force values. However, challenges in defining the fracture plane led to inaccurate predictions of feed force values.

Alternate academic minds posited the inception of a Finite Element (FE) based framework, sculpted around the crucible of anisotropic plasticity theory, tailored for the emulation of unidirectional Fiber-Reinforced Polymers (FRPs) machining dynamics. This innovative approach, marked by conceptual ingenuity, was underpinned by the depiction of the cutting tool as possessing pristine elastic attributes, juxtaposed against the homogeneous fabric of the solid material. Intrinsically woven within this computational edifice, they adroitly harnessed empirically ascertained physical and mechanical parameters to fashion a model that meticulously encapsulated the intricate tableau of chip formation dynamics across an array of composite material substrates. This ambit seamlessly enfolded diverse constituents, including Glass Fiber-Reinforced Plastic materials (GFRPs) distinguished by fiber orientations spanning the cardinal compass points at 0º, 45º, 90º, and 135º (Ramesh et al., 1998). They observed variations in cutting forces based on the fiber orientation angle, similar to those obtained by Bhatnagar (Bhatnagar et al., 1995). They also found that matrix shearing was the predominant cause of fracture. Ribeiro and Coppini proposed an acquisition system to optimize the cutting process and production costs. This system integrates the material, cutting tool, machine tool, and cutting conditions used (Ribeiro & Coppini, 1999). Mahdi and Zhang proposed a 3D adaptive FE algorithm that modeled the fiber and matrix as a composite cell (Mahdi & Zhang, 2001b). They also presented a one to 2D cutting model to forecast the behavior of cutting forces in composite materials with respect to the fiber orientation angle (Mahdi & Zhang, 2001a). Selecting the appropriate cutting tool and parameters can significantly reduce costs. The effect of tool geometry and parameters was evaluated by Wang and Zhang (D. H. Wang et al., 1995a), (D. H. Wang et al., 1995b), analyzing chip formation, surface roughness, and cutting forces. All these aspects are directly influenced by the fiber orientation in the material.

Experimental studies on machining carbon fiber-reinforced plastic materials conducted by various researchers demonstrate that chip formation strongly depends on fiber orientation (D. H. Wang et al., 1995a) (D. H. Wang et al., 1995b) (KANEEDA, 1989). Besides fiber orientation, the material and cutting tool geometry also have a significant influence on chip formation, machined surfaces, tool wear, and cutting forces (Cabrera et al., 2010) (Zweben, 2015). Ramulu et al. conducted studies characterizing the orthogonal cutting of GFRP and analyzed the effects of tool geometry, cutting parameters, fiber orientation and surface quality (Ramulu et al., 2003) (Zweben, 2015).

4. PARAMETERS AFFECTING THE MACHINABILITY OF COMPOSITES

4.1. Material Properties

Preeminent amongst the material's mechanical attributes that wield a commanding influence upon the domain of machining are its intrinsic hardness and robustness (Zheng & Liu, 2013) (El-Sonbaty et al., 2004). Increasing hardness leads to higher abrasive wear and reduced tool life. On the other hand, increasing material strength results in higher cutting forces, cutting temperature, and specific energy, making the material more difficult to machine.

The complex interplay of thermal and physical attributes within the fibers and polymer matrix evinces discernible disparities. The resultant composite characteristics emerge as contingent entities, intricately woven within the fabric of fiber typology, the proportion of fiber content, the nuanced alignment of fibers, and the inherent heterogeneity intrinsic to the matrix medium (Ferreira et al., 1999). Hence, these parameters can significantly influence the tribological behavior of these materials. Reinforced plastic materials exhibit extremely abrasive behavior during chip formation processes. The pronounced mechanical potency intrinsic to the fibers augments the propensity for undue detriment inflicted upon the polymer matrix, concurrently eliciting wear effects upon the employed cutting tool. Within this overarching context, Rahman and co-researchers undertook an insightful inquiry into the realm of Polyetheretherketone (PEEK) composites fortified with carbon fibers. Through discerning experimentation, they delineated an intriguing tapestry wherein the eventual manifestation of a sleek and polished surface morphology in the resultant machined artifact was indisputably ascribed to a symbiotic interplay between the attributes of the polymer matrix itself and the intricate geometrical attributes characterizing the applied cutting tool (Rahman, Ramakrishna, & Thoo, 1999).

4.2. Functional Cutting Parameters

When referring to cutting functional parameters, it generally includes the selection of depth of cut, feed rate, and cutting speed. It is likewise important to consider whether or not to use and appropriately select a lubricant, if necessary. Several works have been devoted to the end results of cutting parameters on turning operations (El-Sonbaty et al., 2004) (Yang & Tarng, 1998) (Varatharajan et al., 2006). These studies have confirmed that there is no universal reference for composite materials. Each material, based on the type and percentage of reinforcing fibers, has its own suitable cutting speed and feed rate values for achieving optimal surface finish and reduced tool wear.

Therefore, it is necessary to design new experiments to ensure satisfactory results that can qualify the machining process (Eriksen, 1999), particularly the experimental relationships that allow correlating cutting parameters with final dimensional characteristics, as these are of central interest to the industry (Lee et al., 2000).

Using a lubricant during the turning process can lead to its absorption by the polymer matrix and result in dimensional changes of the parts and their mechanical properties. Hence, it is preferable to carry out the machining process without coolant. This situation has sparked renewed research activity with significant practical interest to determine whether dry or wet machining is more appropriate (Sreejith et al., 2000). Sardinas et al., using genetic algorithms, presented an optimization study of cutting parameters to optimally machine laminated composites (Sardiñas et al., 2006).

4.3. Cutting Tool

The predominant choices for cutting implements within the domain of machining polymer matrix composite materials embrace the utilization of cemented carbides (K10) and polycrystalline diamond (PCD). An extensive array of research inquiries focusing on the machining of Fiber-Reinforced Polymers (FRPs) has unequivocally converged upon a consensus that PCD tools furnish optimal outcomes within this context. Nevertheless, the conspicuous caveat tethered to PCD tools resides in their elevated cost, oftentimes escalating up to a magnitude tenfold in comparison to their carbide counterparts. Notably, on the crest of recent advancements, diamond-coated tools have heralded a compelling and pragmatic alternative to PCD tools, engendering a landscape of viable solutions.

The machining trajectory intrinsic to fiber-reinforced plastic materials is confronted by intricate challenges, predominantly germinating from the specter of delamination within the composite matrix and the fleeting trajectory characterizing tool longevity (Sardiñas et al., 2006). This necessitates the use of tools made from high wear-resistant materials with high toughness and hot hardness (cemented carbides, polycrystalline diamond, and diamond-coated tools) (Groover, 2012).

The choice of cutting tools and parameters is crucial in the machining process (An et al., 1997) (Benardos & Vosniakos, 2003). The tool material plays a vital role in generating cutting forces (machining force, specific cutting force, cutting power) as it significantly influences the surface finish of the machined workpiece. Cemented carbides are formulated from tungsten carbide and are manufactured using powder metallurgy techniques with cobalt as a binder (Groover, 2012).

The confluence of prodigious thermal conductivity, diminished coefficient of friction, and superlative hardness endow diamond with an exquisite material profile, rendering it an impeccable candidate for cutting tool fabrication. Diamond, revered for these attributes, has garnered extensive application both through its instantiation as polycrystalline diamond inserts adorning cemented carbide tools, and in its unadulterated natural manifestation (Reineck et al., 1996).

Polycrystalline diamond is manufactured by sintering fine dust of granulated diamond crystals at high temperatures and pressures. The use of polycrystalline diamond (PCD) tools has seen significant growth due to their extended tool life and superior surface finish compared to PCD (Bai et al., 2004)

Chambers and Bishop conducted tests on the machining of PEEK/Carbon and Epoxy/Carbon composites using cemented carbide and PCD tools. They inferred that PCD tools provide the best machining quality (Chambers & Bishop, 1995). Another study by Jahanmir supports this result, indicating that PCD cutting tools consistently achieve better surface quality (Jahanmir, s. d.).

The synthesis of diamond at low pressures through Chemical Vapor Deposition (CVD) allows diamond to be used as deposited layers on cutting tools for machining applications (Belmonte et al., 2004) (Faure et al., 1999) (Sheikh-Ahmad et al., 2003). The CVD diamond is synthesized by introducing hydrocarbon gas into a hydrogen environment at temperatures approximately reaching 2000°C, resulting in the deposition of diamond onto a suitable substrate Turning of thin CVD diamond coated Si3N4 and CFRC composites using Si3N4 tools (Bai et al., 2004) and (Olsen et al., 2004). The characteristics of CVD diamond are significantly impacted by its grain structure, consequently, are influenced by the specific synthesis settings (S et al., 1998). There are challenges related to the adhesion of the diamond layer to the substrate and differences in thermal expansion between diamond and substrate materials (Kalpakjian & Schmid, 2002). CVD tool producers have devised diverse techniques for depositing diamond crystals with varying sizes and thicknesses, allowing for the fabrication of diamond coatings ranging from microns to several millimeters, tailored to specific application requirements (S et al., 1998).

CVD diamond implements manifest remarkable attributes, encompassing a diminished coefficient of friction, commendable thermal conductivity, and elevated hardness under escalated thermal conditions (6000-9000 HV), and excellent chemical stability (S et al., 1998). Owing to these superior chemical and thermal properties compared to PCD, CVD diamond tools have a lower risk of fracturing, making them well-suited for finishing operations on modern non-metallic materials and non-ferrous alloys (Chou & Liu, 2005). Moreover, when dealing with composite materials featuring a polymer matrix, PCD diamond proves beneficial due to its remarkable hardness, enabling it to resist abrasive wear effectively, while its smooth surface demonstrates reduced adhesion to polymers (Köpf et al., 2006). Composite material machining involves exceptional adhesion of the diamond coating. If the polymeric material sticks to the tool edge, cutting forces increase, leading to additional stresses in the coating. This results in increased roughness and may even cause delamination of the coating.

Numerous research works have been documented, focusing on the accomplishment of diamond coated cutting tools in various turning operations, such as those involving aluminum alloys, graphite, metal matrix composites, and FRPs. Reineck et al. carried out an investigation to assess the impact of diamond-coated cutting tools, produced using the High Current DC-Arc technique, on diverse machining configurations. They conducted tests on specimens made from different materials, spanning from metal matrix composites to Al-Si alloys and glass fiber-reinforced plastics. These authors demonstrated that the diamond-coated tools exhibited superior performance, characterized by high wear resistance and excellent surface quality, compared to conventional uncoated carbide tools (Reineck et al., 1996).

Moreover, an intriguing experimental investigation was conducted by Davim on the machining of metal matrix composite materials (A356/SiC/20p). The author compared the efficacy of polycrystalline diamond tools with that of diamond-coated tools produced via chemical vapor deposition. The comparisons centered on tool life, cutting forces, surface roughness and the PCD tools demonstrated superior performance (Davim, 2002).

Olsen et al. delved into the mechanical, physical and microstructure properties of thick diamond films produced through CVD for use as coatings on cutting tools. The incorporation of electrically conductive diamond within the CVD coating holds the potential to overcome certain manufacturing limitations. This, in turn, could lead to wider adoption in the market and the possibility of replacing PCD (S et al., 1998).

Turchetta et al. examined the end outcome of diamond coated tools formed by CVD and compared them with traditional tungsten carbide tools. Their findings revealed that the CVD-coated tools significantly reduced cutting force and energy consumption, enabling higher cutting speeds without increasing tool stresses (Turchetta et al., 2005).

Arumugam et al. performed an investigation on the comportment of diamond coated tools produced by CVD in the dry turning of aluminum alloy composites. They compared polished CVD-coated carbide tools with non-polished ones. The polished CVD-coated tools exhibited better performance in terms of improved surface finish of the component (polishing reduced cutting forces and enhanced tool life) and reduced tool wear (Arumugam et al., 2006).

Fukui et al. conducted a similar study to investigate the behavior of Diamond-Like Carbon coated tools in the dry machining of aluminum alloys (Fukui et al., 2004).

Other researchers carried out a work on the machinability of diamond coated tools produced using the Time-Modulated CVD in graphite turning and compared the outcomes with those of commercially available polycrystalline diamond (PCD) tools. Their findings showcased that TMCVD tools exhibited higher wear resistance compared to tools coated through conventional techniques or commercially available PCD tools. The fine particle size of the diamond coating led to reduced surface roughness and lower residual stresses, resulting in enhanced performance for TMCVD when compared to conventional diamond and PCD tools (Cabral et al., 2006). Köpf et al. performed a study on the application of diamond coated tools for machining purposes in fiber-reinforced plastics, aluminum alloys, and graphite. Their findings indicated that using diamond-coated tools can yield satisfactory and promising performance, making them a viable option in the marketplace (Köpf et al., 2006). Diamond coated tools produced by CVD, can undoubtedly serve as a cost-effective substitute for PCD tools in machining advanced materials (Chou & Liu, 2005).

5. MACHINING CRITERIA

An array of diverse criteria have been postulated to delineate the evaluation ambit within composite material machining. Notwithstanding, the quest for a globally ratified machining metric remains unfulfilled. Amidst this continuum, specific cutting force and surface roughness emerge as pivotal touchstones, wielding profound significance in the critical appraisal of composite material machining endeavors (Benardos & Vosniakos, 2003). Surface roughness is ubiquitously employed as a hallmark of product quality, routinely harnessed as a technical benchmark for mechanical components. In parallel, the specific cutting force stands as a discerning gauge, casting insight into the intrinsic efficacy of the machining process. The pursuit of attenuated specific cutting force mandates the utilization of polycrystalline diamond (PCD) tools.

While the construct of machinability conventionally coalesces with the intrinsic attributes of the work material, it is imperative to underscore the pivotal role etched by functional cutting parameters, the chosen tools, and the dynamics of machining operation rates. These factors reverberate as crucial determinants collectively wielding substantial sway upon the machining enterprise (Groover, 2020).

5.1. Force and Power Cutting

The inherent importance of cutting force magnitudes resides in their profound influence on tool longevity. These forces inherently exert a dominion over the mechanical attributes of the machinated material, particularly its intrinsic hardness. In scenarios involving composites, these forces intricately hinge upon multifarious factors, including the volumetric composition of fibers, their specific typology, orientation vectors, and the nuanced attributes characterizing the cutting instrument itself—encompassing material composition, geometric configuration, and prevailing tool health.

Parallelly, the facet of cutting power conveys indispensable insights into the quantum of energy requisitioned during the machining endeavor, an expenditure necessitated for the purpose of material elimination and the concomitant attainment of prescribed finishing criteria—entailing surface topology refinement, dimensional fidelity, and tolerance criteria compliance. Ergo, the imperative to curtail cutting power intricately interlinks with the objective of diminishing cutting forces, thereby mandating systematic modulations in operational cutting parameters and the underlying tool instantiation. This orchestrated drive towards diminished energy consumption invariably culminates in concomitant reduction of production overheads.

5.2. Specific Cutting Force

The parameter of specific cutting force stands forth as an additional discerning metric, derived from the underpinning cutting forces and proffering substantive utility in the appraisal of machining process efficacy. In a notable investigative venture, Sreejith and collaborators embarked upon an empirical exploration into the impact of specific cutting force within the ambit of polymer matrix composite material machining, employing tools fashioned from cemented carbide substrates.

Concurrently, the study harnessed the manifestation of acoustic emission signals engendered during the machining procedures, affording revelatory insights. Amidst their meticulous scrutiny, a pivotal juncture of machining velocity, designated as the "critical machining speed," emerged resplendently. Furthermore, the inquiry unfurled an elucidatory vista concerning the envelope of temperatures wherein the specific cutting pressure steadfastly maintained its equilibrium, thus augmenting the comprehension of thermo-mechanical interplays (Sreejith et al., 2007).

5.3. Tool Wear

Discernible disparages emerge in tool longevity during the process of machining thermoplastic materials. The vital determinant governing this metric resides in the nature of the selected cutting implement, with distinct choices encompassing cemented carbides and polycrystalline diamonds (PCD). Yet, the scientific consensus predominantly resonates around the premise that PCD tools, by and large, epitomize the zenith in prolonged tool lifespan (Jahanmir, s. d.). Tool wear is used as a machining criterion for metals, metal matrix composite materials, and ceramics (Ravindra et al., 1993b) (Ravindra et al., 1993a) (Bonifacio & Diniz, 1994) (Dolinšek & Kopač, 1999). In the case of polymer matrix composite materials, especially polycrystalline diamond tools, determining the tool life requires prolonged testing where significant material removal must occur to observe substantial wear.

In the case of tools with diamond coating or cemented carbides, which are materials with high wear resistance, conducting tests capable of accurately revealing the tool life is challenging. However, wear is easier to detect when using other types of conventional tools.

The distinctive wear patterns arising from the intricate process of machining polymer matrix composite materials can be succinctly characterized as the confluence of two fundamental mechanisms. The abrasive wear component emanates from the fibers, notable for their appreciable hardness, inducing abrasion upon tool surfaces. In parallel, the plastic matrix imparts a complementary facet through its remarkable adhesion attributes, thus engendering the potential for delamination events, a phenomenon that precipitates both the diminution of cutting tool edges and a transformative influence upon their integrity (Köpf et al., 2006).

5.4. Surface Finish

Now, there is an increasing growing need for components manufactured from fiber-reinforced plastics, and increased requirements are also being imposed on the desired dimensional accuracy and surface finish (Caggiano, 2018b), (Parida et al., 2014), (Palanikumar, 2006). Surface finish is a very useful parameter in practice for characterizing machined materials. Surface roughness stands as a pivotal parameter wielding substantial dominion over the operational comportment and efficacy of mechanical components in active service, concurrently exerting a tangible impact upon the economics of production (Abouelatta & Mádl, 2001) (Cai et al., 2021). It is, therefore, a fundamental parameter in quality control. Surface roughness is critical in mechanical contacts, fluid flow, and semiconductor applications.

In most applications today, impeccable dimensional accuracy, high quality machined surfaces, and optimal surface integrity are demanded. As a result, numerous research investigations have been carried out to optimize cutting parameters to attain stated surface finish (Eriksen, 1999), (Abouelatta & Mádl, 2001). These studies have demonstrated that the profile and roughness of machined surfaces depend largely on the type, fiber orientation, and measuring direction adopted for roughness measurement (Jahanmir, s. d.) (Ramulu et al., 1993) (Cenna & Mathew, 1997). The profile texture of machined surfaces is more susceptible to variations in fiber winding angle compared to variations in tool feed rate (Caggiano, 2018b).

To achieve the chosen surface finish, it is vital to comprehend the cutting and the kinetics and material removal mechanisms that affect the behavior of cutting tools (Sreejith et al., 2000). Generally, tool loss occurs through progressive wear or fracture. The degree of tool wear affects the quality of machined surfaces. Jahanmir et al. (Jahanmir, s. d.), Spur and Wunsch (Caggiano, 2018b) studied the turning process of FRP (GFRP with polyester and epoxy resin matrix) and found that roughness increases with increasing feed rate and is not dependent on cutting speed. Conversely, Santhanakrishman et al., Ramulu et al. achieved higher surface quality by raising the cutting speed. In all cases, this remains an area with many questions that need to be addressed (Ramulu, 1997) (Santhanakrishnan et al., 1988).

Kopac and Bahor conducted a research project in which they investigated the impact of machining parameters on surface roughness. To analyze this relationship in metal turning, they employed an DOE and ANOVA. Furthermore, this experimental approach has been adopted by other researchers for studying the machining of FRP (Kopač & Bahor, 1999). Bernardos and Vosniakos presented various practical methodologies they used to predict surface roughness (experimental design, IA, etc.). They considered the effect of machining parameters, cutting tool properties, and workpiece characteristics on surface outcome (Benardos & Vosniakos, 2003). Several studies have been carried out in the field of turning FRPs using polycrystalline diamond tools. These studies have led to the conclusion surface finish rises with an upsurge in feed rate and declines as cutting speed is raised. One particular property of FRPs is their anisotropy. Various authors have inspected how fiber orientation can impact the quality of machined surfaces and its impact on tool wear. In general, when fibers are cut in tension, lower roughness and cutting forces are achieved, although the tool wear ratio increases. Conversely, roughness and cutting forces increase when fibers are cut in shear (Bhatnagar et al., 1995), (Takeyama & Iijima, 1988).

Eriksen undertaken research on the outcome of production parameters on surface finish in the machining of thermoplastics reinforced with short fibers and found that the effect of fiber orientation is practically insignificant on roughness (Eriksen, 1999). However, Wang and Zhang conducted a study through experimentation using orthogonal cutting of unidirectional FRP and determined that subsurface layer defects were present, cutting forces, and surface roughness undergo substantial alterations with fiber orientation. The fiber orientation angle is the main reason that controls the surface outcome of the machined components. The 90° is the most relevant angle beyond which severe damage occurs beneath the machined surface (X. M. Wang & Zhang, 2003).

Chambers and Bishop conducted research on the machining of PEEK/Carbon composite and FRPs employing alternative cutting instruments, such as cemented carbides or PCD. They determined that PCD cutting tools exhibit the most exceptional overall results (Bai et al., 2004).

Davim and Reis conducted research on the influence of cutting parameters (feed rates and cutting speeds) on dimensional accuracy and surface finish in the turning of FRP tubes. They used analysis of variance and orthogonal arrays to examine the cutting behavior of FRPs with K15 cutting tools. The aim was to create correlations between surface roughness, dimensional accuracy and cutting parameters for FRP components. Depending on the cutting parameters used, they were able to accomplish surfaces with adequate dimensional accuracy for mechanical construction, IT 9, and 10 (G. P. Petropoulos et al., 2006).

Some researchers evaluated the behavior of polycrystalline diamond cutting tools during the machining process and analyzed the influence of the material on the final state of machined texture. They detected a serious temperature range from 300-350°C and speed of 300 m/min. On the basis of on tool wear tests, they determined that PCD tools can be employed cost-effectively in the turning process of FRPs, as they have a longer tool life and provide better surface finish on the workpieces (Sreejith et al., 2007).

The multi-factor experimental analysis allowed the development of maps for controlling turned surfaces in the case of metals (G. P. Petropoulos et al., 2006). In other studies, various techniques are applied to predict surface roughness based on cutting forces, using genetic algorithms or neural networks (Lin et al., 2020) (Brezocnik et al., 2004) (Hanafi, Khamlichi, Cabrera, Nuñez López, et al., 2012) (Hanafi et al., 2013) (Hanafi, Khamlichi, Cabrera, & Nuñez López, 2012). Similarly, these methodologies can be employed in FRP to inspect the impact of cutting parameters on surface roughness and forecast specific values under predefined operational conditions.

CONCLUSION

Polymeric matrix-based composites have earned distinction owing to their exceptional performance attributes and inherent lightweight superiority, thereby demarcating them from conventional materials as well as alternative composite formulations. This distinctive assemblage of traits has propelled their pervasive integration within a myriad of industrial domains, instigating a fervent surge in research endeavors. Predominantly, this thrust has been channeled into deciphering their machinability, forming the pivotal pursuit of the current investigation.

In alignment with this research aim, the present endeavor undertakes an expansive survey of extant literature, engendering a compendium that encapsulates both evolved cutting paradigms and diverse technical intricacies germane to the art of machining polymeric matrix composites. Within this purview, a meticulous analysis unfurls concerning the manifold factors that wield dominion over machinability determinants, underscored by a focal spotlight on salient aspects such as surface finish nuances and the intricate orchestration of cutting forces.

Amidst the expansive canvas of cutting tool selection, empirical trials have consistently unveiled the ascendancy of polycrystalline diamond (PCD) tools, invariably yielding outcomes of minimum energy expenditure concomitant with optimal surface refinement. Beyond this, the study casts illumination upon the poignant insight that the optimal manufacturing pathway for these materials veers towards eschewing cooling interventions, an inherent premise that necessarily enkindles concomitant requisites within functional parameter domains.

Furthermore, a discerning imperative surface, advocating the adoption of reduced feed rates coupled with judicious cutting speeds to seamlessly accommodate the multifarious applications encompassed by polymeric matrix-based composites.

Integral to the backdrop of historical investigations in this milieu, the initial trajectory of machinability inquiries predominantly attended to the purview of thermosetting matrix-laden materials, distinguished by the reinforcement of extended fibers rather than their truncated counterparts. As a pivotal tenet of this inquiry, the research imperatives undertaken herein were keenly directed towards proffering insights into the machining dynamics intrinsic to thermoplastic matrix composites, emboldened by the entwined fortitude of short fiber reinforcement. Accordingly, classical tenets akin to the Merchant criterion stood enlisted to formulate cutting models, impeccably aligning with the inherent dynamics of minuscule reinforcement dimensions and the concomitant initiation of plastic deformations during the cutting progression, thereby fostering an enduring milieu of continuous chip formation.

This labyrinthine intellectual pursuit duly paves a pioneering trail, beckoning forthcoming empirical investigations to hone the cutting models via the amalgamation of assimilated insights and the empirical veracity gleaned from this expansive realm of exploration. All these revelations coalesce within a tapestry enriched by an assembly of globally acknowledged scholarly articles, encapsulating the zenith of current scientific thought whilst traversing the realm of fiber-reinforced polymer machining.

REFERENCES

Abouelatta, O. B., & Mádl, J. (2001). Surface roughness prediction based on cutting parameters and tool vibrations in turning operations. *Journal of Materials Processing Technology*, *118*(1), 269–277. doi:10.1016/S0924-0136(01)00959-1

An, S.-O., Lee, E.-S., & Noh, S.-L. (1997). A study on the cutting characteristics of glass fiber reinforced plastics with respect to tool materials and geometries. *Journal of Materials Processing Technology*, *68*(1), 60–67. doi:10.1016/S0924-0136(96)02534-4

Arola, D., & Ramulu, M. (1997). Orthogonal cutting of fiber-reinforced composites : A finite element analysis. *International Journal of Mechanical Sciences*, *39*(5), 597–613. doi:10.1016/S0020-7403(96)00061-6

Arumugam, P. U., Malshe, A. P., & Batzer, S. A. (2006). Dry machining of aluminum–silicon alloy using polished CVD diamond-coated cutting tools inserts. *Surface and Coatings Technology*, *200*(11), 3399–3403. doi:10.1016/j.surfcoat.2005.08.127

Bai, Q. S., Yao, Y. X., Bex, P., & Zhang, G. (2004). Study on wear mechanisms and grain effects of PCD tool in machining laminated flooring. *International Journal of Refractory & Hard Metals*, *22*(2), 111–115. doi:10.1016/j.ijrmhm.2004.01.007

Belmonte, M., Oliveira, F. J., Lanna, M. A., Silva, C. R. M., Corat, E. J., & Silva, R. F. (2004). Turning of CFRC Composites Using Si3N4 and Thin CVD Diamond Coated Si3N4 Tools. *Materials Science Forum*, *455-456*, 609–613. . doi:10.4028/www.scientific.net/MSF.455-456.609

Benardos, P. G., & Vosniakos, G.-C. (2003). Predicting surface roughness in machining : A review. *International Journal of Machine Tools & Manufacture*, *43*(8), 833–844. doi:10.1016/S0890-6955(03)00059-2

Bhatnagar, N., Ramakrishnan, N., Naik, N. K., & Komanduri, R. (1995). On the machining of fiber reinforced plastic (FRP) composite laminates. *International Journal of Machine Tools & Manufacture*, *35*(5), 701–716. doi:10.1016/0890-6955(95)93039-9

Bonifacio, M. E. R., & Diniz, A. E. (1994). Correlating tool wear, tool life, surface roughness and tool vibration in finish turning with coated carbide tools. *Wear*, *173*(1), 137–144. doi:10.1016/0043-1648(94)90266-6

Breidenstein, B., Denkena, B., Bergmann, B., Wolters, P., & Picker, T. (2022). Turning Copper and Aluminum Alloys with Natural Rocks as Cutting Tools. *Materials (Basel)*, *15*(6), 6. Advance online publication. doi:10.3390/ma15062187 PMID:35329638

Brezocnik, M., Kovacic, M., & Ficko, M. (2004). Prediction of surface roughness with genetic programming. *Journal of Materials Processing Technology*, *157-158*, 28–36. doi:10.1016/j.jmatprotec.2004.09.004

Cabral, G., Reis, P., Polini, R., Titus, E., Ali, N., Davim, J. P., & Grácio, J. (2006). Cutting performance of time-modulated chemical vapour deposited diamond coated tool inserts during machining graphite. *Diamond and Related Materials*, *15*(10), 1753–1758. doi:10.1016/j.diamond.2006.03.007

Cabrera, F. M., Hanafi, I., Khamlichi, A., Jabbouri, A., & Bezzazi, M. (2010). Sur l'usinabilité des composites à matrices polymères renforcée par des fibres. *Mechanics & Industry*, *11*(2), 93–103. doi:10.1051/meca/2010021

Caggiano, A. (2018a). Machining of Fibre Reinforced Plastic Composite Materials. *Materials (Basel)*, *11*(3), 3. Advance online publication. doi:10.3390/ma11030442 PMID:29562635

Caggiano, A. (2018b). Machining of Fibre Reinforced Plastic Composite Materials. *Materials (Basel)*, *11*(3), 442. doi:10.3390/ma11030442 PMID:29562635

Cai, C., An, Q., Ming, W., & Chen, M. (2021). Modelling of machined surface topography and anisotropic texture direction considering stochastic tool grinding error and wear in peripheral milling. *Journal of Materials Processing Technology*, *292*, 117065. doi:10.1016/j.jmatprotec.2021.117065

Caprino, G., Santo, L., & Nele, L. (1998). Interpretation of size effect in orthogonal machining of composite materials. Part I : Unidirectional glass-fibre-reinforced plastics. *Composites. Part A, Applied Science and Manufacturing*, *29*(8), 887–892. doi:10.1016/S1359-835X(98)00028-1

Caprino, G., & Tagliaferri, V. (1995). Damage development in drilling glass fibre reinforced plastics. *International Journal of Machine Tools & Manufacture*, *35*(6), 817–829. doi:10.1016/0890-6955(94)00055-O

Cenna, A. A., & Mathew, P. (1997). Evaluation of cut quality of fibre-reinforced plastics—A review. *International Journal of Machine Tools & Manufacture*, *37*(6), 723–736. doi:10.1016/S0890-6955(96)00085-5

Chambers, A., & Bishop, G. (1995). The drilling of carbon fiber polymer matrix composites. *Tenth International Conference on Composite Materials. III. Processing and Manufacturing*, 565-572.

Chang, C.-S. (2006). Turning of glass–fiber reinforced plastics materials with chamfered main cutting edge carbide tools. *Journal of Materials Processing Technology*, *180*(1), 117–129. doi:10.1016/j.jmatprotec.2006.05.011

Chaubey, S. K., & Gupta, K. (2022). A Review on Fabrication of Cylindrical and Rotating Parts by Thermoelectric-Erosion Based Turning Processes. *Metals*, *12*(7), 7. Advance online publication. doi:10.3390/met12071227

Chegdani, F., & Mansori, M. E. (2018). Mechanics of material removal when cutting natural fiber reinforced thermoplastic composites. *Polymer Testing*, *67*, 275–283. doi:10.1016/j.polymertesting.2018.03.016

Choi, I., Lee, S.-J., Shin, D., & Suh, J. (2021). Green Picosecond Laser Machining of Thermoset and Thermoplastic Carbon Fiber Reinforced Polymers. *Micromachines*, *12*(2), 2. Advance online publication. doi:10.3390/mi12020205 PMID:33671185

Chou, Y. K., & Liu, J. (2005). CVD diamond tool performance in metal matrix composite machining. *Surface and Coatings Technology*, *200*(5), 1872–1878. doi:10.1016/j.surfcoat.2005.08.094

Davim, J. P. (2002). Diamond tool performance in machining metal–matrix composites. *Journal of Materials Processing Technology*, *128*(1), 100–105. doi:10.1016/S0924-0136(02)00431-4

Dolinšek, S., & Kopač, J. (1999). Acoustic emission signals for tool wear identification. *Wear*, *225-229*, 295–303. doi:10.1016/S0043-1648(98)00363-9

El-Sonbaty, I., Khashaba, U. A., & Machaly, T. (2004). Factors affecting the machinability of GFR/epoxy composites. *Composite Structures*, *63*(3), 329–338. doi:10.1016/S0263-8223(03)00181-8

Eriksen, E. (1999). Influence from production parameters on the surface roughness of a machined short fibre reinforced thermoplastic. *International Journal of Machine Tools & Manufacture*, *39*(10), 1611–1618. doi:10.1016/S0890-6955(99)00017-6

Everstine, G. C., & Rogers, T. G. (1971). A Theory of Machining of Fiber-Reinforced Materials. *Journal of Composite Materials*, *5*(1), 94–106. doi:10.1177/002199837100500109

Faure, C., Hänni, W., Schmutz, C. J., & Gervanoni, M. (1999). Diamond-coated tools. *Diamond and Related Materials*, *8*(2-5), 830–833. doi:10.1016/S0925-9635(98)00363-X

Fernando, R., Gamage, J., & Karunathilake, H. (2022). Sustainable machining : Environmental performance analysis of turning. *International Journal of Sustainable Engineering*, *15*(1), 15–34. doi:10.1080/19397038.2021.1995524

Ferreira, J. R., Coppini, N. L., & Levy Neto, F. (2001). Characteristics of carbon–carbon composite turning. *Journal of Materials Processing Technology*, *109*(1), 65–71. doi:10.1016/S0924-0136(00)00776-7

Ferreira, J. R., Coppini, N. L., & Miranda, G. W. A. (1999). Machining optimisation in carbon fibre reinforced composite materials. *Journal of Materials Processing Technology*, *92-93*, 135–140. doi:10.1016/S0924-0136(99)00221-6

Fukui, H., Okida, J., Omori, N., Moriguchi, H., & Tsuda, K. (2004). Cutting performance of DLC coated tools in dry machining aluminum alloys. *Surface and Coatings Technology*, *187*(1), 70–76. doi:10.1016/j.surfcoat.2004.01.014

Groover, M. P. (2012). Fundamentals of Modern Manufacturing : Materials, Processes, and Systems 5th Edition (5th Edition Binder Ready Version). Wiley.

Groover, M. P. (2020). *Fundamentals of Modern Manufacturing : Materials, Processes, and Systems*. John Wiley & Sons.

Hanafi, I., Cabrera, F. M., Khamlichi, A., Garrido, I., & Manzanares, J. T. (2013). Artificial neural networks back propagation algorithm for cutting force components predictions. *Mechanics & Industry*, *14*(6), 6. Advance online publication. doi:10.1051/meca/2013084

Hanafi, I., Khamlichi, A., Cabrera, F. M., & Nuñez López, P. J. (2012). Prediction of surface roughness in turning of PEEK cf30 by using an artificial neural network. *Journal of the Chinese Institute of Industrial Engineers*, *29*(5), 337–347. doi:10.1080/10170669.2012.702690

Hanafi, I., Khamlichi, A., Cabrera, F. M., Nuñez López, P. J., & Jabbouri, A. (2012). Fuzzy rule based predictive model for cutting force in turning of reinforced PEEK composite. *Measurement*, *45*(6), 1424–1435. doi:10.1016/j.measurement.2012.03.022

Henry Ononiwu, N., Akinlabi, E. T., & Ozoegwu, C. G. (2021). Optimization techniques applied to machinability studies for turning aluminium metal matrix composites : A literature review. *Materials Today: Proceedings*, *44*, 1124–1129. doi:10.1016/j.matpr.2020.11.228

Hocheng, H., & Tsao, C. C. (2005). The path towards delamination-free drilling of composite materials. *Journal of Materials Processing Technology*, *167*(2), 251–264. doi:10.1016/j.jmatprotec.2005.06.039

Hocheng, H., & Tsao, C. C. (2006). Effects of special drill bits on drilling-induced delamination of composite materials. *International Journal of Machine Tools & Manufacture*, *46*(12), 1403–1416. doi:10.1016/j.ijmachtools.2005.10.004

Jahanmir, S. (n.d.). Machining of ceramics and composites. https://cir.nii.ac.jp/crid/1130282269159410816

Jawahir, I. S., & Balaji, A. K. (2018). *Machining*. Routledge Handbooks Online. doi:10.1201/9781351045636-140000423

Kalpakjian, S., & Schmid, S. R. (2002). *Manufactura, ingeniería y tecnología*. Pearson Educación.

Kaneeda, T. (1989). CFRP cutting mechanism. *Proc. of 17th North American Manufacturing Research Conf.*, 220.

Kopač, J., & Bahor, M. (1999). Interaction of the technological history of a workpiece material and the machining parameters on the desired quality of the surface roughness of a product. *Journal of Materials Processing Technology*, *92-93*, 381–387. doi:10.1016/S0924-0136(99)00239-3

Köpf, A., Feistritzer, S., & Udier, K. (2006). Diamond coated cutting tools for machining of non-ferrous metals and fibre reinforced polymers. *International Journal of Refractory & Hard Metals*, *24*(5), 354–359. doi:10.1016/j.ijrmhm.2005.11.013

Koplev, A. (1980). Cutting of CFRP with Single Edge Tools : 3rd International Conference of Advances in Composite Materials. *Advanced Composite Materials*, *2*, 1597–1605.

Koplev, A., Lystrup, A., & Vorm, T. (1983). The cutting process, chips, and cutting forces in machining CFRP. *Composites*, *14*(4), 371–376. doi:10.1016/0010-4361(83)90157-X

Lee, B. Y., Tarng, Y. S., & Lii, H. R. (2000). An investigation of modeling of the machining database in turning operations. *Journal of Materials Processing Technology*, *105*(1), 1–6. doi:10.1016/S0924-0136(00)00535-5

Lin, Y.-C., Wu, K.-D., Shih, W.-C., Hsu, P.-K., & Hung, J.-P. (2020). Prediction of Surface Roughness Based on Cutting Parameters and Machining Vibration in End Milling Using Regression Method and Artificial Neural Network. *Applied Sciences (Basel, Switzerland)*, *10*(11), 11. Advance online publication. doi:10.3390/app10113941

Lu, Z., & Yoneyama, T. (1999). Micro cutting in the micro lathe turning system. *International Journal of Machine Tools & Manufacture*, *39*(7), 1171–1183. doi:10.1016/S0890-6955(98)00092-3

Mahdi, M., & Zhang, L. (2001a). A finite element model for the orthogonal cutting of fiber-reinforced composite materials. *Journal of Materials Processing Technology*, *113*(1), 373–377. doi:10.1016/S0924-0136(01)00675-6

Mahdi, M., & Zhang, L. (2001b). An adaptive three-dimensional finite element algorithm for the orthogonal cutting of composite materials. *Journal of Materials Processing Technology*, *113*(1), 368–372. doi:10.1016/S0924-0136(01)00676-8

Mathew, J., Ramakrishnan, N., & Naik, N. K. (1999a). Investigations into the effect of geometry of a trepanning tool on thrust and torque during drilling of GFRP composites. *Journal of Materials Processing Technology*, *91*(1), 1–11. doi:10.1016/S0924-0136(98)00416-6

Mathew, J., Ramakrishnan, N., & Naik, N. K. (1999b). Trepanning on unidirectional composites : Delamination studies. *Composites. Part A, Applied Science and Manufacturing*, *30*(8), 951–959. doi:10.1016/S1359-835X(99)00012-3

Mittal, R. K., Kulkarni, S. S., & Singh, R. K. (2017). Effect of lubrication on machining response and dynamic instability in high-speed micromilling of Ti-6Al-4V. *Journal of Manufacturing Processes*, *28*, 413–421. doi:10.1016/j.jmapro.2017.04.007

Mohamed, S. B., Rashid, R. A., Muhamad, M., & Ismail, J. (2019). Composite Materials and Types of Machining. In *Down Milling Trimming Process Optimization for Carbon Fiber-Reinforced Plastic* (pp. 1–14). Springer. doi:10.1007/978-981-13-1804-7_1

Olsen, R. H., Dewes, R. C., & Aspinwall, D. K. (2004). Machining of electrically conductive CVD diamond tool blanks using EDM. *Journal of Materials Processing Technology*, *149*(1), 627–632. doi:10.1016/j.jmatprotec.2003.11.052

Palanikumar, K. (2006). Cutting Parameters Optimization for Surface Roughness in Machining of GFRP Composites using Taguchi's Method. *Journal of Reinforced Plastics and Composites*, *25*(16), 1739–1751. doi:10.1177/0731684406068445

Parida, A. K., Das, R., Sahoo, A. K., & Routara, B. C. (2014). Optimization of Cutting Parameters for Surface Roughness in Machining of gfrp Composites with Graphite/fly Ash Filler. *Procedia Materials Science*, *6*, 1533–1538. doi:10.1016/j.mspro.2014.07.134

Petropoulos, G., & Pandazaras, C. (2003). Evaluating the real profile length in turning of carbon steels. *Industrial Lubrication and Tribology*, *55*(3), 128–136. doi:10.1108/00368790310470967

Petropoulos, G. P., Pandazaras, C. N., Vaxevanidis, N. M., & Antoniadis, A. (2006). Multi-parameter identification and control of turned surface textures. *International Journal of Advanced Manufacturing Technology*, *29*(1), 118–128. doi:10.1007/s00170-004-2483-y

Picard, Y. N., Adams, D. P., Vasile, M. J., & Ritchey, M. B. (2003). Focused ion beam-shaped microtools for ultra-precision machining of cylindrical components. *Precision Engineering*, *27*(1), 59–69. doi:10.1016/S0141-6359(02)00188-5

Rahman, M., Ramakrishna, S., Prakash, J. R. S., & Tan, D. C. G. (1999). Machinability study of carbon fiber reinforced composite. *Journal of Materials Processing Technology*, *89-90*, 292–297. doi:10.1016/S0924-0136(99)00040-0

Rahman, M., Ramakrishna, S., & Thoo, H. C. (1999). Machinability Study of Carbon/Peek Composites. *Machining Science and Technology*, *3*(1), 49–59. doi:10.1080/10940349908945682

Ramesh, M. V., Seetharamu, K. N., Ganesan, N., & Sivakumar, M. S. (1998). Analysis of machining of FRPs using FEM. *International Journal of Machine Tools & Manufacture*, *38*(12), 1531–1549. doi:10.1016/S0890-6955(98)00008-X

Ramulu, M. (1997). Machining and surface integrity of fibre-reinforced plastic composites. *Sadhana*, *22*(3), 449–472. doi:10.1007/BF02744483

Ramulu, M., Kim, D., & Choi, G. (2003). Frequency analysis and characterization in orthogonal cutting of glass fiber reinforced composites. *Composites. Part A, Applied Science and Manufacturing*, *34*(10), 949–962. doi:10.1016/S1359-835X(03)00203-3

Ramulu, M., Wern, C. W., & Garbini, J. L. (1993). Effect of fibre direction on surface roughness measurements of machined graphite/epoxy composite. *Composites Manufacturing*, *4*(1), 39–51. doi:10.1016/0956-7143(93)90015-Z

Rance, J., Hall, S., de Bartolomeis, A., & Shokrani, A. (2019). Future Direction of the Sustainable Turning of Difficult-to-Machine Materials. *Sustainable Design and Manufacturing*, *2019*, 111–120. doi:10.1007/978-981-13-9271-9_11

Ravindra, H. V., Srinivasa, Y. G., & Krishnamurthy, R. (1993a). Modelling of tool wear based on cutting forces in turning. *Wear*, *169*(1), 25–32. doi:10.1016/0043-1648(93)90387-2

Ravindra, H. V., Srinivasa, Y. G., & Krishnamurthy, R. (1993b). Tool wear monitoring in turning using a pattern recognition technique. *Journal of Materials Processing Technology*, *37*(1), 731–740. doi:10.1016/0924-0136(93)90132-P

Reineck, I., Sjöstrand, M. E., Karner, J., & Pedrazzini, M. (1996). Diamond coated cutting tools. *International Journal of Refractory & Hard Metals*, *14*(1), 187–193. doi:10.1016/0263-4368(96)83433-0

Ribeiro, M. V., & Coppini, N. L. (1999). An applied database system for the optimization of cutting conditions and tool selection. *Journal of Materials Processing Technology*, *92-93*, 371–374. doi:10.1016/S0924-0136(99)00233-2

S, S. R., R, B. J., E, C. S., J, P. C. S., G, S. C., A, W., H, W. C. J., & N, D. C. (1998). CVD diamond : A new engineering material for thermal, dielectric and optical applications. *Industrial Diamond Review, 58*(578), 69-71.

Sabet, F. A., Su, F. Y., McKittrick, J., & Jasiuk, I. (2018). Mechanical Properties of Model Two-Phase Composites with Continuous Compared to Discontinuous Phases. *Advanced Engineering Materials*, *20*(10), 1800505. doi:10.1002/adem.201800505

Sakuma, K., & Seto, M. (1983). Tool wear in cutting glass-fiber-reinforced plastics : The relation between fiber orientation and tool wear. *Bulletin of the JSME*, *26*(218), 1420–1427. doi:10.1299/jsme1958.26.1420

Santhanakrishnan, G., Krishnamurthy, R., & Malhotra, S. K. (1988). Machinability characteristics of fibre reinforced plastics composites. *Journal of Mechanical Working Technology*, *17*, 195–204. doi:10.1016/0378-3804(88)90021-6

Sardiñas, R. Q., Reis, P., & Davim, J. P. (2006). Multi-objective optimization of cutting parameters for drilling laminate composite materials by using genetic algorithms. *Composites Science and Technology*, *66*(15), 3083–3088. doi:10.1016/j.compscitech.2006.05.003

Selvakumar, S., Sreebalaji, V. S., & Ravikumar, K. (2021). Machinability Analysis and Optimization in Micro turning on tool wear for Titanium Alloy. *Materials and Manufacturing Processes*, *36*(7), 792–802. doi:10.1080/10426914.2020.1866198

Shastri, A., Nargundkar, A., & Kulkarni, A. J. (2021). Optimization of Micro-turning Process. In *Socio-Inspired Optimization Methods for Advanced Manufacturing Processes* (pp. 119–126). Springer. doi:10.1007/978-981-15-7797-0_9

Sheikh-Ahmad, J. Y., Stewart, J. S., & Feld, H. (2003). Failure characteristics of diamond-coated carbides in machining wood-based composites. *Wear*, *255*(7), 1433–1437. doi:10.1016/S0043-1648(03)00179-0

Singh, R., Gupta, M. K., Sarikaya, M., Mia, M., & Garcia-Collado, A. (2021). Evaluation of machinability-based sustainability indicators in the eco-benign turning of Ti3Al2.5V alloy with textured tools. *International Journal of Advanced Manufacturing Technology*, *116*(9), 3051–3061. doi:10.1007/s00170-021-07667-2

Somireddy, M., & Czekanski, A. (2020). Anisotropic material behavior of 3D printed composite structures – Material extrusion additive manufacturing. *Materials & Design*, *195*, 108953. doi:10.1016/j.matdes.2020.108953

Sousa, V. F. C., & Silva, F. J. G. (2020). Recent Advances in Turning Processes Using Coated Tools—A Comprehensive Review. *Metals*, *10*(2), 2. Advance online publication. doi:10.3390/met10020170

Spencer, A. J. M. (1972). *Deformations of Fibre-reinforced Materials*. Clarendon Press.

Sreejith, P. S., Krishnamurthy, R., & Malhotra, S. K. (2007). Effect of specific cutting pressure and temperature during machining of carbon/phenolic ablative composite using PCBN tools. *Journal of Materials Processing Technology*, *183*(1), 88–95. doi:10.1016/j.jmatprotec.2006.10.003

Sreejith, P. S., Krishnamurthy, R., Malhotra, S. K., & Narayanasamy, K. (2000). Evaluation of PCD tool performance during machining of carbon/phenolic ablative composites. *Journal of Materials Processing Technology*, *104*(1), 53–58. doi:10.1016/S0924-0136(00)00549-5

Takeyama, H., & Iijima, N. (1988). Machinability of Glassfiber Reinforced Plastics and Application of Ultrasonic Machining. *CIRP Annals*, *37*(1), 93–96. doi:10.1016/S0007-8506(07)61593-5

Turchetta, S., Carrino, L., & Polini, W. (2005). CVD diamond insert in stone cutting. *Diamond and Related Materials*, *14*(3), 641–645. doi:10.1016/j.diamond.2004.10.031

Uhlmann, E., Sammler, F., Richarz, S., Reucher, G., Hufschmied, R., Frank, A., Stawiszynski, B., & Protz, F. (2016). Machining of Carbon and Glass Fibre Reinforced Composites. *Procedia CIRP*, *46*, 63–66. doi:10.1016/j.procir.2016.03.197

Ullen, N. B., Hasak, D. S. M. A., & Dirikolu, M. H. (2020). Factors influencing the machinability during turning sinter-hardened Cu-Ni-Mo based steel : Dependency on cutting speed, feed rate and cutting depth. *Journal of Engineering Research*, *8*(4), 4. Advance online publication. doi:10.36909/jer.v8i4.9031

Varatharajan, R., Malhotra, S. K., Vijayaraghavan, L., & Krishnamurthy, R. (2006). Mechanical and machining characteristics of GF/PP and GF/Polyester composites. *Materials Science and Engineering B*, *132*(1), 134–137. doi:10.1016/j.mseb.2006.02.010

Voss, H., & Friedrich, K. (1987). On the wear behaviour of short-fibre-reinforced peek composites. *Wear*, *116*(1), 1–18. doi:10.1016/0043-1648(87)90262-6

Wang, D. H., Ramulu, M., & Arola, D. (1995a). Orthogonal cutting mechanisms of graphite/epoxy composite. Part I : Unidirectional laminate. *International Journal of Machine Tools & Manufacture*, *35*(12), 1623–1638. doi:10.1016/0890-6955(95)00014-O

Wang, D. H., Ramulu, M., & Arola, D. (1995b). Orthogonal cutting mechanisms of graphite/epoxy composite. Part II : Multi-directional laminate. *International Journal of Machine Tools & Manufacture*, *35*(12), 1639–1648. doi:10.1016/0890-6955(95)00015-P

Wang, X. M., & Zhang, L. C. (2003). An experimental investigation into the orthogonal cutting of unidirectional fibre reinforced plastics. *International Journal of Machine Tools & Manufacture*, *43*(10), 1015–1022. doi:10.1016/S0890-6955(03)00090-7

Xu, J., & El Mansori, M. (2017). Wear characteristics of polycrystalline diamond tools in orthogonal cutting of CFRP/Ti stacks. *Wear*, *376-377*, 91–106. doi:10.1016/j.wear.2016.11.038

Yang, W. H., & Tarng, Y. S. (1998). Design optimization of cutting parameters for turning operations based on the Taguchi method. *Journal of Materials Processing Technology*, *84*(1), 122–129. doi:10.1016/S0924-0136(98)00079-X

Zheng, H., & Liu, K. (2013). Machinability of Engineering Materials. In *Handbook of Manufacturing Engineering and Technology* (pp. 1–34). Springer. doi:10.1007/978-1-4471-4976-7_2-1

Zweben, C. (2015). Composite Materials. In *Mechanical Engineers' Handbook* (pp. 1–37). John Wiley & Sons, Ltd. doi:10.1002/9781118985960.meh110

Chapter 5
Efficient Machining Strategies for Polymer Matrix Composites

Chikesh Ranjan
https://orcid.org/0000-0003-1197-601X
National Institute of Technology, Rourkela, India

Ritesh Singh
https://orcid.org/0000-0001-8055-1256
Birla Institute of Technology, Mesra, India

Kaushik Kumar
https://orcid.org/0000-0002-4237-2836
Birla Institute of Technology, Mesra, India

ABSTRACT

Efficient machining strategies for polymer matrix composites (PMCs) are integral to modern manufacturing, navigating the intricate challenges posed by these advanced materials. This chapter delves into a comprehensive exploration of machining methodologies tailored specifically for PMCs, emphasizing optimization techniques crucial for precision and productivity. It delves into the nuances of tool selection, cutting parameters, and novel approaches to mitigate damage while enhancing machining efficacy. Innovative advancements in tooling technologies and adaptive machining processes are scrutinized to address the unique characteristics of PMCs. The chapter underscores the significance of sustainable practices within machining, focusing on waste reduction, energy efficiency, and environmentally conscious methodologies to align with contemporary manufacturing standards.

DOI: 10.4018/978-1-6684-9927-6.ch005

INTRODUCTION

This chapter delves into the realm of efficient machining strategies tailored for polymer matrix composites. It covers aspects such as tool selection, the risk of damage, and the key parameters involved in this process.

Machining polymer matrix composites (PMCs) poses a significant challenge in terms of cutting tool prerequisites. In contrast to metal cutting, where chip formation is primarily due to plastic deformation, cutting PMCs involves compression shearing and the fracture of fiber reinforcement and matrix (Kaw 2006). This necessitates stringent criteria for cutting edge geometry and material selection. Achieving a sharp cutting edge and a substantial positive rake angle becomes crucial for effectively shearing fibers, while the tool material must possess high hardness and toughness to withstand the fibers' abrasiveness and the intermittent loads induced by their fracture (Daniel, Ishai 2006). The options for tool materials that fulfill these demanding requirements are quite limited.

Urresti (2022) identified that the tool wear and, consequently, the surface integrity in the machining of CFRP composites are predominantly influenced by the fiber orientation, highlighting it as a crucial parameter in the process.

In their research, Bi et al. (2022) carried out both theoretical analysis and practical experiments on CFRP machining, utilizing multi-tooth milling cutters featuring a segmented right-hand helical cutting edge (SRHCE) and a segmented left-hand helical cutting edge (SLHCE). Their findings indicated that the SLHCE tool exhibited greater wear resistance compared to the SRHCE tool. Additionally, they observed that a reduced feed per tooth resulted in increased tool wear and diminished surface quality.

According to Haddad et al. (2014), the use of a multi-tooth cutter in CFRP milling results in reduced damage and lower cutting forces, consistent with the findings reported by Chatelain and Zaghbani (2012). Additionally, Slamani et al. (2015) observed that, during CFRP trimming, the normal cutting force surpassed the feed force.

Lin and Shen (1999) investigated the high-speed drilling characteristics of unidirectional (UD) fiberglass-reinforced composites, exploring a cutting speed range from 210 to 850 m/min. Two types of drills, a twist drill and a multi-face drill, were subjected to testing. Their observations revealed that the axial distance between the outer corner of the cutting edge and the drill tip increased with tool wear. Consequently, the authors proposed a variation in tip height (axial distance) as a parameter for quantifying wear in multi-facial (C-shaped) drills. This discovery was further affirmed by Lin and Chen (1996) in the context of composites drilled with carbon fiber reinforcement. They noted significant wear in tungsten carbide drills as the cutting speed escalated.

Recently, Abd-Elwahed (2022) employed a modeling and optimization strategy grounded in response surface analysis and artificial neural networks. This approach aimed to predict drilling process parameters and optimize cutting conditions for woven glass fiber-reinforced epoxy composites with varying laminate thicknesses. Their findings indicated that achieving optimal results, such as maximizing drilling torque and minimizing the delamination factor, involves employing a low feed rate and high spindle speed during the drilling process.

Mohan et al. (2005) determined that the thickness and drill size collectively constitute the most crucial factors impacting torque, while cutting speed and drill size are the primary influencers of cutting thrust. Erturk et al. (2021) delved into the impact of machining parameters on the drilling performance of GFRP composite, highlighting the influence of tool coating on tribo-mechanical behavior. Their study also emphasized the significant role of drill bit type in determining material temperature.

Caprino and Tagliaferri (1995) reported that feed rate strongly influences the type of damage in GFRP drilling. Inoue et al. (1997) recommended a high feed rate for drilling a high number of holes with consistent quality, while a low feed rate was preferable for achieving high-quality holes. For small-diameter drills, they emphasized the ratio of drill radius to yarn width as a key parameter in estimating tool life.

Liu et al. (2018) investigated the impact of cutting tool geometry on delamination factor and thrust force during GFRP drilling. Comparing a twist drill with three candlestick drills featuring different tip geometries, they concluded that optimizing the geometric angles of the candlestick drill significantly improved peel-up and push-down delamination while reducing thrust force. Palanikumar et al. (2008) explored the influence of cutting parameters and drill geometry on delamination in GFRP composite drilling, utilizing a response surface model to predict the delamination factor and highlighting the limited effect of the drill point angle.

This chapter provides an overview of cutting tool materials and their respective wear characteristics during PMC machining. It delves into various aspects such as types of tool wear, methods for measuring tool wear, and the underlying mechanisms governing wear. Additionally, the chapter offers recommendations for best practices in selecting tool materials tailored to specific applications.

The Machining Strategies of Composites

Machining strategies for composites involve a range of approaches aimed at efficiently shaping and processing composite materials, such as Polymer Matrix Composites (PMCs), Metal Matrix Composites (MMCs), and Ceramic Matrix Composites (CMCs). Some key machining strategies include (Davim and Mata 2007):

- **Tool Selection and Material:** Choosing appropriate cutting tools based on the composite's composition and properties is crucial. Tools with specialized geometries, coatings, and materials tailored to withstand the composite's hardness, abrasiveness, and thermal characteristics are essential.
- **Optimized Tool Geometry:** Tailoring tool geometry, including rake angle, clearance angle, cutting edge sharpness, and nose radius, to suit the specific composite's properties influences chip formation, cutting forces, and surface finish.
- **Cutting Parameters Optimization:** Adjusting cutting speeds, feed rates, and depths of cut to optimize material removal rates while minimizing tool wear and potential damage to reinforcing fibers or the matrix (Koplev et al. 1999).
- **Cooling and Lubrication:** Implementing effective cooling and lubrication methods to manage heat generation during machining, reducing thermal damage to the composite and prolonging tool life.
- **Adaptive Machining Techniques:** Incorporating adaptive machining strategies utilizing real-time monitoring, sensing systems, and predictive analytics to dynamically adjust cutting parameters and tool paths for enhanced precision and efficiency.
- **Safety Protocols:** Prioritizing safety measures such as dust control, proper ventilation, use of personal protective equipment, and adherence to safety guidelines to mitigate hazards associated with machining composite materials.
- **Specialized Cutting Techniques:** Exploring specialized cutting methods like abrasive water jet machining, laser cutting, or ultrasonic machining for specific applications or challenging composite compositions.

Efficient machining strategies for composites involve a nuanced understanding of the material's properties, tooling requirements, cutting parameters, and safety considerations. Continuous research and advancements in machining technologies are vital for developing tailored strategies that maximize productivity while preserving the structural integrity of composite materials.

Davim et al. (2004) demonstrated that Brad and Spur drills outperform Stub Length drills in terms of delamination and surface roughness. Jessy et al. (2015) focused on cooling conditions during GFRP drilling on a CNC lathe, discovering that internal coolant significantly reduced temperature by 76% compared to dry drilling and 66% compared to external coolant methods. This reduction in temperature corresponded to decreased flank wear and increased tool life.

Figure 1 indicates that the enhancement in research on machining of polymer matrix composites has gone up by approximately 200 times in last two decades which effectively also shows the demand on the same.

Figure 1. Machining of polymer matrix composites
(a) Year wise (https://www.webofscience.com/wos/woscc/analyze-results/1d801ef0-85ea-4cf8-ba48-3a392601d548-c453d416);
(b) Subject wise (https://www.webofscience.com/wos/woscc/analyze-results/1d801ef0-85ea-4cf8-ba48-3a392601d548-c453d416);
(c) Category wise (https://www.webofscience.com/wos/woscc/analyze-results/1d801ef0-85ea-4cf8-ba48-3a392601d548-c453d416)
(Accessed on January 11, 2024)

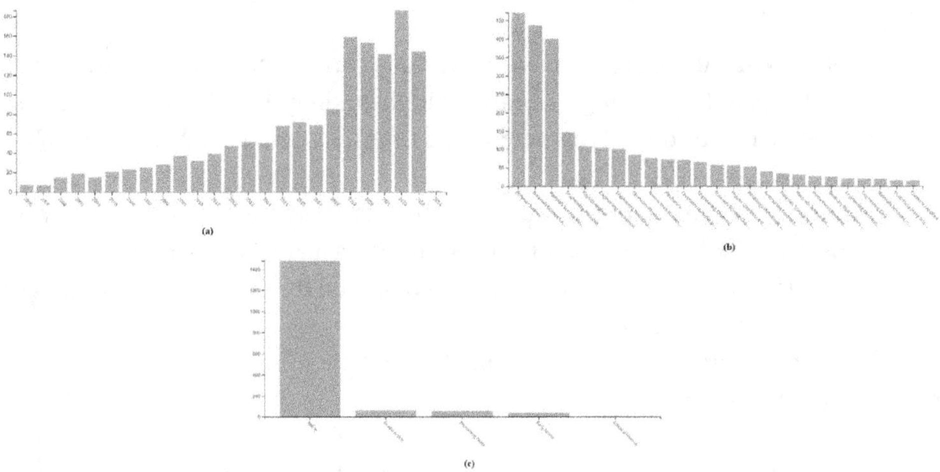

Tool Materials for Machining Composites

Tool materials utilized in machining composite parts must satisfy multiple criteria. These tools require robust resistance to friction and abrasion at key interfaces such as chip/rake face and machined surface/clearance face (Ferriera et al. 1999). Notably, excessive heating during work piece manipulation can lead to thermal degradation of the matrix, necessitating tools with high-temperature hardness and efficient heat dissipation capacities to counter friction-generated heat. Crucially, the cutting edge's resistance to abrasion wear is vital, as any loss of sharpness can result in significant damage within the composite material, including issues like fiber pullout and delamination.

The prevailing industrial tool materials, in ascending order of hardness, include:

- **High-speed steels:** These are high-alloy steels incorporating tungsten, chromium, molybdenum, and vanadium. While these tools can be coated through Physical Vapor Deposition (PVD) with thin films (ranging from 1 to 5µm) of hard materials like TiN, TiCN, TiAlN, and AlCrN to enhance performance, they are occasionally used for small series or low-quality requirements due to their limited resistance to abrasion.

- **Carbides:** These materials, produced via Powder Metallurgy, offer higher hardness and abrasion resistance than HSS, attributed to the presence of fine particles such as CW, TiC, and/or TaC. Further enhancement is achieved by coating with materials like TiN, TiCN, TiAlN, Al2O3, AlCrN, or diamond through processes like PVD or CVD, exhibiting good adhesion on the carbide tool.
- **Cubic boron nitride:** Available in various forms, including electrolytically deposited platelets on tungsten carbide bodies, sintered CBN blanks brazed on carbide plates, or directly sintered on a tungsten carbide substrate (Hocheng et al. 1992).
- **Diamond:** As the hardest material, it exists in synthetic or natural crystal forms deposited on HSS or carbide bodies with an electrolytic binder (Ni), or in polycrystalline diamond (PCD) form obtained by sintering diamond grains with a Co binder on a carbide substrate.
- **Ceramics:** Ceramics are unsuitable due to their susceptibility to mechanical and thermal shocks.

Ross et al. (2005) identified the cutting tool's condition by employing pre-trained networks, including AlexNet, VGG-16, ResNet, MobileNet, and Inception-V3. The optimal network for tool condition monitoring was suggested, taking into account the impact of hyperparameters such as batch size, learning rate, solver, and train-test split ratio.

In summary, research on PMCs machining predominantly advocates for carbides or PCD tools. PCD tools are preferred for their superior wear resistance, while carbide tools are chosen for their cost-effectiveness.

Carbides Tools

Carbide tools are widely used and preferred for machining composites, including Polymer Matrix Composites (PMCs) and other composite materials. Here's why:

- **Material Composition:** Carbide tools, typically made of tungsten carbide (WC) or a combination of tungsten carbide and cobalt (WC-Co), offer exceptional hardness, strength, and wear resistance (Konig et al. 1999).
- **Advantages:** Carbide tools excel in cutting abrasive and tough materials like composites due to their superior hardness. They maintain sharp cutting edges for longer durations, resulting in extended tool life and reduced downtime for tool changes.

- **Coatings:** Carbide tools can be further enhanced with specialized coatings (e.g., TiN, TiCN, TiAlN) using methods like Physical Vapor Deposition (PVD) or Chemical Vapor Deposition (CVD). These coatings augment the tool's wear resistance, reduce friction, and improve cutting performance when machining composites.
- **Performance:** Their excellent wear resistance makes carbide tools well-suited for handling the abrasive nature of composite materials, including the challenge posed by reinforcing fibers.
- **Applications:** Carbide tools find widespread use in various machining operations involving composites, ranging from drilling and milling to cutting and shaping, due to their ability to maintain cutting edge integrity under demanding conditions.
- **Considerations:** While carbide tools exhibit remarkable wear resistance, careful selection of cutting parameters (speeds, feeds, depths of cut) is crucial to optimize performance and mitigate potential tool damage in composite machining applications.

In summary, carbide tools are favored in the machining of composites due to their inherent hardness, wear resistance, and ability to withstand the abrasive nature of composite materials, ensuring prolonged tool life and improved machining efficiency.

Carbides Tools Coatings

Carbide tool coatings have become a prevalent method for enhancing tool performance. Figure 2 illustrates the categorically organized coating on tools for machining composites. In this representation, it is evident that in 2020, there was an increased amount of work conducted in this specific area.

Figure 2. Year wise publication on coating on tools for machining of composites (https://www.webofscience.com/wos/woscc/analyze-results/a82bf 258-ce5f-4c49-9e7d-df5c53ec7dc2-c454776b) (Accessed on January 11, 2024)

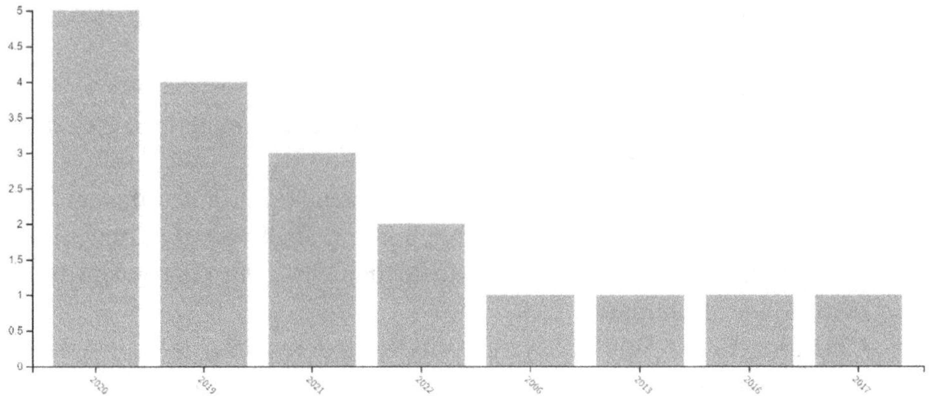

These coatings fall into four main categories based on the materials utilized:

- **Titanium-based coatings**

Titanium-based coatings for tools involve applying compounds like titanium carbide (TiC), titanium nitride (TiN), titanium diboride (TiB2), titanium carbonitride (TiCN), among others, onto the surface of tools. These coatings enhance tool performance by improving hardness, wear resistance, and reducing friction during cutting operations. Aluminum is often added to these coatings, resulting in combinations like TiAlN, TiAlON, and AlCrN, which further augment the coating's properties, including heat resistance and overall durability. These coatings are commonly used in various machining applications to extend tool life and enhance cutting performance in different materials, including metals, composites, and alloys.

Titanium-based coatings refer to thin layers of materials primarily composed of titanium or titanium alloys that are applied onto the surface of tools or components. These coatings are known for their exceptional properties such as high hardness, wear resistance, and corrosion resistance. Titanium-based coatings are commonly used in various industries, including aerospace, automotive, and manufacturing, to enhance the performance and longevity of cutting tools, molds, and other components subjected to harsh operating conditions.

- **Methods of Tools Coatings**

- **Physical Vapor Deposition (PVD):** PVD is a common method for depositing thin coatings onto tools. In this process, the coating material is vaporized in a vacuum chamber and then condensed onto the tool surface, creating a thin film. Popular titanium-based coatings produced by PVD include titanium nitride (TiN) and titanium aluminum nitride (TiAlN).
- **Chemical Vapor Deposition (CVD):** CVD involves a chemical reaction between gaseous precursors and the tool surface, leading to the deposition of a solid coating. Titanium carbide (TiC) and titanium carbonitride (TiCN) coatings are examples of titanium-based coatings produced through CVD.

Application

Titanium-based coatings find application in machining strategies for Polymer Matrix Composites (PMCs) due to their favorable properties. When machining PMCs, challenges such as tool wear, delamination, and surface finish need to be addressed. Here's how titanium-based coatings contribute:

- **Wear Resistance:** Titanium-based coatings, especially TiN and TiAlN, provide enhanced wear resistance to cutting tools. This is crucial in machining PMCs, where abrasive fillers in the composite material can lead to accelerated tool wear.
- **Reduced Friction:** Coated tools often exhibit reduced friction during machining, leading to lower cutting forces and improved chip evacuation. This is beneficial in preventing delamination and ensuring a smoother machining process in PMCs.
- **Extended Tool Life:** Titanium-based coatings increase the longevity of cutting tools, reducing the frequency of tool changes. This not only improves efficiency but also contributes to cost savings in the machining of PMCs.
- **Improved Surface Finish:** Coated tools contribute to achieving a better surface finish on machined parts, which is essential in applications where surface aesthetics and integrity are critical, such as in aerospace components made from polymer matrix composites.

- **Ceramic coatings**

Ceramic coatings for carbide tools involve applying ceramic materials onto the surface of the tools. These coatings, which can include compounds like aluminum oxide (Al2O3) or titanium nitride (TiN), are known for their exceptional hardness and resistance to wear. They are used to enhance the performance and longevity of cutting tools by providing a hard, protective layer that withstands high temperatures and abrasive forces during machining operations. Ceramic coatings contribute to reducing friction, improving tool life, and enabling efficient cutting across a variety of materials, including metals, alloys, and some composites.

Ceramic coatings are thin layers of ceramic materials applied to the surface of tools or components. These coatings are known for their excellent hardness, wear resistance, and high-temperature stability. They are widely used in various industries to enhance the performance and durability of cutting tools, molds, and components subjected to extreme operating conditions.

- **Methods of Tools Coatings:**
    - **Physical Vapor Deposition (PVD):** PVD is a common method for depositing ceramic coatings onto tools. In this process, ceramic materials are vaporized in a vacuum chamber and then condensed onto the tool surface, creating a thin, hard coating. Common ceramic coatings produced by PVD include titanium nitride (TiN), titanium carbonitride (TiCN), and aluminum oxide (Al2O3).
    - **Chemical Vapor Deposition (CVD):** CVD involves a chemical reaction between gaseous precursors and the tool surface, resulting in the deposition of a solid ceramic coating. Popular ceramic coatings produced through CVD include silicon carbide (SiC) and titanium carbide (TiC).

Application

Ceramic coatings play a vital role in machining Polymer Matrix Composites (PMCs), addressing specific challenges associated with tool wear, delamination, and surface finish. Here's how ceramic coatings contribute:

- **Enhanced Hardness and Wear Resistance:** Ceramic-coated tools exhibit high hardness and exceptional wear resistance, making them well-suited for machining PMCs. This is particularly important as PMCs often contain abrasive fillers that can lead to accelerated tool wear.
- **Reduced Friction:** Ceramic coatings contribute to reduced friction during machining, leading to lower cutting forces and improved chip evacuation. This helps prevent delamination and ensures a smoother machining process in PMCs.
- **High-Temperature Stability:** Ceramics are known for their thermal stability. In machining PMCs, where elevated temperatures may occur due to friction,

the high-temperature resistance of ceramic coatings helps maintain tool integrity and performance.

- **Improved Surface Finish:** The hardness and wear resistance of ceramic coatings contribute to achieving a better surface finish on machined parts. This is crucial in applications where surface quality is essential, such as aerospace components made from polymer matrix composites.
- **Extended Tool Life:** Ceramic coatings significantly extend the lifespan of cutting tools, reducing the need for frequent tool changes. This improves efficiency and cost-effectiveness in the machining of PMCs.

• Ultra-hard coatings employing diamond

Ultra-hard coatings utilizing diamond for carbide tools involve depositing diamond-based materials onto the surface of these tools. Diamond is one of the hardest known materials, and its application as a coating enhances the tools' hardness, wear resistance, and durability to an exceptional degree. These coatings, often referred to as diamond-like coatings (DLC) or polycrystalline diamond (PCD) coatings, offer superior performance in extreme cutting conditions, providing excellent abrasion resistance and reducing friction during machining processes. By integrating diamond into the tool surface, these coatings significantly extend tool life and enable efficient cutting across various materials, including metals, composites, and abrasive substances.

Ultra-hard coatings employing diamond refer to thin layers of materials that incorporate diamond particles or utilize diamond-like carbon (DLC) structures. Diamonds, being one of the hardest known materials, contribute to the exceptional hardness and wear resistance of these coatings. These coatings are utilized to enhance the durability and performance of cutting tools, particularly in applications where extreme hardness and wear resistance are crucial.

• Methods of Tools Coatings:

- **Chemical Vapor Deposition (CVD):** CVD is a common method for depositing ultra-hard coatings employing diamond. In this process, a gas mixture containing carbon is introduced into a vacuum chamber, and under controlled conditions, diamond crystals form and adhere to the tool surface.
- **Physical Vapor Deposition (PVD):** PVD methods, such as pulsed laser deposition or magnetron sputtering, can be employed to create ultra-hard coatings with diamond-like carbon structures. These methods involve depositing carbon-containing species onto the tool surface in a vacuum environment, resulting in a hard and wear-resistant coating.

Application

Ultra-hard coatings employing diamond offer specific advantages when machining Polymer Matrix Composites (PMCs):

- **Extreme Hardness:** The incorporation of diamond or diamond-like structures provides ultra-hardness to the coating, making it suitable for machining PMCs containing abrasive fillers. This hardness helps resist wear and abrasion during cutting.
- **Enhanced Wear Resistance:** The ultra-hard nature of these coatings significantly improves wear resistance, making them effective in extending tool life when machining PMCs. This is particularly beneficial in applications where tool wear is a critical concern.
- **Reduced Friction:** The hardness and lubricity of diamond coatings contribute to reduced friction during machining. This characteristic helps in minimizing heat generation and preventing issues like delamination in PMCs.
- **Improved Surface Finish:** Ultra-hard diamond coatings contribute to achieving a superior surface finish on machined PMC components. This is crucial in applications where surface quality is a key requirement.
- **Extended Tool Life:** The combination of extreme hardness and wear resistance leads to an extended tool life. This reduces the need for frequent tool changes, enhancing the efficiency and cost-effectiveness of machining PMCs.

- **Self-lubricating Coatings**

Self-lubricating coatings for carbide tools incorporate materials such as molybdenum disulfide (MoS2) or a combination of tungsten carbide (CW) and cobalt (C) as an outer layer. These coatings are designed to reduce friction and heat generation during cutting operations by providing a lubricating layer on the tool surface. MoS2, known for its lubricating properties, forms a low-friction layer that helps in reducing tool wear and prolonging tool life. Similarly, coatings with a combination of CW and C offer lubricating characteristics, enhancing the tool's ability to glide smoothly during machining processes, particularly in high-temperature and high-pressure conditions. These self-lubricating coatings contribute to improved cutting performance, reduced wear, and increased efficiency when machining various materials, including metals and alloys.

Tool coatings, whether a single layer or a combination of various materials, typically measure up to 15μm in total thickness. Whether employing CVD (chemical vapor deposition) or PVD (Physical Vapor Deposition), the coating process aims for a strong bond between the coating and the substrate. Adhesion to the substrate is crucial in selecting coatings, which are also chosen for their high-temperature properties like hardness, toughness, and thermal conductivity. For instance, materials like alumina (Al2O3) within multilayer coatings serve to thermally insulate, preserving the mechanical properties of the substrate. These coatings are an integral part of the tool in direct contact with the work piece material, tailored to impart specific properties required for optimal performance, such as wear resistance, reduced friction, and chemical stability.

In machining composite materials, ultra-hard coatings yield the most favorable outcomes. Diamond, particularly, possesses ideal properties for this purpose. Its exceptional hardness renders it highly resistant to abrasive wear compared to other cutting materials. Additionally, its strong chemical stability and limited affinity with non-ferrous materials prevent the formation of built-up edges, maintaining consistently high cutting performance. Unlike PCD coatings, diamond coatings exhibit adaptability to conform to various tool geometries, allowing for their application on tools like WC carbide mills, drills, and inserts. Manufacturers' experiences indicate that the most optimal results are achieved with submicron carbides (ranging from 0.5 to 3μm) containing less than 10% cobalt content (Ramulu et al. 1989).

Tool for Polymer Matrix Composites

Tools used for machining Polymer Matrix Composites (PMCs) vary based on the specific requirements of the composite material and the machining process. Some common tool types used for PMCs include (Krishnamurthy et al. 1992):

- **Sintered Carbide Tools:** These tools are made through a sintering process, where carbide powder is compressed and heated to create a solid tool. Sintered carbide tools offer high hardness and durability, making them suitable for machining PMCs due to their wear resistance (Boothroyd et al. 1989).
- **Solid Carbide End mill Coated:** Solid carbide tools are composed entirely of carbide material and are coated with specialized materials like TiAlN or TiCN. The coating enhances their performance by providing increased hardness, wear resistance, and reduced friction during machining of PMCs.
- **Electroplated Diamond Tools:** Electroplated diamond tools involve depositing a layer of diamond grit onto the tool's surface using an electroplating process. These tools offer exceptional abrasion resistance and are efficient for cutting PMCs due to the diamond's hardness.
- **Multi-Pyramidal-Like Edge Tool:** This tool design features multiple left-hand and right-hand helical edges forming small pyramidal edges along the cutting length. Such a design aids in clean shaving of fibers in PMCs, reducing damage and improving the quality of the machined surface.

Each of these tool types presents specific advantages in machining PMCs, offering varying degrees of wear resistance, cutting efficiency, and precision. The selection of the appropriate tool type depends on the machining requirements, the specific properties of the composite material, and the desired outcomes of the machining process.

Cutting Mechanisms

Cutting mechanisms in composite materials involve various processes due to their heterogeneous nature. When a tool engages with a composite material, several mechanisms come into play (Shaw 2005):

- **Fiber Fracture:** Fiber fracture refers to the breaking or severing of the reinforcing fibers within a composite material. In fiber-reinforced composites, such as carbon fiber composites or glass fiber composites, these fibers provide the material with strength and stiffness. When a cutting tool engages with the composite, it applies forces that surpass the strength of these fibers, causing them to break. The cutting tool's action creates stress on the fibers, leading to their fracture. This process is vital during machining but needs careful consideration, as the integrity of the composite's mechanical properties relies on the arrangement and strength of these fibers. Proper machining techniques aim to minimize excessive fiber fracture, ensuring the composite retains its structural strength and performance after machining.
- **Matrix Shear:** Matrix shear occurs in composite materials when the cutting tool applies force that exceeds the strength of the matrix material holding the reinforcing fibers together. In fiber-reinforced composites, like carbon fiber or glass fiber composites, the matrix material serves to bond and support the fibers.

During cutting or machining processes, the tool's action imposes stress on the matrix, causing it to deform and shear. This shearing action allows the fibers to separate from the matrix, enabling the cutting tool to penetrate and shape the material.

Matrix shear is a critical aspect of composite machining as it affects the way the material responds to cutting forces. Managing this process is essential to avoid excessive damage to the matrix, ensuring that the structural integrity of the composite is preserved after machining.

- **Delamination:** Delamination refers to the separation or splitting of layers within a composite material along its laminate structure. In laminated composites, which consist of multiple layers or plies bonded together, each layer contributes to the material's overall strength.

During cutting or machining operations, the tool's action can induce tensile and shear stresses in the composite layers. These stresses may cause adjacent layers to separate or detach, leading to delamination. This separation compromises the structural integrity of the material, affecting its strength and stiffness.

Delamination is a critical concern in composite machining as it can significantly impact the quality and performance of the final machined part. Techniques and tooling strategies are employed to minimize delamination and preserve the integrity of the composite structure during cutting processes.

- **Tool Wear:** Tool wear refers to the gradual deterioration or loss of material from a cutting tool's surface during machining operations. When a tool engages with a work piece, especially in materials like composites, metals, or alloys, various factors contribute to wear over time.

In composite machining, the abrasive nature of the material, combined with the cutting forces and temperatures generated during the process causes the tool's cutting edges to degrade. This degradation can take the form of flank wear (wear along the cutting edge), crater wear (depression or erosion on the tool face), or edge chipping.

Tool wear affects machining accuracy, surface finish quality, and the tool's lifespan. Managing and monitoring tool wear are crucial to maintaining machining precision, optimizing tool performance, and ensuring cost-effective production in composite machining processes. Strategies such as using appropriate coatings, tool materials, and cutting parameters help mitigate and control the effects of wear on cutting tools.

Effect of Fiber Orientation in Tool Wear

The influence of fiber orientation on tool wear refers to how the alignment and arrangement of reinforcing fibers within a composite material affect the degradation of cutting tools during machining processes. In fiber-reinforced composites, the orientation and distribution of fibers play a significant role in determining the cutting forces and tool wear characteristics. When machining these materials, the cutting tool interacts with both the reinforcing fibers and the matrix.

If the fibers are oriented in the direction of cutting or perpendicular to the cutting edge, they can act as abrasives, leading to increased tool wear. The tool's contact with these aligned fibers causes higher cutting forces and greater abrasive action, resulting in more rapid degradation of the cutting edges.

Conversely, if the fibers are randomly oriented or aligned at angles to the cutting direction, they might offer less resistance to the tool's motion, potentially reducing abrasive wear on the cutting edges.

Understanding the fiber orientation and its impact on tool wear is essential in selecting appropriate cutting strategies, tool materials, and machining parameters to optimize tool performance and minimize wear when machining fiber-reinforced composites.

The primary cause of abrasive wear on the tool material stems from friction at the interfaces between the chip and the tool, as well as the tool and the work piece. This friction is predominantly caused by the hard components present in the machined material, particularly the fibers. Abrasive wear manifests as the development of ribbed zones or the loss of sharpness in the cutting edge, leading to an increase in the cutting edge radius.

When the cutting edge radius is comparable to the diameter of the fibers, the fibers are cleanly cut. However, if the cutting edge radius grows to approximately five to ten times the diameter of the fibers, the fibers are no longer cleanly cut but rather repelled and poorly severed by the tool.

Understanding these cutting mechanisms is crucial for selecting appropriate tooling, optimizing machining parameters, and ensuring the integrity of the final machined composite component.

Machining of Polymers

Machining polymers involves the shaping, cutting, or drilling of polymer-based materials like plastics. Unlike metals, polymers have different properties that influence machining techniques (Bhatnagar et al. 1995):

- **Material Characteristics:** Polymers vary in hardness, thermal properties, and behavior when subjected to cutting forces. Softer polymers might deform easily, while harder ones can be more brittle.
- **Tool Selection:** Specialized tools, often made from carbide or high-speed steel, are used due to polymers' lower heat resistance compared to metals. Diamond-coated tools or those with specialized coatings can also be employed to reduce friction and improve surface finish.
- **Machining Techniques:** Techniques like milling, turning, drilling, and sawing are used, but parameters such as cutting speed, feed rate, and depth of cut must be adjusted to avoid excessive heat buildup or material deformation.
- **Cooling and Lubrication:** Cooling methods, such as using coolants or air jets, are employed to dissipate heat generated during machining, preventing material melting or distortion.

- **Surface Finish:** Achieving a smooth surface without burrs or irregularities is crucial in polymer machining, often requiring careful tool selection and precise control of cutting parameters.

Machining polymers demands a nuanced approach due to their unique properties, necessitating adjustments in tooling, techniques, and cooling methods to ensure accurate cuts, prevent material damage, and maintain the desired quality of the finished product (Jain. et al. 2022)

The machining of polymers has been extensively explored, notably by Mahakur et al. (2022), whose findings are well-documented in a renowned monograph. Recent literature (Kandavalli et al. 2023) also delves into this area, driven by increased polymer usage in optical, electronic, and consumer products.

In machining fiber-reinforced composites, the polymer matrix, despite being less resistant to cutting than the reinforcing fibers due to lower strength and stiffness, significantly influences chip formation and chip types during machining. Moreover, the polymer matrix, with its distinct thermal properties, especially thermal conductivity, plays a pivotal role in determining cutting zone temperatures (Malik et al. 2022, Rajamurugan et al. 2022)

The behavior of polymers in machining is impacted by various process parameters such as tool material, rake angle, cutting-edge radius, depth of cut, and cutting speed. Additionally, the rheological and thermal properties of polymers profoundly affect chip formation. Thermosets and thermoplastics exhibit contrasting mechanical responses to applied loads thermoset are brittle with minimal strain to fracture, while thermoplastics tend to be ductile, exhibiting considerable strain to fracture.

The material's viscous behavior influences its ultimate strength and elongation to fracture, both dependent on the rate of load application. Polymers shift from ductile to brittle behavior as the deformation rate increases. These differences lead to varied chip formation and machined surface quality across different polymer types based on cutting parameters (Jain et al. 2022).

Adjusting the rake angle and reducing the depth of cut lessens material deformation in chip formation. However, increasing cutting speed has dual effects: accelerating strain rates, potentially inducing brittle failure, and simultaneously elevating temperatures, enhancing material ductility through increased molecular chain mobility. The dominance of these effects depends on the polymer's molecular weight and glass transition temperature (Ross et al. 2023, Saikrupa et al. 2023).

Tian et al. (2019) explored the effects of machining parameters and clearance angles of outer cutting edges during GFRP composite drilling using three candlestick drills. Utilizing the response surface approach, they analyzed the delamination factor and identified that damage at the hole exit results from a simultaneous combination of cutting force, thrust force, and drilling temperature. The study cautioned against high-speed drilling with relatively small clearance angles due to increased temperature at the hole exit, leading to resin failure and enhanced exit damage.

Chip Formation in Polymer Matrix Composites

Chip formation in Polymer Matrix Composites (PMCs) involves the process of material removal during machining. Several factors influence chip formation in PMCs:

- Effect of Tool Geometry:

The effect of tool geometry on machining, especially in Polymer Matrix Composites (PMCs), is substantial. Tool geometry encompasses various aspects like rake angle, clearance angle, cutting edge sharpness, and nose radius (Arola et al. 1996). Each element influences chip formation, cutting forces, and surface quality:

- **Rake Angle:** Affects the direction and magnitude of cutting forces. Positive rake angles reduce cutting forces but can lead to increased heat generation, while negative rake angles enhance tool stability but may increase cutting forces.
- **Clearance Angle:** Impacts the tool's ability to move through the material without rubbing against it. An inadequate clearance angle might cause friction, leading to heat buildup and tool wear.
- **Cutting Edge Sharpness:** A sharper cutting edge allows for better penetration and chip formation but may be more susceptible to wear. Blunt edges might increase cutting forces and produce poorer surface finishes.
- **Nose Radius:** Influences stress distribution during cutting. A larger nose radius distributes forces over a wider area, reducing localized stress and potential damage to fibers. Smaller nose radii can provide better accuracy but might risk higher stress concentrations.

Optimizing tool geometry is crucial in PMC machining to ensure efficient material removal, minimize tool wear, control cutting forces, and prevent damage to reinforcing fibers or the matrix material. It's about finding the right balance to achieve optimal machining performance and preserve the composite's structural integrity.

- Effect of Nose Radius:

The nose radius of a cutting tool has a significant impact on machining, particularly in Polymer Matrix Composites (PMCs):

- **Stress Distribution:** A larger nose radius helps distribute cutting forces over a broader area, reducing localized stress on the material being cut. This can mitigate the risk of damaging the composite's fibers or matrix by lowering stress concentrations.
- **Cutting Forces:** A larger nose radius generally decreases cutting forces during machining. This reduction can help prevent excessive tool wear, minimize vibrations, and improve the overall machining process stability.
- **Surface Finish:** A larger nose radius tends to produce smoother surface finishes on the machined part due to reduced localized stress. However, excessively large nose radii might compromise cutting precision or accuracy.
- **Fiber Damage:** In PMCs, a larger nose radius might lower the risk of damaging reinforcing fibers during cutting by exerting less concentrated force on the fibers, reducing the likelihood of breakage or pull-out.

Balancing the nose radius is crucial in PMC machining. While a larger radius may offer benefits in reducing stress and improving surface finish, an overly large radius could compromise cutting accuracy or precision. Selecting an appropriate nose radius is essential for achieving efficient material removal while preserving the composite's integrity.

- Effect of Matrix Material:

The matrix material in Polymer Matrix Composites (PMCs) significantly influences the machining process:

- **Mechanical Properties:** The mechanical properties of the matrix, such as hardness, brittleness, and toughness, affect how it deforms and responds to cutting forces during machining. Softer matrices might deform more readily, while harder or more brittle matrices may generate higher cutting forces.
- **Thermal Properties:** Matrix materials with distinct thermal conductivities and heat resistance impact the temperature generated at the cutting zone. Materials with poor thermal conductivity may lead to higher localized temperatures, affecting tool wear and material deformation.
- **Chip Formation:** The behavior of the matrix material during machining affects chip formation. Softer matrices might shear more easily, contributing to chip formation, while harder matrices may exhibit more complex deformation modes.
- **Tool Wear:** Matrix materials with abrasive components or high hardness levels may accelerate tool wear. Tools used for machining PMCs need to

withstand the abrasiveness of the matrix material without compromising their cutting efficiency.

Understanding the properties of the matrix material is crucial in selecting appropriate cutting parameters, tool materials, and machining techniques to achieve efficient material removal while minimizing tool wear and preserving the integrity of the composite structure. Balancing these factors ensures optimal machining performance in PMCs.

- Effect of Fibers:

The fibers within Polymer Matrix Composites (PMCs) have a significant impact on the machining process:

- **Hardness and Abrasiveness:** Reinforcing fibers, such as carbon, glass, or aramid, are typically harder than the matrix material. During machining, these hard fibers can cause tool wear due to their abrasive nature, especially if the tool's cutting edges are not properly designed or if cutting parameters are not optimized.
- **Fiber Orientation and Distribution:** The alignment and distribution of fibers within the matrix influence chip formation and the mode of material removal. Properly aligned fibers can either aid or hinder the cutting process based on their orientation relative to the cutting tool.
- **Tool Interaction:** Interaction between the cutting tool and fibers determines chip formation. If the tool engages properly with the fibers, clean cutting can occur. However, improper tool geometry or excessive force can lead to fiber pull-out, breakage, or poor chip formation.
- **Structural Integrity:** Preserving the integrity of the fibers is crucial in maintaining the composite's structural properties. Damage to fibers during machining can compromise the overall strength and performance of the composite material.

Balancing the challenges posed by the fibers in PMCs is essential for successful machining. Optimizing tool geometry, selecting appropriate cutting parameters, and considering fiber orientation are crucial factors in achieving efficient material removal while minimizing damage to the reinforcing fibers.

Optimizing tool geometry, selecting appropriate cutting parameters, understanding the properties of the matrix and fibers, and ensuring suitable machining techniques are crucial in controlling chip formation in PMCs. Balancing these factors is essential to achieve efficient material removal, minimize damage to reinforcing fibers, and maintain the integrity of the composite structure (Spur and Lachmund, 1999).

Innovative Cutting: Harnessing AI and Machine Learning for Advanced Techniques

The development of novel cutting techniques leveraging emerging technologies like artificial intelligence (AI) and machine learning (ML) represents a significant advancement in the field of machining and manufacturing processes. This innovative approach involves integrating AI and ML algorithms into various aspects of cutting processes to enhance efficiency, precision, and overall performance. Here's a brief overview:

- **Process Optimization:** AI and ML algorithms analyze large datasets generated during cutting processes to identify patterns and trends. By leveraging this information, these technologies optimize cutting parameters such as speed, feed rate, and tool selection, leading to improved efficiency and reduced production costs.
- **Predictive Maintenance:** AI and ML can predict tool wear and potential machine failures by monitoring real-time data from sensors and machinery during cutting operations. This allows for proactive maintenance, minimizing downtime and extending the lifespan of cutting tools and equipment.
- **Adaptive Control Systems:** AI-driven adaptive control systems continuously adjust cutting parameters based on real-time feedback. This dynamic responsiveness enhances precision and adaptability, allowing the cutting process to automatically adapt to variations in material properties and environmental conditions.
- **Quality Control:** Machine learning algorithms can analyze the quality of finished products by assessing parameters such as surface finish, dimensional accuracy, and integrity. This ensures that products meet stringent quality standards, reducing defects and waste.
- **Tool Path Optimization:** AI and ML contribute to the optimization of tool paths by considering factors like material properties, machine capabilities, and tool wear. This results in more efficient and precise cutting, reducing cycle times and improving overall productivity.
- **Intelligent Tooling Systems:** Smart tools equipped with sensors and AI capabilities can dynamically adjust their cutting parameters based on the material being processed. This leads to enhanced tool life, reduced energy consumption, and improved overall sustainability.

- **Learning from Experience:** By accumulating data from past cutting processes, AI and ML models "learn" from experience. This accumulated knowledge can be applied to future operations, allowing for continuous improvement and refinement of cutting techniques over time.

The integration of AI and ML in cutting processes represents a paradigm shift towards intelligent, data-driven manufacturing. These technologies offer the potential to revolutionize the way cutting operations are planned, executed, and optimized, ultimately leading to more efficient, precise, and cost-effective machining solutions.

Safety Aspects in Machining of Polymer Matrix Composites

Safety is paramount when machining Polymer Matrix Composites (PMCs) due to potential risks associated with the materials and machining processes:

- **Dust and Particulate Hazard:** Machining PMCs can generate fine dust and particles, which may contain hazardous components. Proper ventilation, dust collection systems, and personal protective equipment (PPE) such as respirators, goggles, and gloves are crucial to prevent inhalation or skin contact.
- **Tooling Hazards:** Machining tools operating at high speeds pose a risk of injury. Ensuring proper tool guarding, secure fixturing, and following safety protocols when handling, installing, or changing tools mitigates the risk of accidents.
- **Heat and Thermal Hazards:** High-speed machining can generate heat, especially when cutting through composites. Adequate cooling systems and thermal protection measures are essential to prevent burns or material degradation.
- **Chemical Exposure:** Some composite materials might release harmful fumes or gases during machining. Understanding the specific material's composition and using appropriate machining techniques in well-ventilated areas or with proper exhaust systems minimizes exposure risks.
- **Material Handling:** Handling large or heavy composite work pieces requires proper lifting equipment and techniques to prevent strain or injury. Workers should be trained in safe material handling practices.
- **Fire Hazard:** Some machining processes involving composites might create sparks or heat buildup, increasing the risk of fire. Fire safety measures, such as fire extinguishers and proper fire protocols, are crucial.

- **Training and Education:** Adequate training for operators and personnel on safe machining practices, handling hazardous materials, emergency procedures, and the use of safety equipment is essential to mitigate risks.

Prioritizing safety measures, implementing proper training protocols, utilizing appropriate PPE, and adhering to safety guidelines and regulations are vital to ensuring a safe working environment during the machining of Polymer Matrix Composites (Toug et al. 2023).

CONCLUSION

The investigation into "Efficient Machining Strategies for Polymer Matrix Composites" underscores the intricate challenges and possibilities associated with machining these composite materials. The complexity of PMCs necessitates a thorough approach, covering aspects from tool selection to process optimization and safety considerations. This in-depth analysis emphasizes the crucial role played by tool geometry, cutting parameters, and material properties in achieving efficient material removal while preserving the integrity of the composite structure. Amid ongoing research and technological progress, the focus on developing more sustainable and effective machining strategies for Polymer Matrix Composites persists, offering the promise of heightened productivity and material processing quality while ensuring machining operations' safety. Our research findings indicate that incorporating nanostructured coatings and composite tool materials significantly enhances tool durability, leading to prolonged tool life and diminished wear during machining operations. The application of machine learning algorithms for real-time adaptive machining, along with digital twin technology, proves successful in dynamically adjusting cutting parameters, optimizing efficiency, and providing a virtual representation of the machining environment. Furthermore, our study demonstrates that innovative cooling and lubrication techniques, such as cryogenic machining and minimum quantity lubrication, contribute to superior thermal management, minimizing thermal damage, and enhancing overall surface finish. The implementation of in-process monitoring and control systems, supported by strategically placed sensors, proves instrumental in early issue detection and ensuring consistent product quality. Tailoring customized machining strategies to specific polymer matrix composites, considering factors like fiber orientation and matrix composition, leads to improved precision and reduced defects.

FUTURE SCOPE

The future of "Efficient Machining Strategies for Polymer Matrix Composites" holds immense potential for transformative advancements in the manufacturing landscape. Embracing advanced cutting tool materials, including nanostructured coatings and composite tools, can significantly enhance tool durability and reduce wear. The integration of machine learning for real-time adaptive machining, along with digital twin technology, enables dynamic adjustments to cutting parameters, optimizing efficiency and supporting predictive maintenance. Innovations in cooling and lubrication techniques, such as cryogenic machining or minimum quantity lubrication, promise improved thermal management and enhanced surface finish. Further exploration of hybrid machining processes, like laser-assisted machining, offers versatile solutions to overcome challenges associated with tool wear. Developing sophisticated in-process monitoring and control systems, strategically employing sensors for data capture, ensures early issue detection and consistent product quality. Customizing machining strategies for specific polymer matrix composites, considering fiber orientation and matrix composition, remains a pivotal focus. Additionally, collaborative industry-academia initiatives can expedite the practical implementation of these strategies, fostering knowledge exchange and technology transfer for sustainable and efficient manufacturing practices.

ACKNOWLEDGMENT

The authors sincerely acknowledge the comments and suggestions of the reviewers that have been instrumental for improving and upgrading the paper in its final form.

REFERENCES

Abd-Elwahed, M. S. (2022). Drilling process of GFRP composites: Modeling and optimization using hybrid ANN. *Sustainability (Basel)*, *14*(11), 6599. doi:10.3390/su14116599

Arola, D., Ramulu, M., & Wang, D. H. (1996). Chip formation in orthogonal trimming of graphite/epoxy composite. *Composites. Part A, Applied Science and Manufacturing*, *27A*(2), 121–133. doi:10.1016/1359-835X(95)00013-R

Bhatnagar, N., Ramakrishnan, N., Naik, N. K., & Komanduri, R. (1995). On the machining of fiber reinforced plastic (FRP) composite laminates. *International Journal of Machine Tools & Manufacture*, *35*(5), 701–716. doi:10.1016/0890-6955(95)93039-9

Bi, G., Wang, F., Fu, R., & Chen, P. (2022). Wear characteristics of multi-tooth milling cutter in milling CFRP and its impact on machining performance. *Journal of Manufacturing Processes*, *81*, 580–593. doi:10.1016/j.jmapro.2022.07.008

Boothroyd, G., & Knight, W. (1989). *Fundamentals of Machining and Machine Tools* (2nd ed.). Marcel Dekker.

Caprino, G., & Tagliaferri, V. (1995). Damage development in drilling glass fibre reinforced plastics. *International Journal of Machine Tools & Manufacture*, *35*(6), 817–829. doi:10.1016/0890-6955(94)00055-O

Chatelain, J.-F., & Zaghbani, I. (2012). Effect of tool geometry special features on cutting forces of multilayered CFRP laminates. *Journal of Composite Materials*, *85*, 90.

Daniel, I. M., & Ishai, O. (2006). *Engineering Mechanics of Composite Materials* (2nd ed.). Oxford University Press.

Davim, J. P., & Mata, F. (2007). New machinability study of glass fiber reinforced plastics using polycrystalline diamond and cemented carbide (K15) tools. *Materials & Design*, *28*(3), 1050–1054. doi:10.1016/j.matdes.2005.09.019

Davim, J. P., Reis, P., & Antonio, C. C. (2004). Experimental study of drilling glass fiber reinforced plastics (GFRP) manufactured by hand lay-up. *Composites Science and Technology*, *64*(2), 289–297. doi:10.1016/S0266-3538(03)00253-7

Erturk, A. T., Vatansever, F., Yarar, E., Guven, E. A., & Sinmazcelik, T. (2021). Effects of cutting temperature and process optimization in drilling of GFRP composites. *Journal of Composite Materials*, *55*(2), 235–249. doi:10.1177/0021998320947143

Ferriera, J.R., Coppini, N.L., & Miranda, G.W.A. (1999). *Machining optimization in carbon fibre reinforced composites materials.* Journal of Materials Processing Technology, 92, 135–140.

Haddad M, et al. (2014) *Study of the surface defects and dust generated during trimming of CFRP: influence of tool geometry, machining parameters and cutting speed range. Compos* A Appl Sci Manuf., 66, 142–54.

Hocheng, H., Puw, H. Y., & Yao, K. C. (1992).*Experimental aspects of drilling of some fiber-reinforced plastics. Proceedings of the Machining of Composites Materials Symposium, ASM Materials Week,* 127–138.

Inoue, H., Aoyama, E., Hirogaki, T., Ogawa, K., Matushita, H., Kitahara, Y., & Katayama, T. (1997). Influence of tool wear on internal damage in small diameter drilling in GFRP. *Composite Structures*, *39*(1–2), 55–62. doi:10.1016/S0263-8223(97)00068-8

Jain, A., Kumar, C. S., & Shrivastava, Y. (2022) *Machining of composite materials using different conventional and unconventional machining processes: a short review.* Adv Mech Energy Technol Select Proc ICMET, 75–83.

Jessy, K., Dinakaran, D., & Seshagiri Rao, V. (2015). Influence of different cooling methods on drill temperature in drilling GFRP. *International Journal of Advanced Manufacturing Technology*, *76*(1), 609–621. doi:10.1007/s00170-014-6280-y

Kandavalli, S. R., Khan, A. M., Iqbal, A., Jamil, M., Abbas, S., Laghari, R. A., & Cheok, Q. (2023). Application of sophisticated sensors to advance the monitoring of machining processes: Analysis and holistic review. *International Journal of Advanced Manufacturing Technology*, *125*(3-4), 989–1014. Advance online publication. doi:10.1007/s00170-022-10771-6

Kaw, K. (2006). *Mechanics of Composite Materials* (2nd ed.). Taylor & Francis.

Konig, W., Wulf, Ch., Grab, P., & Willerscheid, H. (1985). Machining of fibre reinforced plastics. *CIRP Annals*, *34*(2), 537–547. doi:10.1016/S0007-8506(07)60186-3

Koplev, A., Lystrup, A., & Vrom, T. (1983). The cutting process, chips, and cutting forces in machining CFRP. *Composites*, *14*(4), 371–376. doi:10.1016/0010-4361(83)90157-X

Krishnamurthy, R., Santhanakrishnan, G., & Malhotra, S. K. (1992).*Machining of polymeric composites.* Proceedings of the Machining of Composites Materials Symposium, ASM Materials Week, 139–148.

Lin, S., & Chen, I. (1996). Drilling carbon fiber-reinforced composite material at high speed. *Wear*, *194*(1–2), 156–162. doi:10.1016/0043-1648(95)06831-7

Lin, S.-C., & Shen, J.-M. (1999). Drilling unidirectional glass fiber-reinforced composite materials at high speed. *Journal of Composite Materials*, *33*(9), 827–851. doi:10.1177/002199839903300903

Liu, L., Qi, C., Wu, F., Zhang, X., & Zhu, X. (2018). Analysis of thrust force and delamination in drilling GFRP composites with candle stick drills. *International Journal of Advanced Manufacturing Technology*, *95*(5), 2585–2600. doi:10.1007/s00170-017-1369-8

Mahakur, V. K., Bhowmik, S., & Patowari, P. K. (2022). Machining parametric study on the natural fiber reinforced composites: A review. *Proceedings of the Institution of Mechanical Engineers. Part C, Journal of Mechanical Engineering Science*, *236*(11), 6232–6249. doi:10.1177/09544062211063752

Malik, K., Ahmad, F., & Gunister, E. (2022). Drilling performance of natural fiber reinforced polymer composites: A review. *Journal of Natural Fibers*, *19*(12), 4761–4779. doi:10.1080/15440478.2020.1870624

Mohan, N., Ramachandra, A., & Kulkarni, S. (2005). Influence of process parameters on cutting force and torque during drilling of glass–fiber polyester reinforced composites. *Composite Structures*, *71*(3–4), 407–413. doi:10.1016/j.compstruct.2005.09.039

Palanikumar, K., Rubio, J. C., Abrão, A., Esteves, A., & Davim, J. P. (2008). Statistical analysis of delamination in drilling glass fiber-reinforced plastics (GFRP). *Journal of Reinforced Plastics and Composites*, *27*(15), 1615–1623. doi:10.1177/0731684407083012

Rajamurugan, T., Rajaganapathy, C., Jani, S. P., Gurram, C. S., Allasi, H. L., & Damtew, S. Z. (2022). Analysis of drilling of coir fiber-reinforced polyester composites using multifaceted drill bit. *Advances in Materials Science and Engineering*, *2022*, 1–9. doi:10.1155/2022/9481566

Ramulu, M., Faridnia, M., Garbini, J. L., & Jorgensen, J. E. (1989). Machining of graphite/epoxy materials with polycrystalline diamond (PCD) tools. In M. Ramulu & M. Hashish (Eds.), *Machining Characteristics of Advanced Materials* (Vol. 16, pp. 33–40). ASME.

Ross, N. S., Sheeba, P. T., & Shibi, C. S. (2023). A novel approach of tool condition monitoring in sustainable machining of Ni alloy with transfer learning models. *Journal of Intelligent Manufacturing*. Advance online publication. doi:10.1007/s10845-023-02074-8

Saikrupa, C. H., ChandraMohan Reddy, G., & Venkatesh, S. (2023). *Experimental investigation of cutting conditions in turning of Al-MMC.* Materials Today: Proceedings, 76, 488–494. doi:.matpr.2022.11.154 doi:10. 1016/j

Shaw, M. C. (2005). *Metal Cutting Principles* (2nd ed.). Oxford University Press.

Slamani, M., Gauthier, S., & Chatelain, J.-F. (2015). A study of the combined effects of machining parameters on cutting force components during high-speed robotic trimming of CFRPs. *Measurement*, *59*, 268–283. doi:10.1016/j.measurement.2014.09.052

Spur, G., & Lachmund, U. (1999). Turning of fiber-reinforced plastics. In Machining of Ceramics and Composites. Marcel Dekker.

Tian, J., Wu, F., Zhang, P., Lin, B., Liu, T., & Liu, L. (2019). The coupling effect and damage analysis when drilling GFRP laminates using candlestick drills. *International Journal of Advanced Manufacturing Technology*, *102*(1), 519–531. doi:10.1007/s00170-018-03216-6

Toug, L., Mouritz, A., & Bannister, M. (2002). *3D fiber reinforced polymer composites*. Elsevier.

Urresti, I., Llanos, I., de Lacalle, L. N. L., & Zelaieta, O. (2022). Study of the surface integrity during CFRP trimming: Tool material and geometry, fiber orientation and tool wear effect analysis. *Procedia CIRP*, *108*, 660–664. doi:10.1016/j.procir.2022.03.102

Wang, F. (2022). Effects of cutting speed and fiber orientation on tool wear and machining quality in milling CFRP with PCD cutter. *Proceedings of the Institution of Mechanical Engineers. Part B, Journal of Engineering Manufacture*.

Chapter 6
Advanced Polycrystalline Diamond Coatings on Cutting Tools for CFRP Composites:
A Review

Senthilkumar Chandrasekaran
https://orcid.org/0000-0001-5375-1754
SNS College of Technology, India

Devaraj E.
https://orcid.org/0009-0000-2217-2872
CMR University, India

ABSTRACT

This review chapter focuses on enhancing the cutting performance of diamond-coated tools when machining CFRP composite materials. The enhancement is achieved by utilizing advanced polycrystalline diamond coatings on WC-Co-based cutting tools. Additionally, a proposed reconditioning method for these diamond-coated tools is discussed. The chapter elaborates on a hot filament chemical vapor deposition (HFCVD) technique for preparing diamond coatings with various film morphologies, achieved through boron doping. It covers the fundamentals, equipment used, deposition parameters, and substrate pretreatment techniques involved in the HFCVD method. Furthermore, it explores the strategy of improving interface adhesion by doping chemical elements into diamond coatings.

DOI: 10.4018/978-1-6684-9927-6.ch006

1. INTRODUCTION

Throughout human history, composite materials have been crucial to the housing of early civilizations and the advancement of future technologies. Indeed, for thousands of years, people have used composite materials in a variety of contexts. When early Egyptians and Mesopotamian settlers built sturdy, long-lasting structures out of a mixture of mud and straw, composite materials were first utilized around 1500 BC. A brick's ability to withstand bending, tearing, and squeezing is derived from the combination of mud and straw. Boats and other ancient composite goods were kept sturdy by the straw (Johnson, 2018). The first composite bow was created by the Mongols in 1200 AD by combining "animal glue," bone, and wood. Birch bark was crushed and wrapped around the bows.

The composite Mongolian bows contributed to the military supremacy of Genghis Khan. Wartime requirements led to many of the great breakthroughs in composites because of their advantages, such being strong and lightweight. A Japanese sword, for instance, that is composed of soft iron and steel has good flexural and impact resistance (Gay et al., 2003). The creation of carbon and boron fibers in the 1960s has been identified as the turning point in the contemporary decade in the history of the composite. The first carbon fiber was patented in 1961 and made commercially available a few years later. The use of composites in the aerospace, defense, nuclear, automotive, and marine (boat and yacht) industries has increased significantly since the 1970s. Wind energy systems, pressure vessels, sports goods (golf clubs, tennis rackets, bicycles, and surfers), and biomedical items are among the products that use composite constructions or components (Das et al., 2016).

Composites are materials made of two or more constituent pieces combined to produce a product that is better than the sum of its parts (Brent, 2008; Callister & Rethwisch, 2008). One is often a softer bulk phase termed the matrix encased in a hard, stiff, and strong distributed phase known as reinforcement. While the matrix is either a polymer, metal, or ceramic, the reinforcement, which can take the shape of particles, fibers, whiskers, etc., is typically formed of glass, carbon, aramid, or metallic compounds. The matrix distributes a load, maintains the fibers in place, shields the reinforcement from abrasion and chemical reactions, and establishes the composite material's service temperature. These components affect the final properties of the material even though they keep their original qualities when combined to produce the composite. In more recent times, materials made from sheets of pre-impregnated fibers in an appropriate resin matrix have been referred to as "advanced composites" (Brent, 2008; Callister & Rethwisch, 2008; Soutis, 2005; UMECO, 2013). These sheets are usually between 100 and 150 µm thick; they are often referred to as plies or prepregs. Advanced composites offer several benefits, including near-net-shape formability, improved rigidity and damping, high strength/stiffness to weight ratios that result in significant weight reduction, and customizable properties to meet loading requirements. They are especially appealing as components for the automobile and aerospace industries.

Figure 1. Mechanical properties of various metals and composites (UMECO, 2013)

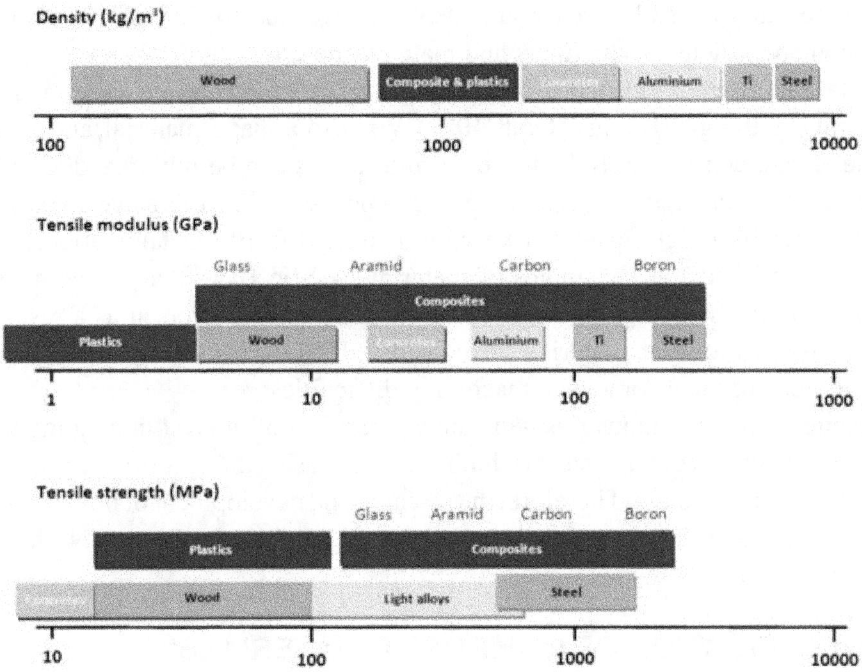

In the civilian aircraft sector, for instance, composites are being used more often. For the new Boeing 787 Dreamliner, for instance, around 50% of the entire weight is made of composites, compared to less than 5% for the Boeing 757/767 manufactured in the 1980s (Sheikh-Ahmed, 2009). Comparably, CFRP makes up 22-25% (35 tons) of the new Airbus A380's structural weight, which is predicted to result in a 12% decrease in fuel usage.

Figure 1 compares the characteristics of several materials, including composites and traditional metals. The benefits of composites include increased stiffness, increased strength, and decreased weight. Even though the composite-based component fabrication process yields nearly net-shaped products, a variety of machining processes, such as edge trimming, drilling, milling, turning, etc., are sometimes still required to meet assembly requirements and part quality criteria.

Furthermore, there are other ways in which the machinability of composites varies from that of normal materials and their alloys, including chip formation, the need for specific cutting tools, and operating parameters (Bhatnagar et al., 1995; Karataş & Gökkaya, 2018; König et al., 1985). This is mainly because the mechanical and thermal properties of the reinforcement and matrix differ, and composites are inherently non-homogeneous and anisotropic (Bhatnagar et al., 1995; Ferreira et al., 1999; Teti, 2002). For instance, thermal stresses in composites can result in deformation and possibly even component damage due to the varying thermal expansion coefficients of the fibers and matrix materials.

Surface layer separation can also be brought on by strong cutting forces and low interlaminate strength (König & Graß, 1989). When tool shape, material, and cutting parameters are chosen carefully, these harmful forces can be minimized (Sheikh-Ahmed, 2009). Furthermore, rigid fibers in composites have an abrasive impact that invariably results in significant tool wear and a limited tool life. Delamination, burrs, and ripping of fibers are examples of material-specific flaws that the composite workpiece frequently displays when cutting tool conditions worsen as a result of wear and tear. Composite materials are considered to be challenging to cut, as seen by the aforementioned composite machining difficulties.

In spite of this, the industries generate a great deal of interest in creating new, affordable methods for composite machining and increasing their profitability across a range of industrial sectors. Therefore, the research and development of novel ways to enhance composite cutting performance is the main emphasis of this review chapter.

2. CLASSIFICATION OF COMPOSITE MATERIALS

Composites can be generically categorized as metal matrix composites, ceramic matrix composites, or polymer matrix composites based on the matrix system that is used to classify them. As seen in Figure 2 (Callister & Rethwisch, 2008), they can also be divided into groups according to the format or arrangement of their reinforcement, which usually consists of particles, whiskers, fibers (continuous, discontinuous, aligned, and random), and structural arrangements.

Compared to exclusively metal-based alternatives, metal matrix composites (MMC) can offer higher temperature working limitations (John & Blackmon, 1990). Particle reinforcement such as silicon and boron carbides or continuous carbon, boron, and silicon fibers can be used to reinforce metals like aluminum, copper, nickel, magnesium, steel, titanium, and their alloys. These composites are made by either infiltrating a preform with liquid metal under pressure or soaking fibers in molten metal (Brent, 2008).

Figure 2. Classification for various composites according to reinforcement format (Callister & Rethwisch, 2008)

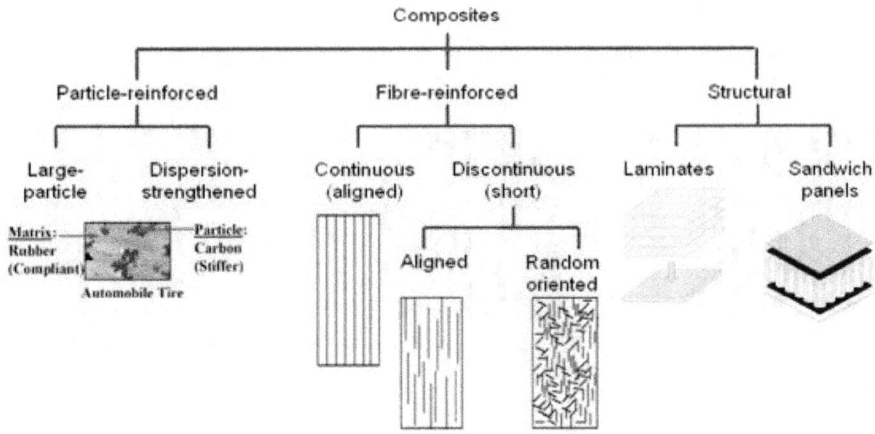

The advantages of MMCs over CMCs and PMCs are their high thermal and electrical conductivity, resistance to organic fluid degradation, and nonflammability. Corrosion-induced interfacial deterioration at the fiber/matrix contact is the primary drawback of MMCs.

Fibers serve as the reinforcement in polymer matrix composites (PMCs), which are made of a polymer resin matrix. In dispersion strengthened composites, the matrix material is combined with small particles (diameters ranging from 10^{-5} mm to 10^{-4} mm). By assisting the matrix in resisting deformation, these particles help create a stronger and harder material. The particle diameter of large-particle reinforced composites is usually several micrometers, and it bears a significant portion of the load. This arrangement, with relatively large carbon particles contained within a polymer matrix, is exemplified by car tires. Usually created using chemical vapor deposition (CVD), whiskers are anisotropic single crystals with length/diameter ratios ranging from 10 to 100. Their diameter is normally within the range of 0.1 to 1 μm. With MMCs and CMCs, particles and whiskers are typically utilized because their random distribution preserves the isotropic nature of the composite.

Present-day engineering fibers often have a circular shape with a diameter of several tens of micrometers, a relatively long axis, and stronger mechanical characteristics along the length axis. The part that follows will provide a more thorough explanation of fiber reinforced composites.

Laminates, a homogenous material made of many layers of fiber-reinforced plastic sheets, are commonly used in the fabrication of structural composites. Both the individual layers and the fiber orientation of the various structural layers affect the laminate's characteristics. Occasionally, sandwich panels with a honeycomb structure are made using laminates.

Figure 3. The most commonly used fabric types (UMECO, 2013)

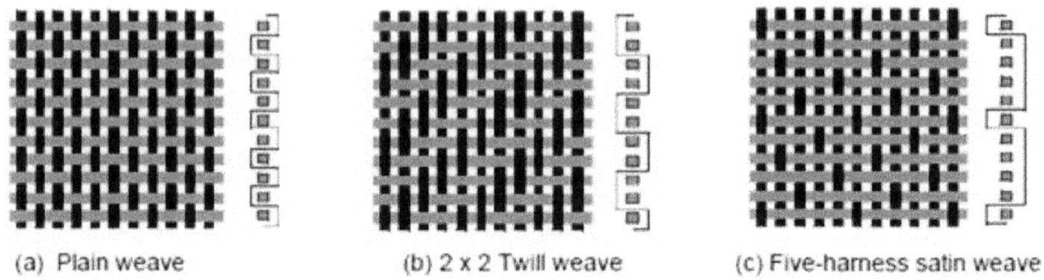

Figure 4. Example of stacking sequence in multidirectional layup (Liu et al., 2012)

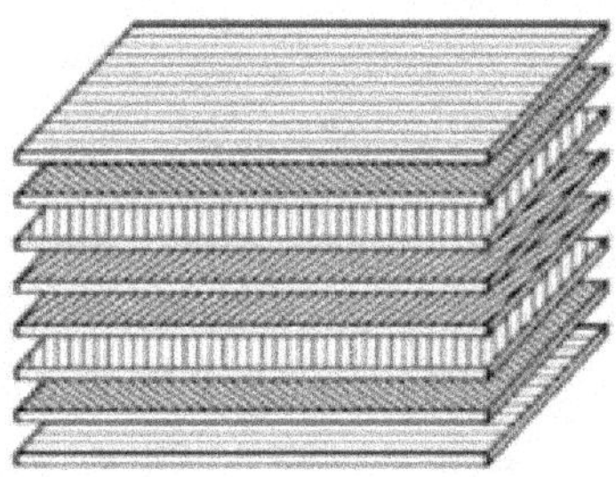

2.1 Fiber-Reinforced Plastic (FRP) Composites

When brittle and strong fibers are mixed into a soft and ductile polymer matrix, PMCs are known as fiber-reinforced plastics (FRPs) (Teti, 2002). The main fibrous patterns in fiber-reinforced polymer (FRP) composites are continuous fibers in the form of woven or unidirectional (UD) fabrics. While woven constructions use fibers that meet perpendicularly, unidirectional materials have their fibers organized in a single direction. Figure 3 illustrates the many types of woven fabrics, such as plain weaves, twill weaves, and satin weaves, in which each fiber yarn crosses over and under every other perpendicular thread. As seen in Figure 4, multidirectional composite laminates can be created by laying up both UD and woven plies.

The most popular fiber types used in PMC are carbon, glass, and aramid; the type of reinforcing fibers is indicated by the letters C, G, or A that come before the acronym FRP (Teti, 2002). Although to a lesser degree, boron and polyethylene fibers are also utilized in FRP composites. Figure 5 displays the general mechanical characteristics and relative costs for each particular fiber.

Usually, a liquid substance is drawn through an aperture or a precursor is pulled through it to create fibers. They are then provided in the form of tows, which are made up of single continuous fiber filaments and often include a bundle of 3,000–30,000 filaments.

Since their initial introduction in the 1930s, glass fibers have been a common reinforcement material for polymer matrix composites. The most popular kinds are C-glass, S-glass, and E-glass, which are electrical, high strength, and stiff.

Figure 5. Relative properties for various individual fibers (UMECO, 2013)

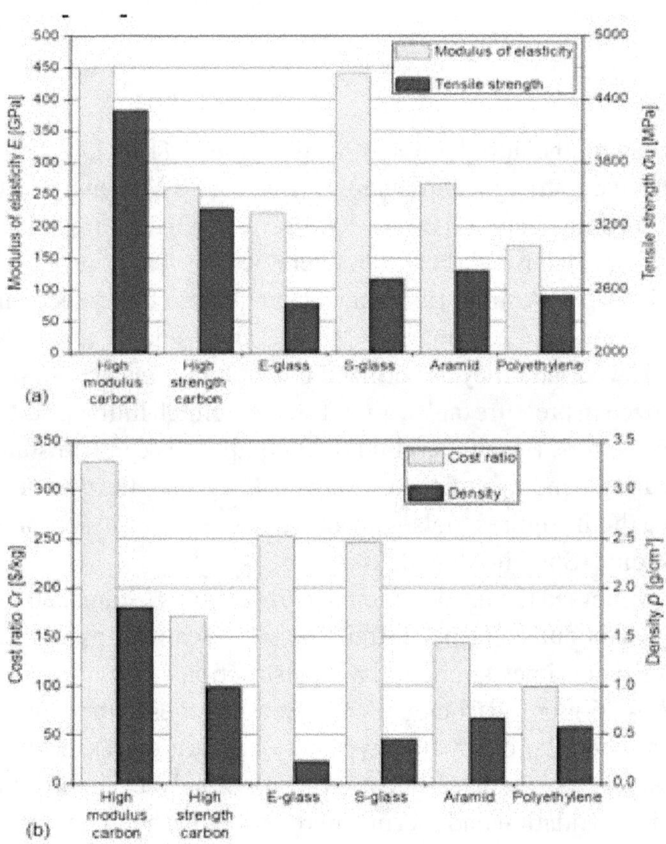

Whereas C-glass fibers are utilized in industries needing corrosion resistance, such the chemical and marine sectors, S-glass fibers are utilized in aerospace applications. Glass fibers are less expensive than other fibers and have superior chemical resistance, although they can only be used at temperatures below 200 °C (Callister & Rethwisch, 2008). In addition, compared to carbon fibers, they have a lower tensile strength and a higher density. Furthermore, the abrasive nature of glass fibers negatively impacts the machinability of glass fiber reinforced polymers.

Figure 6. Manufacturing process for making carbon/graphite fibers (Brent, 2008)

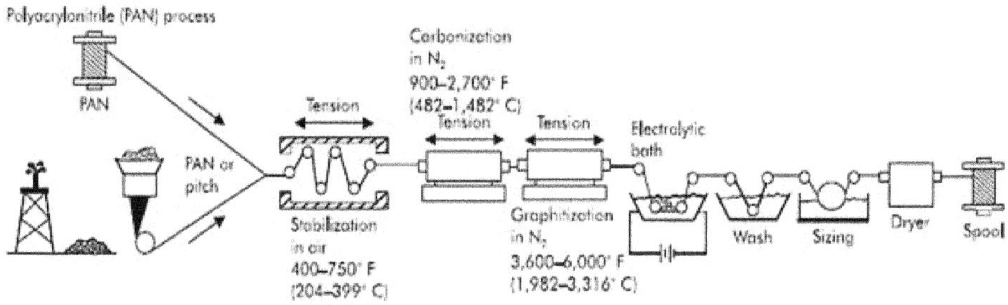

The 1970s saying the introduction of aramid fibers. Their low density (~1.44 g/cm3) and high tensile strength and modulus (σu = 3720 MPa and E = 63-143 GPa) are their distinguishing features (Brent, 2008). Although they are quite strong under tension, their compression and shear characteristics are very poor, and their stiffness can approach 125 GPa. Aramid fibers can also maintain their mechanical qualities at temperatures between -200 and 200 °C. The most widely used trade name for these fibers is Kevlar, and they are utilized in a polymer matrix. Aramid fibers are commonly utilized in pressure tanks, aircraft wings, clutch linings, bulletproof vests, and battlefield shelters. Because aramid fiber composites can withstand significant deformation during bending and then spring back, causing fuzzing of the machined surface, cutting them requires tools with a high degree of edge sharpness and a tiny cutting-edge radius (Sheikh-Ahmed, 2009).

The 1960s witnessed the introduction of carbon fibers (Soutis, 2005). They have a broad variety of moduli and tensile strengths, which can be up to three times stiffer than steel and fifteen times stronger than construction steel, in addition to having a low density (~1.6 g/cm^3) (Chung, 1994). Because these fibers have the highest specific strength/modulus of any fiber type, they are regarded as the most significant material for reinforcing. Pitch, cellulose (Rayon), or polyacrylonitrile fibers (PAN) are carbonized by oxidation and thermal pyrolysis at high temperatures to produce commercial carbon fibers. The standard procedures for producing carbon fiber based on PAN are depicted in Figure 6. In order to crosslink the fibers, they are first pulled and oxidized in air before being carbonized in a nitrogen atmosphere. By doing this, non-carbon atoms are guaranteed to be eliminated, resulting in fibers made entirely of carbon. To increase the fibers' purity and crystallinity, more graphitization is done. Size and surface treatment are applied after graphitization. Because carbon fibers are anisotropic, the degree to which the graphite layers are oriented with respect to the fiber axis influences the fiber's characteristics. Graphite layers are composed of carbon atoms bonded together by a strong covalent connection to form hexagonal rings. The graphite layers are held together by secondary bonds, which allow for slide along the hexagonal planes (Brent, 2008). A greater temperature during the graphitization process encourages the graphite layers to align with the direction of the fiber, which raises the tensile modulus.

The general purpose (GP), high performance (HP), and activated carbon fibers (ACF) are the three major categories into which carbon fibers fall (Chung, 1994). The structure of the GP and HP kinds is isotropic and amorphous. On the other hand, the abundance of open micropores in activated carbon fibers might serve as adsorption sites, which restricts their application. They are used to make gas masks and water treatment units, among other environmental protection gear. The characteristics of the various carbon fiber types are listed in Table 1, and the attributes of carbon fibers categorized based on their strength/modulus ratio and tensile strength are shown in Table 2.

The matrix in fiber-reinforced polymer (FRP) composites serves to transmit and distribute externally imposed loads to the reinforcement, support and link the reinforcement phase, regulate the chemical and electrical properties of the composite, and stop cracks from spreading (Brent, 2008; Callister & Rethwisch, 2008; Das et al., 2016; Gay et al., 2003). Furthermore, the matrix establishes the composite part's maximum functioning temperature because it often melts, softens, or degrades at a lower temperature than the fibers. The glass transition temperature (Tg), at which many of PMCs' physical qualities change or deteriorate, meaning that PMCs cannot be employed close to or above this temperature (Brent, 2008). To increase the matrix phase's mechanical characteristics and permitted operating temperature, it can be combined with other additives or used pure.

For FRP composites, thermoset polymers (like polyester and epoxy) and thermoplastic polymers (like polyamide and PEEK) are the two main types of matrix materials that are typically used.

Table 1. Properties of various carbon fiber types (Chung, 1994)

Type	Tensile modulus (GPa)	Strength to modulus ratio %
UHM (ultra high modulus)	> 500	-
HM (high modulus)	> 300	< 1
IM (intermediate modulus)	< 300	> 1
LM (low modulus)	< 100	Isotropic structure
HT (high strength)	Tensile strength > 3 GPa	1.5 - 2

Table 2. Classification of carbon fibers based on their tensile strength and modulus (Chung, 1994)

Type	Diameter (μm)	Tensile strength (MPa)	Tensile modulus of elasticity (GPa)
General purpose (GP)	(7 – 15)	700	40
High performance (HP pitch)	(9 – 11)	5000	350
High performance polyacrylonitrile (HP PAN)	(4 – 8)	2000	200-800
Activated carbon fibres (ACF)	(7 – 15)	200	1500

Thermosets are substances that often change from a liquid to a solid by a chemical reaction or curing. The substance is made up of unlinked monomers in liquid form. The chemical reaction will be started by the presence of heat, the inclusion of a second material as a catalyst, or both. The substance solidifies as a result of the molecules' formation of much longer molecular chains and cross-link networks during this reaction. The thermoset state has undergone an irreversible and permanent alteration. Thermosets are very strong and stiff and do not melt when heated (Teti, 2002). Instead, exposure to high heat will cause the material to deteriorate. Although it is less expensive, the polyester matrix is mostly utilized in glass fiber composites.

There is a market for thermoset and thermoplastic materials alike. Compared to thermoplastics, thermosets make it easier to wet reinforcing fibers and shape final part geometries. However, thermoplastic polymers provide a simple way to recycle composite parts. About 80% of the materials utilized in the composite sector are thermoset (Das et al., 2016).

The mechanical characteristics of the most widely used CFRP composites are compared to comparable Aramid fiber reinforced polymer (AFRP) and Glass fiber reinforced polymer (GFRP) composites in Table 3. The data presented leads to the conclusion that CFRP composites, in comparison to other FRP composites, have advantageous mechanical qualities such low density, strength, and modulus of elasticity. Nonetheless, it's possible to anticipate the high price of CFRP products.

However, CFRP-based composites are currently among the most cutting-edge materials available. For the majority of its existence, aerospace projects have been the main forces behind the CFRP sector. In the development of electric vertical take-off and landing (eVTOL) aircraft and unmanned aerial vehicles (UAVs) equipped with electric propulsion systems, CFRP composites have recently emerged as a fundamental structural material. As a result, CFRP composite used in aerospace applications is the material of choice for the workpiece in this evaluation of cutting performance.

Table 3. Mechanical properties for the common CFRP laminates against GFRP and AFRP composites (Teti, 2002)

FRP material	Tensile strength, σu (MPa)	Elastic modulus, E (MPa)	Strain to failure, εu (%)	Density, ρ (g/cm^3)
CFRP				
Unidirectional (Vf=60%) High strength	1,200	145,000	0.9	1.6
Unidirectional (Vf=60%) High modulus	800	220,000	0.3	1.6
GFRP				
Unidirectional (Vf=60%)	1,000	45,000	2.3	2.1
Woven cloth*	100-300	10,000-20,000	-	1.5-2.1
Chopped roving* (short fibers)	50-200	6,000-12,000	-	1.3-2.1
Sheet molding compound*(short fibers)	10-20	500-2,000	-	1.3-1.9
AFRP				
Unidirectional (Vf=60%)	1,000	75,000	1.6	1.4
* For these materials Vf=20-50%				

2.2 Description of CFRP Composite Components

A single, orthotropically behaving, unidirectional layer, or ply, is the fundamental component of every continuous fiber composite item (Brent, 2008). Typically, the material provider is the one to identify the characteristics of each individual layer. Depending on the intended use, laminates are usually created by stacking layers of woven or unidirectional fabric at various fiber orientations (Gay et al., 2003).

Every ply in an N-layer laminate has a unique fiber orientation, which is numbered starting from the top surface. The angles of the reinforcing fibers within the plies are used to code the laminate, with the orientation of a single ply on the stack being represented by either a (+) or (-) sign from the reference axis (0°). The notation that is utilized to indicate the order in which a FRP laminate is stacked consists of brackets, parentheses, slashes, and subscripts. Subscript s (outside the brackets) indicates a symmetric laminate; slashes "/" are used to separate layers or groups of layers in a certain direction; and brackets "()" are used to indicate all the fiber orientations contained within a laminate in line with the stacking sequence. As an example, the UD laminate layup (0/+45/-45/90/-45/+45/0) that begins at the top surface can be explained as follows (Figure 7).

Figure 7. Different possible ply orientation for UD laminate layup

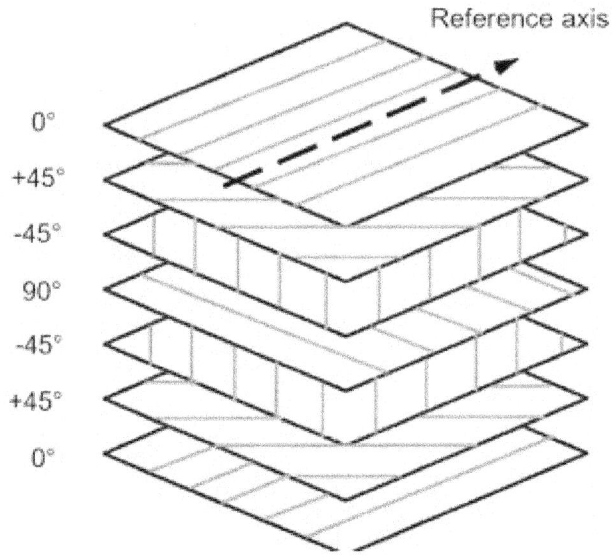

Figure 8. Quasi-isotropic vs. unidirectional lay-up (Hexcel, 2013)

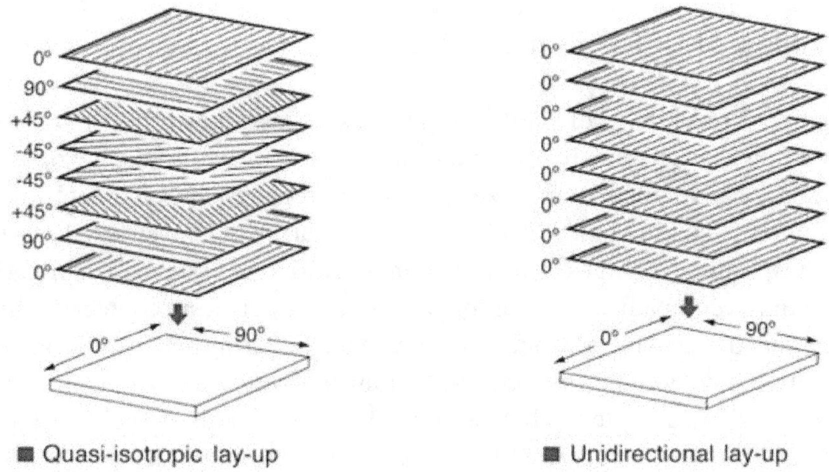

Only half of the layers specified should have an even number of plies in a symmetric laminate. For instance, the notation (02/90/90/02)s designates a laminate with 12 plies, the first of which is oriented in 0° and is repeated twice. The laminate is symmetric with respect to a center plane. A laminate structure is referred to as quasi-isotropic if it has an equal number of plies at each primary fiber orientation, such as (0/45/-45/90)s, as seen in Figure 8.

2.3 Manufacturing of CFRP Composites

Advanced composite components can be made using a variety of techniques, such as vacuum bag molding, autoclave molding, filament winding, press molding, pressure bag molding, thermal expansion molding, and pultrusion, as was previously discussed (Sheikh-Ahmed, 2009). For the creation of high-performance composites, prepreg lay-up is utilized. The procedure can be either automatic—using a robot or CNC machine, as in the case of the automatic tape laying (ATL) process—or manual—involving hand lay-up (HLU). To harden the matrix, the prepregs need to be cured in an autoclave and placed in a vacuum bag. For maritime and railroad applications, the curing procedure can be completed in an oven; however, an autoclave is the preferred method for the highest-performance aerospace composites (Sheikh-Ahmed, 2009). Fiber arrangement and flaws, such as matrix cracks (e.g., voids, porosities), fiber cracks, interface cracks, delamination, or inclusions that may materially alter its properties, are usually the main areas of attention for inspection of fiber reinforced composite parts (Chung, 1994). Metallographic analysis, X-ray analysis, infrared thermal imaging, and C-scan procedures are examples of inspection techniques.

2.4 Drilling of CFRP Composites

Conventional machining of CFRP composites necessitates tools equipped with sharp cutting edges to effectively handle abrasive fibers, while keeping operational parameters relatively low to meet surface quality requirements and control cutting temperatures (Sheikh-Ahmed, 2009). Among the array of conventional machining techniques, drilling emerges as a paramount operation in composite fabrication.

Drilling, accompanied by subsequent procedures such as countersinking and reaming, stands out as the predominant method for preparing composite components for assembly and joining, particularly within the aerospace sector. It is widely acknowledged that a single small aircraft may require upwards of a hundred thousand mounting holes, while larger aircraft can demand over a million (Karataş & Gökkaya, 2018). Furthermore, manufacturers have indicated that approximately 40% of material removal volume in the aerospace industry is attributed to drilling operations. Hence, hole production assumes a pivotal role in composite manufacturing, with drilling typically serving as the concluding operation.

In comparison, milling sees less use as a material removal process in the production of composite parts compared to its application in metal machining. This disparity arises because CFRP components are typically manufactured to near-net shape, with subsequent milling primarily employed for deburring and edge trimming to achieve precise contouring (Karataş & Gökkaya, 2018), considerable attention should be paid to all factors affecting the drilling of the composite material. Milling is less commonly employed in the manufacturing of composite parts compared to metals, primarily due to the fact that CFRP components are typically produced near net shape, with any subsequent milling mainly reserved for deburring and edge trimming to ensure contour shape accuracy.

In addition to traditional drilling methods, several advanced drilling techniques and unconventional machining processes have shown promising outcomes.

Vibration-assisted drilling (VAD) has garnered considerable interest in recent years from both academia and industry. In VAD, a low (<1000 Hz) or high (>1000 Hz) frequency and low amplitude vibration are applied to the drill along the feed direction during drilling (Brehl & Dow, 2008). Consequently, VAD operates as a pulsed intermittent cutting process, differing from the continuous cutting process of conventional drilling. This results in reduced thrust force and improved chip evacuation during VAD, leading to decreased delamination compared to conventional drilling. In fact, under similar drilling conditions, thrust force induced by VAD is reduced by 20–30% compared to conventional drilling (Arul et al., 2006; Linbo et al., 2003).

Additionally, VAD has shown a reduction of approximately 50% in delamination area, along with improvements in circularity and surface roughness (Makhdum et al., 2014). Despite its promise as a CFRP drilling technique, the industrial application of VAD requires machining centers equipped with oscillation systems, which are currently unavailable on the market.

Researchers have explored the feasibility of water jet machining for cutting FRPs. This method utilizes water jet erosion under high pressure with abrasive slurry to cut the target composite material. To date, abrasive water jet (AWJ) machining has been the predominant method used in industry for trimming CFRP composite materials among non-conventional machining processes due to its minimal sensitivity to material properties and flexibility (Azmir & Ahsan, 2008). However, severe delamination and chipping persist at the hole exit (Shanmugam et al., 2002; Wang, 1999).

Laser machining, a non-contact process, holds promise for CFRP drilling due to its advantages of low stress, minimal damage, and high precision. However, the sensitivity of the CFRP resin matrix to heat poses a significant limitation to laser cutting, leading to the formation of a heat-affected zone (HAZ) around the hole, which in turn affects matrix recession, fiber distortion, and delamination (Cheng et al., 1998; Pan & Hocheng, 1996). Additionally, machining taper can affect the depth drilling ability in CFRP components. A recent promising advancement is the proposal of a picosecond laser "double rotation" cutting technique, which has shown no interlaminar tearing, cracks, or burrs on the cutting section, and significantly reduces the HAZ (Ouyang et al., 2021).

Despite the myriad advancements in various machining techniques, this study primarily focuses on the cutting performance of CFRP, with an emphasis on conventional drilling to underscore current research trends.

2.4.1 Drilled Hole Quality

Since hole quality affects the composite parts' service life after assembly, it is one of the most crucial factors to consider when evaluating drilling performance. First, as seen in Figure 9, König and Grab (1989) presented a system for evaluating hole quality by categorizing faults frequently connected to drilled holes in FRP. The workpiece surface roughness, hole roundness error, dimensional mistakes, damage to surface layers, damage inside a peripheral zone, and cracks are among the quality requirements in this case. Similar to this, Abrão et al. (2007) examined the primary hole quality flaws that occur when drilling CFRP, such as surface delamination, fiber pullout, and inadequate surface roughness. There are more categories for damage to surface layers, including fuzzing, delamination, edge chipping, and spalling. Composite component strength and fatigue life are greatly impacted by tool wear, the presence of powdery chips, workpiece surface and internal delamination, fiber pullout, and reduced flexural strength.

Figure 9. Quality criteria when drilling FRPs (König & Graß, 1989)

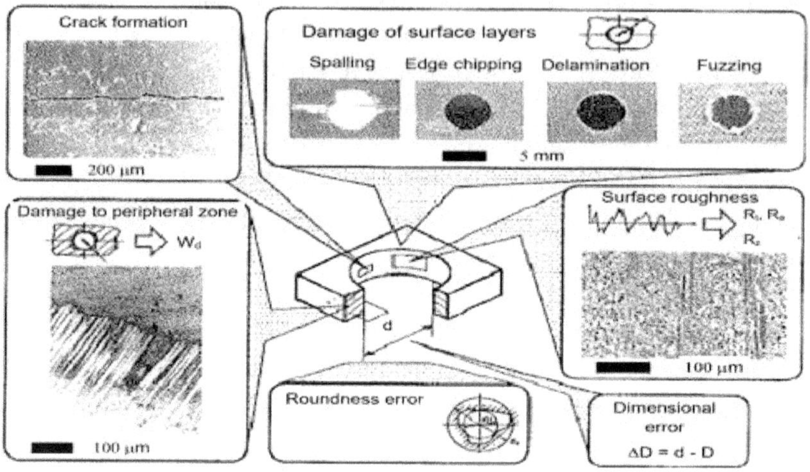

Figure 10. Delamination forms: (a) peel-up at entrance and (b) push-out at exit (Liu et al., 2012)

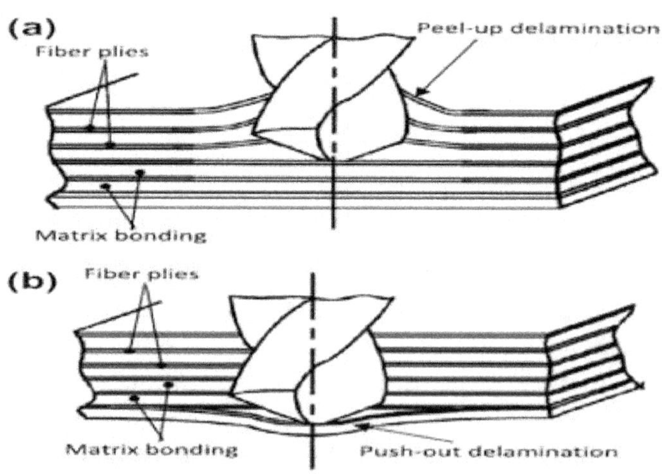

While Zhang et al.'s study (Zhang et al., 2001) found that the main hole defect was spalling, several other publications (Feito et al., 2015; Geng et al., 2019; Karataş & Gökkaya, 2018; Shyha et al., 2009; Xu et al., 2013; Zhang et al., 2001) revealed that surface delamination was the main damage criterion when drilling CFRP. The primary cause of stiffness/strength decrease in composite constructions, which results in the laminate's catastrophic failure under bending or shear loading conditions, is widely recognized to be delamination (Abrão et al., 2007). To regulate delamination and achieve the necessary hole quality, it is crucial to evaluate drilling-induced delamination.

Figure 10 illustrates the two possible forms of delamination: push-out at the hole exit and/or peel-up at the hole entry. At the hole entry, peel-up happens when the tool pulls away from the composite laminate's upper layer, causing separation and the creation of an entry delamination zone. The uncut laminate thickness cannot withstand deformation as the drill gets closer to the bottom of the hole; as a result, the tool works like a punch, separating the thin uncut layers from the rest of the laminate. Then, when the thrust force is greater than the material's interlaminar bond strength, push-out delamination starts at the hole exit.

Figure 11. Diagram of the delamination damage: (a) uniform damage, (b) cracks, (c) uniform damage with cracks, and (d) uniform damage with fine cracks (Geng et al., 2019)

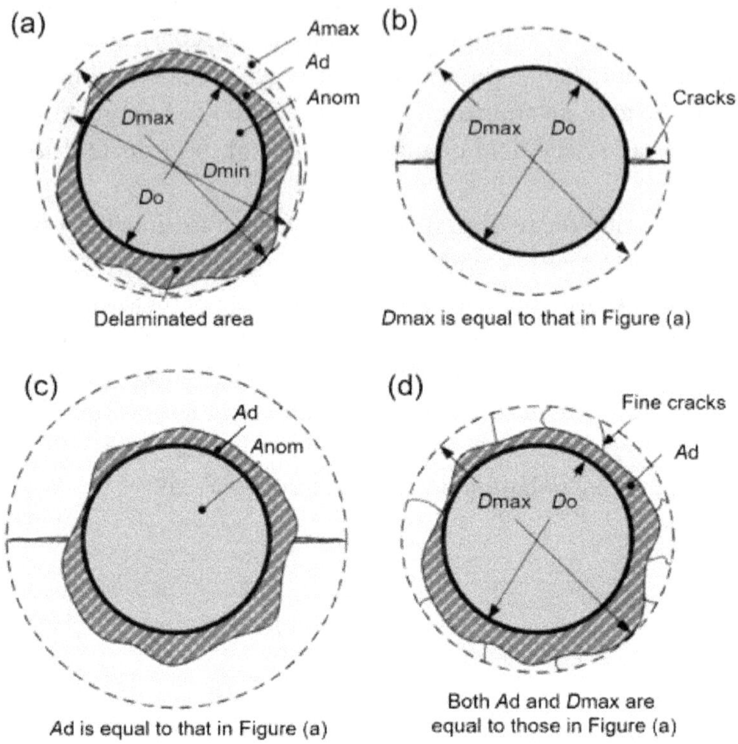

Chen (1997) proposed the delamination evaluation method that is currently in use. Two circles are concentric, as shown in Figure 11(a), and the delamination factor Fd is defined as the ratio of the maximum diameter of the damaged area (Dmax) to the nominal diameter of the hole (Do). When there is a regular pattern in the delamination, this delamination factor yields good results. However, if the delamination exhibits small cracks and ruptures, the delamination area may not be indicated during calculation, making the delamination factor inappropriate. In fact, it is evident from comparing Figure 1.11(a) and (b) that the same Fd is recorded for two different situations with the same Dmax. It is evident, therefore, that the prior state is more harmed. Later, Da Silva (Geng et al., 2019) suggested the minimal delamination factor (Fdmin), a comparable evaluation metric. The two circles in Figure 11(a) may not be concentric, but this delamination factor is determined by dividing the diameter of the minimal delamination area (Dmin) by the Do. When the delaminated area is not uniformly positioned around the drilled hole, this method can accurately determine the delamination factor findings. Nevertheless, it also disregards the damage area's contribution to the delamination assessment.

Faraz et al. (2009) proposed a Two-dimensional delamination factor (*F*a). It is defined as the ratio of the delamination area (*A*d) to the nominal hole area (*A*nom) and is expressed as a percentage, as shown below:

$$F_a = \left(\frac{A_d}{A_{nom}}\right)\%$$

The *F*a seems to be better than mentioned above delamination factors due to the consideration of the delamination area. However, it is still not accurate when the delamination damage occurs accompanied by long cracks, as shown in Figure 11(c).

In order to improve the accuracy of delamination assessment, several combination parameters were proposed (Geng et al., 2019). One of them is the adjusted delamination factor (*F*da) which is proposed by Davim et al. (2007) and calculated as shown below:

$$F_{da} = F_d + \frac{A_d\left(F_d^2 - F_d\right)}{A_{\max} - A_{nom}}$$

where the area corresponding to Dmax is denoted by Amax. The maximum crack length and damage area contributions are both taken into consideration in the construction of the adjusted delamination factor. In fact, it seems that the adjusted delamination factor is more accurate than the delamination factors that were previously discussed. That might still be off, though, in certain unique situations where a lot of tiny cracks appear, as Fig. 11(d) shows.

The previously mentioned delamination factors are only able to measure the extent of in-plane delamination. The delaminated damages inside the inner composite layers can be simply and precisely evaluated thanks to advancements in non-destructive delamination detection technology (such as ultrasonic scanning by an acoustic microscope). A new delamination factor (Fv) based on the Fa was recently proposed by Xu et al. (2018) in order to measure the interlaminar delamination within a composite laminate. The definition of this new delamination factor, which is independent of composite thickness, is the cumulative volume (Vd) to the nominal hole volume (Vnom) of the delaminated layers.

Figure 12. Acceptable damage (Bombardier Aerospace, 2011)

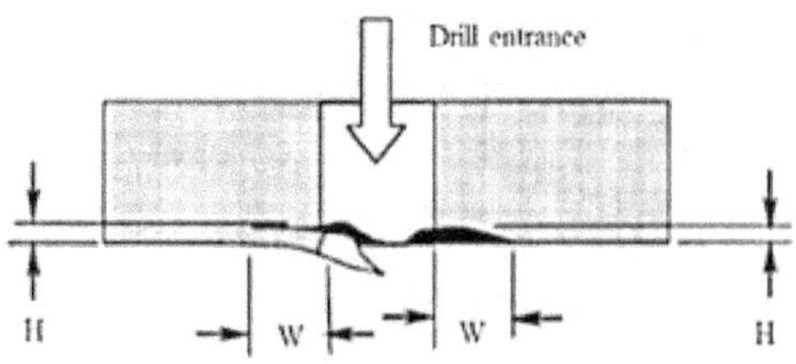

	Fabrics		UD tape	
Hole Diameter (Nominal)	H (mm)	W (mm)	H (mm)	W (mm)
3/16" (4.76 mm)	0.18	0.75	0.35	2.5
1/4" (6.35 mm)	0.18			2.5
5/16" (7.93mm), 3/8" (9.52mm)	0.18	1.0	0.35	3.0
7/16" (11.11mm)	0.18	1.0	0.35	4.0

The *F*v can be determined as follows:

$$F_v = \frac{V_d}{V_{nom=}} = \frac{1}{p}\sum_{k=1}^{p}\frac{A_d^k}{A_{nom}}$$

Where *p* is the total number of the delaminated plies and *k* is the number of specific delaminated ply.

However, there is currently no consensus on how to evaluate the delamination. Additionally, the evaluation of drilling-induced delamination in research fields mostly makes use of the previously established delamination parameters.

Determining interlaminar failure and interfacial debonding between neighboring plies is crucial from an industrial perspective because of the potential for product rejection. Because of its innate ability to spread across interlayer plies, this kind of damage is frequently caused by the thermo-mechanical coupling effects that follow a drilling operation. It can occur both inside the interior layers of a composite and in its outermost layers, forming an out-of-plane imperfection (Xu et al., 2018). As a result, non-destructive diagnostic systems are frequently employed for manufacturing product quality control, particularly in the aerospace sector. Breakout damage cannot go over certain limits in the evaluation of peel-up and push-out delamination described above. Drill hole damage limits, for instance, were shown in Figure 12 and are used in the production of commercial products.

Because of their elastoplastic recovery and bouncing-back properties, CFRP composites can have uncut fibers and fiber pull-out on their machined edges, while fibers severed from a hole's periphery exhibit ripping. These two types of defects exhibit symmetrical distribution patterns and arise based on the fiber cutting angle of the composite laminate under examination (Turki et al., 2014; Xu et al., 2013). Uncut fibers typically have no detrimental effects on the mechanical characteristics of CFRP sections that have been machined; nevertheless, removing them would cost more money and necessitate extra machining procedures.

Figure 13. Uncut fibers at the exit of drilled holes in UD-CFRPs (Geier et al., 2019)

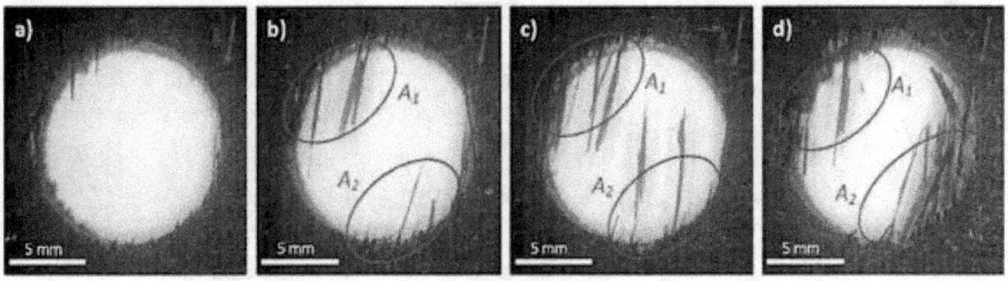

Figure 13. Uncut fibers at the exit of drilled holes in UD-CFRPs [42]

According to Xu et al. (2014), the appearance of uncut fibers surrounding the hole's edge often exhibits two burr areas, which are visible at the hole's exit where uncut fibers occur. Figure 13, for example, shows such uncut fibers near the exit of holes and associated burr regions, designated A1 and A2.

3 TOOL SHAPES, MATERIALS, AND COATINGS APPLIED FOR DRILLING OF CFRP COMPOSITES

3.1 Tool Shapes

One of the major influences on the CFRP damages previously discussed is the geometry of the cutting tool. The production of composite products involves the use of a large variety of drilling instruments. Figure 14 shows the most common drilling instruments used in CFRP drilling.

Figure 14. Schematic drawings of different drilling tools for CFRPs: (a) conventional twist drill, (b) double point angle twist drill, (c) brad and spur drill (candlestick drill), (d) dagger drill (one-shot drill), (e) step drill, (f) core drill, and (g) the special core drill (Geier et al., 2019)

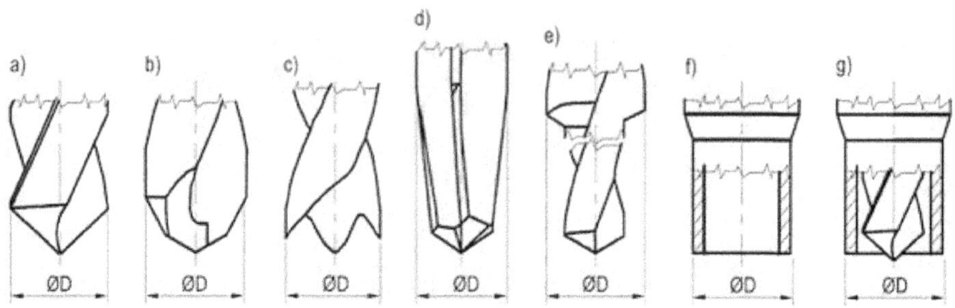

As seen in Figure 14(a), point angle is one of the crucial angles for traditional twist drills. A larger point angle causes the laminates' uncut layers to be extruded, which raises the thrust force and may cause delamination and the appearance of uncut fibers on machined surfaces. Consequently, it is believed that the twist drill's tip angle mostly influences the quality of the drilled hole.

According to Gaitonde et al. (2008), employing a lower point angle of 85° reduced the delamination factor at the hole exit by up to 45%. Additionally, Heisel and Pfeifroth (2012) showed that for CFRP laminates, a minor tip angle of a twist drill can reduce push-out delamination. On peel-up delamination, point angle has a different tendency to affect the process. Lower delamination at hole entry was achieved by Shyha et al. (2009) using an angle point of 140° as opposed to 118°. Lower delamination drill holes are achievable by the use of low feed rates and optimum process parameters, such as cutting speed, etc. Due to the thorough analysis of drilling performance and tool cost, traditional twist drills continue to be the industry standard.

Because double point angle twist drills have a second point angle, they create less thrust force and delamination than ordinary twist drills (see Figure 14(b)). Because less cutting power is applied to the secondary cutting edges, the second point angle may reduce the axial force and improve the outer cutting corners' resistance to wear in comparison to earlier ones.

Brad and spur drills, as seen in Figure 14(c), have good performance outcomes in terms of FRP material delamination. Su et al. (2018) claim that this is because of the cut-push effect of sharp flank cutting edges, which occurs when the drill pushes off the last layer of composite after first cutting it. As a result, delamination and the quantity of uncut fibers can be reduced.

Figure 14(d) illustrates the dagger drill, also known as the one-shot drill, which is a unique double-point angle drill with straight flutes. The ultimate surface of the hole is reammed by the secondary edges' modest point angle. In comparison to other drills, the axial force of this one is low. Hence, push-out delamination and uncut fibers at the hole's exit are satisfactorily demonstrated by one-shot drills. Several researches looked into the cutting capacity of these drills (Feito et al., 2015; Fernandes & Cook, 2006; Murphy et al., 2002; Xu et al., 2013; Xu et al., 2014; Xu et al., 2018). Based on the geometry of a traditional one-shot drill, Jia et al. (2016) recently presented a unique intermittent-saw tooth drill. This intermittent saw tooth structure produced an upward cutting that successfully decreased hole exit delamination. The number of high-quality holes drilled by this novel drill is nearly 6 times of that by conventional one-shot drill.

Step drills are built utilizing a step-control method, as seen in Figure 14(e): a smaller diameter drill is used to drill a pilot hole first, and a greater diameter tool is subsequently applied to reamed the final surface (Geng et al., 2019). Within this process, the first step's objective is to swiftly and effectively build a pilot hole, while the second step's objective is to make the hole's ultimate surface with the least amount of geometrical damage possible. Tsao and Hocheng (2003) state that the modest thrust force of the second drilling stage makes the latter goal possible. The impact of the chisel edge on delamination can be reduced by using step drills. Additionally, a single step drill can be used to drill holes of various diameters and countersunk shapes. The step drill is a useful tool for reducing entry and exit delamination, although it has a restricted first-to-second drilling step diameter ratio.

As shown in Figure 14(f), a core drill is a hollow grinding drill with a bonded diamond cutting surface. It uses abrasive cutting to directly manufacture the hole's ultimate surface. As a result of the little chip section that is left over after cutting, there is very little thrust force, which also means that there is very little entry and exit delamination. However, in contrast to helical flute drills, chip removal is a challenging issue. To resolve this problem of chip removal, conventional compound core-special drills are designed to avoid the chip removal clog in drilling, as shown in Figure 14(g).

Conventional compound core-special drills can be considered as a combination of an outer drill, a directly core drill, and an inner drill, including various twist drills, saw drills or candlestick drills, etc. The main problem of special core drills is the kinematic constraint of the inner and the outer drills. The application of diamond grinds for the outer drill requires a higher cutting speed for effective machining compared with the inner drill ability. Such cutting speed leads to fast inner drill wear. As a solution to the problem, Tsao and Chiu (2011) developed a novel experimental environment to combat the problem of relative motion and chip removal between the inner and outer drills. Nevertheless, special core drills based approaches for CFRP machining are not applied for the manufacturing of commercial products.

3.2 Tool Materials

The choice of tool material is governed by several factors including workpiece specification, productivity requirements, product design, machine tool condition, and tooling cost. According to Sheikh-Ahmed (2009), tool materials more suitable for machining composites include cemented carbides and polycrystalline diamond (PCD).

Cemented carbide tools comprising hard carbide particles, mainly WC, are joined by a cobalt (Co) binder, which is generally in a liquid phase at the sintering temperature of above 1573 K (Sheikh-Ahmed, 2009; Trent & Wright, 2001). The cobalt content is typically between 4-12%, depending on the grade of carbide. The mechanical and physical properties of WC-Co tool materials are highly dependent on the cobalt content and carbide grain size. In general, a smaller grain size results in higher hardness and wear resistance while both hardness and compressive strength also improve with lower cobalt content. Therefore, submicron/fine grain-sized carbides are preferred for the machining composites in order to withstand abrasion attacks from the reinforcement. On the other hand, it has been reported that WC-Co based cutting tools will be rapidly worn out, and perform very short tool lifetime. Their wear resistance still cannot meet the requirements of the long-duration machining of CFRP materials (Lei et al., 2013).

Sintering diamond grits (2–50 μm) at temperatures and pressures as high as 1773 K and 60 GPa, respectively, yields synthetic polycrystalline diamond (Trent & Wright, 2001). Cobalt is the primary metal catalyst used to encourage diamond intergrowth and guarantee an appropriately dense result. Compared to genuine diamond, the resultant material usually has more homogeneous mechanical characteristics and a high hardness (8,000–10,000 HV). Polycrystalline diamond blanks are often made as discs up to 75 mm in diameter, bonded onto a WC substrate and containing a diamond layer that is 0.5–0.7 mm thick. By employing segments/sectors of PCD cut from the original disc or blank, fabricating tools are used to create the various insert shapes (Heath, 2001). The size of the diamond grains, the degree of intergrowth, and the amount and distribution of catalyst material all affect the PCD tool's characteristics. Larger diamond grit goods are harder than tungsten carbide tools, but when edge sharpness is important, smaller grains are chosen. Cutting temperatures must not be permitted to get over ~ 1073 K since this is the temperature at which diamond turns into graphite (Trent & Wright, 2001). PCD tools can be used successfully since CFRP materials often have cutting temperatures lower than this value. The primary drawback of PCD tools is their restricted range of tool forms.

3.3 Tool Coatings

It is commonly recognized that the use of tool coatings improves the productivity and wear resistance of machining a variety of materials. The types of tool coatings used today are separated by the processes used for coating deposition, which include chemical (CVD) and physical (PVD) vapor depositions. To create a hard, wear-resistant surface that can improve performance and guarantee chemical stability, one or more layers are deposited and attached to the substrate. The coating material's adherence to the substrate determines the coating's performance. As a result, choosing the coating thickness is a judgment call based on experimental findings. Although a thicker coating does delay tool wear, it also increases the risk of brittle fracture and flacking.

The PVD method covers a broad family of vacuum coating processes in which the employed material is physically removed from a source by evaporation or sputtering. Then, it is transported by the energy of the vapor particles, and condensed as a film on the surfaces of appropriately placed parts under vacuum. Chemical compounds are deposited by either using similar source material, or by introducing reactive gases, such as nitrogen, oxygen, or simple hydrocarbons, that is reacted with metal from the PVD source (Bouzakis et al., 2012).

The CVD procedure, which is usually carried out at temperatures between 1073 and 1373 K, involves a gas-phase chemical reaction, such as that of methane, nitrogen, etc., in an excess of hydrogen above a solid surface, such as a carbide cutting tool (Asmussen & Reinhard, 2002). To increase wear resistance, a single layer or multilayer coating, consisting of polycrystalline diamond or hard metals like TiC or TiN, is placed on the tool surface in microns (Sheikh-Ahmed, 2009). With its high compressive strength, low coefficient of friction, minimal chemical reactivity (apart from ferrous workpieces), and high hardness level (10,000 HV), CVD diamond coating is the perfect material for machining non-ferrous materials. The properties of the diamond or hardmetal layer are determined to a large extent by the gas mixing ratio and the process temperature of the tool surface. However, the main limitation of the CVD diamond coating process is the poor adhesion of diamond film due to the presence of cobalt in the carbide tool substrate (Xu et al., 2018).

Diamond-like carbon (DLC) is a mixture of hard carbon atoms, commonly referred to as sp^3 and sp^2 hybrid carbon-based bonded arrangements respectively, which are linked together in an amorphous structure (Lifshitz, 1999; Vandevelde et al., n.d.). Unlike CVD diamond products, DLC coatings are produced using the PVD technique involving sputter deposition or an ion-beam system operating at lower temperatures (up to 773 K) (Lifshitz, 1999), which does not cause any difficulties with the cobalt binder in carbide tools. The hardness of DLC depends on the sp^3

element concentration in the mixture. Where the element concentration approaches 100%, the coating hardness approximates that of natural diamond (8,000-10,000 HV), but drops by ~50% (3,000-5,000 HV) when the sp^3 diamond concentration is reduced to ~ 15-20% (Vandevelde et al., n.d.). DLC layer thickness is typically of the order of several microns, whereas diamond coatings can be up to 25 μm.

Since CFRP composites gained popularity, a wide range of coatings with varying kinds and deposition methods have been applied on cutting tools to provide wear-resistant protection during machining. For example, the tool performances of CVD and PVD multilayer coatings were examined by Mkaddem et al. (2013). On turning inserts, three different coating types—PVD-TiAlN/AlCrO, PVD-TiN/TiAlN, and CVD-TiCN/Al2O3/TiN—were applied. The coatings were then examined in dry cutting CFRP. The findings demonstrated that wear mechanisms are dominated by fiber abrasiveness, independent of coating type. Wang et al. (2013) have investigated the tool wear of uncoated, CVD diamond-coated and AlTiN coated carbide drills when drilling the CFRP. It was observed that the diamond coating significantly reduced the tool wear, while the AlTiN coating performs no visible improvement over the uncoated drill. To date, the use of DLC-coated tools for the machining of CFRP composites is still limited. Murphy et al. (2002) found limited benefits in using DLC coatings over uncoated tools in terms of tool wear when drilling CFRP.

Figure 15. Flank wear results of different tool materials when turning CFRP (Ferreira et al., 1999)

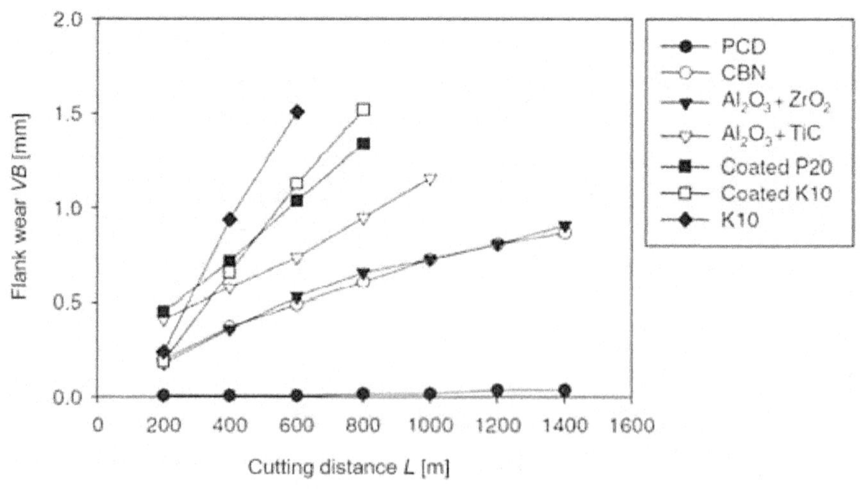

Furthermore, they stated that coatings with TiN and DLC had no advantageous effect on lowering thrust forces or delamination. When used to turn CFRP, as illustrated in Figure 15, tools with PCD inserts outperformed all competitors, including sintered cubic boron nitride (CBN), ceramics, coated and uncoated WC-Co based tools, according to Ferreira et al. (1999). In summary, it is clear that all of the examined coatings aside from diamond provide only modest improvements.

Many researchers have recently assessed CVD diamond-coated tools for cutting composite materials, with favorable outcomes (Davim & Mata, 2008; Faraz et al., 2009; Wang et al., 2018; Xu et al., 2013; Xu et al., 2014; Xu et al., 2018). When turning composites, Davim and Mata (2008) observed reduced cutting forces when using CVD diamond coated tools as opposed to WC and PCD inserts, even if the surface roughness produced was about same. Additional advantages of CVD diamond coatings over comparable solid ultra-hard tool materials are their reduced cost (about 30% less than brazed PCD) and relative simplicity of application on intricate tool geometries. It is well known that hot filament CVD (HFCVD) offers a good deal of practicality when it comes to mass producing complex-shaped diamond-coated cutting tools and uniformly depositing diamond films on intricate surfaces. Zhang et al. (2013) conducted research on new types of doped diamond coated cutting tools based on the benefits of HFCVD. Their findings show that developed diamond films can significantly improve the cutting performance of the drill for CFRP machining; in particular, the best performance of the Si-doped diamond coated drill is achieved by combining nano-scale grain size, lower surface roughness, and better adhesion.

Figure 16. Factors affecting machinability of CFRP composites

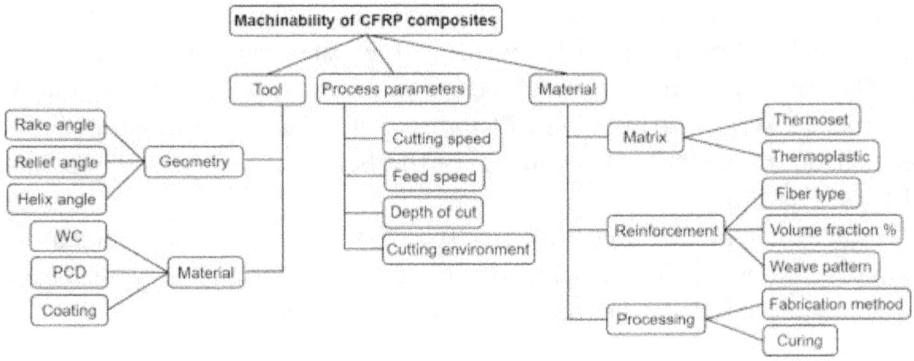

Further, Wang et al. (2017) compared the cutting behaviors of different types of diamond coated drills in CFRP drilling, and indicated that boron-doped microcrystalline and nanocrystalline composite diamond coating performs better than any monolayer diamond coating.

In summary, influence of the tool coating on the cutting performance of the coated drills is obvious. Moreover, although varieties of high-performance diamond coatings, including some composite diamond films, have been extensively used on drills for CFRP machining, the tool failure should be further clarified, because the wear, fracture and film removal still exist.

4. MACHINABILITY OF CFRP COMPOSITES

Machining performance when cutting CFRPs is dependent on the composite specification including fiber and matrix properties, fiber orientation, type of weave, and curing conditions, in addition to the appropriate selection of cutting parameters and tool geometry, as summarized in Figure 16 (Bhatnagar et al., 1995; Ferreira et al., 1999; König & Graß, 1989; König et al., 1985; Wang et al., 1995).

4.1 Effect of Material Properties

Fiber orientation (denoted here by θ) has been found to be the main factor when cutting FRP. It significantly affects cutting forces, torque, machined surface quality, and chip formation (Bhatnagar et al., 1995; König et al., 1985; Wang et al., 1995). Unlike metal work piece, material removal in FRP work piece is typically the result of a series of mini-fractures owing to the brittleness of the material elements.

The primary fiber orientation of the workpiece with respect to the cutting direction and tool rake angle determined the five categories into which Wang et al. (1995) categorized the different types of cutting mechanisms involved in chip formation observed when machining FRP composites. They were classified as Type I delamination, Type II fiber buckling, Type III fiber cutting, Type IV deformation, and Type V shearing.

Cutting parallel to the fiber direction ($\theta = 0^0$) with positive rake geometry was shown to result in Type I, as illustrated in Figure 17(a). The damage spreads along the fiber/matrix interface after the tool tip initiates the break.

Figure 17. Cutting mechanisms in the orthogonal machining of FRP composites (Wang et al., 1995)

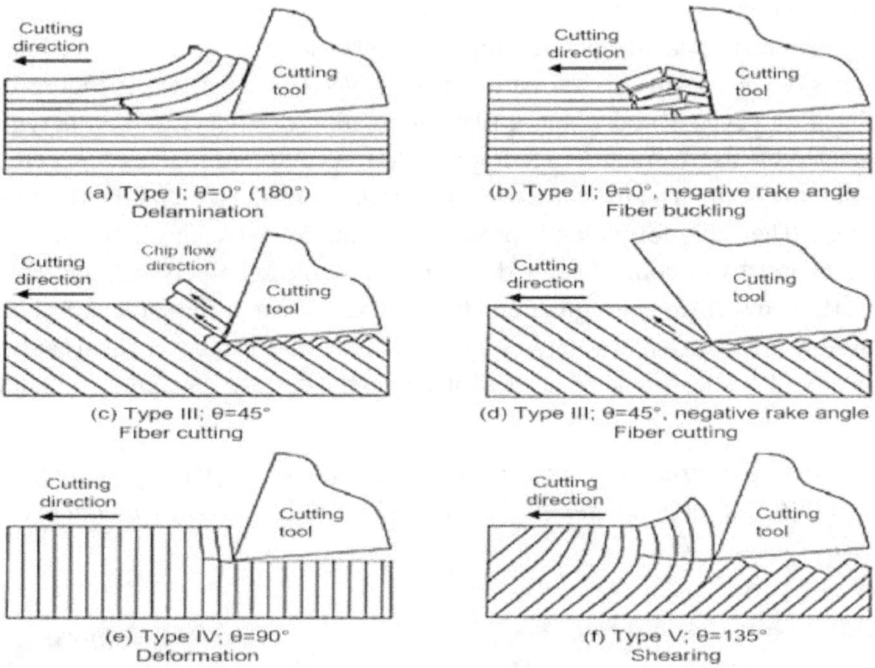

The peeled layer moves up the rake face as the tool progresses into the workpiece, bending it like a cantilever. Then, ahead of the cutting edge and perpendicular to the fiber path, bending-induced fracture takes place. These circumstances usually result in the production of small, uneven, and discontinuous chips. Furthermore, with the repeated cycles of peeling, fiber bending, and fracture, the cutting forces typically vary. Fiber buckling (Type II) becomes prevalent when zero or negative rake angle tools are employed to cut along the fiber direction, as shown in Figure 17(b). Here, fibers are subjected to compressive loading along their axis resulting in buckling loads. Progress of the cutting-edge causes cracks at the fiber/matrix interface and eventually fractures the fibers perpendicular to their axis under intense bucking. This also results in small discontinuous chips. While the fluctuation in cutting forces is smaller than in Type I, the machined surface is comparable.

Regardless of the tool rake angle, fiber cutting Type III mechanism happens when θ is between 0° and 90°. In this instance, fracture resulting from compression-induced shear across the fiber axis and fiber/matrix interfacial shearing along the fiber direction combine to generate the chip. As seen in Figures 17(c) and (d), during compression, cracks form in the fibers above and below the cutting plane, with the latter staying within the machined surface. The chip morphology could be continuous or discontinuous under these circumstances.

When cutting perpendicular to fiber orientation ($\theta = 90^0$), increasing interlaminar shear leading to fracture of the chip segments along the fiber/matrix interface becomes the dominant chip formation mode (Type IV), as shown in Figure 17(e).

Shearing (Type V) occurs at a fiber orientation of 105^0–150^0 and is typically dominated by macro-fracture, see Figure 17(f). As cutting progresses, severe fiber deformation occurs resulting in excessive interlaminar shear along the fiber/matrix interface. Then chip formation is produced by the extensive elastic bending which results in cracks in both fibers and matrix producing relatively long chips. Here, the machined surface is irregular and the fiber ends have different lengths because fracture occurs at different points along their length. Elastic recovery also takes place following fiber shearing and leaves an irregular cut surface, as shown in Figure 18.

Figure 18. Microstructure when orthogonal cutting of graphite/epoxy composite: (a) $\theta = 150^0$ and 0.05 mm depth of cut and (b) $\theta = 120^0$ and 0.1 mm depth of cut (Sheikh-Ahmed, 2009)

Proposed above the cutting mechanisms related to the UD laminates. Multidirectional (MD) CFRP composites are commonly used in many applications contrary to the UD ones.

When drilling MD CFRP composites, variations in cutting forces are common as the tool edge moves through different fiber angles throughout the composite workpiece causes to discontinuous machining and detrimentally influence the cutting tools through their edges micro-chipping. On the other hand, laminate design depends on the required CFRP product properties. Thus, the combination of the tool shape, material and coating should minimize the wear effects of this mechanism.

4.2 Effect of Process Parameters

The drilling thrust force and torque are affected by the cutting speed and feed rate. Both thrust force and torque increase significantly with increasing feed rate due to its direct influence on uncut chip size (König et al., 1985). A high correlation has been reported between feed rate, thrust force, and delamination when drilling high-strength CFRP (Xu et al., 2013). Whereas the effect of cutting speed on thrust force and torque is not significant. Also, the feed rate had a detrimental effect on surface roughness as an increase from 2.0 to 5.0 µm Ra was measured when the feed rate was raised from 0.05 to 0.25 mm/rev (Ferreira et al., 1999).

Most studies that have been published have shown that there is a direct relationship between thrust force and delamination, especially at the hole exit. When drilling UD and MD CFRPs, Chen (1997) found that feed rate affected both thrust force and torque, but cutting speed had no effect on either. Tsao (2008) also achieved similar results. The results of the estimated thrust force closely matched the experimentally measured data. Thrust force and torque were found to increase with cut distance in work by Fernandes and Cook (2006). It is clear that the thrust force rose as the number of drilled holes increased, and this was linked to the cutting tool's chipping and wear. In another research, Durão et al. (2010) studied the influence of different tool geometries, such as twist drills with various point angles, brad & spur drill, dagger drill, and step drill, on thrust force, surface roughness, and delamination. They reported that a low feed rate can decrease thrust force and delamination. Also, they noted that thermal degradation may be considerable at low feed rates.

The effect of process parameters on the characteristics of uncut fibers was investigated by researchers to develop methods for machining composites without leaving uncut fibers. For instance, Xu et al. (2013) observed that the burr area increases with the increase in thrust force. Furthermore, Xu et al. (2018) conducted detailed drilling experiments on CFRPs using different tools, such as dagger, brad & spur, and twist drills, and concluded that feed rate has the most significant influence on the extent of defects. Also, it was observed that the brad & spur drill produced the best performance on damage reduction.

According to cutting force optimization, the minimum peel-up and push-out delamination occurred at a minimum feed. On the other hand, drilling CFRP composites with very low feed seems ineffective due to increased machining time and cost. Also, thermal damage to the CFRP material is expected under increased tool/workpiece contact time. Therefore, the selection of process parameters should match both the delamination control and process efficiency. Usually, a high cutting speed (150-200 m/min), as well as a low feed rate (0.01-0.05 mm/rev), is recommended for CFRP drilling in the industry practice to prevent the drilling-induced delamination and guarantee drilling efficiency at the same time (Sheikh-Ahmed, 2009).

4.3 Effect of Tool Materials and Geometries

Tool materials and geometries play an important role in relation to tool life and machined surface quality when cutting CFRP composites. Feito et al. (2015) studied three different cutting drill geometries, as brad center, step and reamer, in terms of drilling forces and delamination for woven CFRP. The authors found that the reamer drill showed the best results. Shyha et al.

Shyha et al. (2009) analyzed the drilling of small holes in quasi-isotropic CFRP laminate with two different tool geometries, conventional twisted and stepped. Increases in the feed rate lead to thrust force enhancement. Delamination at the hole entrance was higher for the stepped drill than for the twisted drill. However, the increment of point angle, from 118° to 140°, leads to a decrease in delamination. In another research, Heisel and Pfeifroth (2012) conducted drilling experiments in CFRPs, and analyzed the influence of different point angles (155°, 175°, 185°, 178°) on thrust force and delamination. They discovered that an increase in point angle increases delamination at the exit of the holes. It was also observed that delamination decreases at the entry of the holes simultaneously with an increase of the point angle of twist drills.

Wang et al. (2018) reported that a diamond coating of the cutting tool can reduce thrust force and delamination. Also, the diamond coating increases wear resistance and tool life. At the same time, they found that the number of uncut fibers increases due to increased edge roundness caused by coating growth. The study of Iliescu et al. (2010) shows that the tool life of diamond- coated carbide drills can be 10-12 times higher compared the tool life of uncoated carbide drills. Furthermore, related cutting speed can be 3 times higher when using diamond-coated tools than uncoated ones. Also, the researchers suggested a tool specification for drilling delamination- free in CFRPs, recommends point angle of 125°-130°, helix angle of 35°-40°, clearance angle of 10°-25°, and diamond coating as well. Qiu et al. (2019) studied a drilling characteristics of the dagger, the double point angle, and the candle stick drills when drilling CFRPs. The results show that the side edge has poor ability to remove burrs, and the tip structure of the outer corner of the candle stick drill can greatly reduce the entry damage. However, the outer corner small tip structure of the candle stick drill cannot remove the uncut fiber around $\theta=0°$ of hole exit. The cutting edges of the dagger drill makes a small thrust force at the drilling exit stage. Then, a novel drill is developed for drilling CFRP based on the advantages of the three kinds of drills. The cutting test result shows that the new compound drill can effectively remove fibers and reduce thrust force at the drilling exit stage, form burr-free and small delamination hole.

4.4 Tool Wear Phenomena

The most commonly used criterion to determine the end of tool life is maximum flank wear (VBBmax) which is typically between 0.3-0.6 mm for metals, however, there is no established tool wear criterion concerning cutting FRP composites (Sheikh-Ahmed, 2009). When drilling CFRP, both flank and chisel edge wear is evident, which increases with the number of drilled holes. The principal tool wear mechanism experienced when cutting CFRP composites is abrasion with micro-chipping often present as a secondary wear mode. Abrasion occurs when hard workpiece particles progressively indent and erode tool surfaces and remove tiny amounts of material from the cutting tool. In the case of CFRP, this is attributed to the highly abrasive fibers within the material. The presence of micro-chipping is generally a result of force oscillations due to the composites' inhomogeneous nature. Unlike metal machining, diffusion wear is rarely seen when cutting CFRP due to the relatively low temperatures and pressures involved. Gross fracture and edge chipping can also occur under interrupted cutting conditions due to the variation of mechanical properties between the reinforcement and matrix phase. The typical wear pattern observed however when cutting CFRP is severe edge rounding of the tool (Sheikh-Ahmed, 2009) as shown in Figure 19. The considered above tool wear phenomenon is typical for non-coated WC-Co-based tools. In the case of the application of CVD diamond-coated tools, the tool wear mechanism has an inherent phenomenon. During cutting, the fracture modes associated with diamond-coated tools may involve cyclic flaking and delamination at the cutting edges. This phenomenon usually begins with build-up along the cutting edges, which unbalances forces along the cutting edges, and is followed by fracturing along the chisel edge and peripheral corners. Differences in the thermal expansion coefficient along the interface between the coating and substrate produce unbalanced internal stress, resulting in shear force. Shear stress exceeding the threshold of fracture toughness in the coating layer can trigger flaking and delamination. Low frequency load can also be the source of fatigue failure (Kuo et al., 2014). Cyclic loading at the cutting edge can lead to the initiation of cracks in the rim area, which tend to propagate rapidly as micro-cracks coalesce into striations, leading to catastrophic failure.

Figure 19. A cross section of a replica from a cutting edge when machining FRPs (α_o is the rake angle and γ_o is the clearance angle) (Sheikh-Ahmed, 2009)

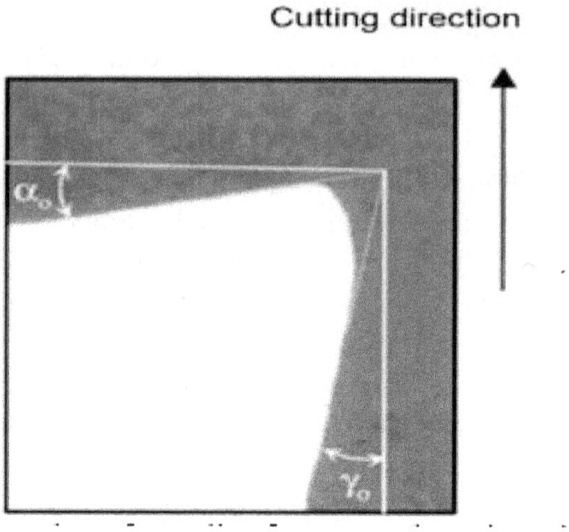

Figure 20. Schematic illustration showing tool wear patterns in (a) adhesion and abrasion wear, (b) extended abrasion wear on the flank face, (c) flaking and delamination, and (d) breakage of substrate material (Kuo et al., 2018)

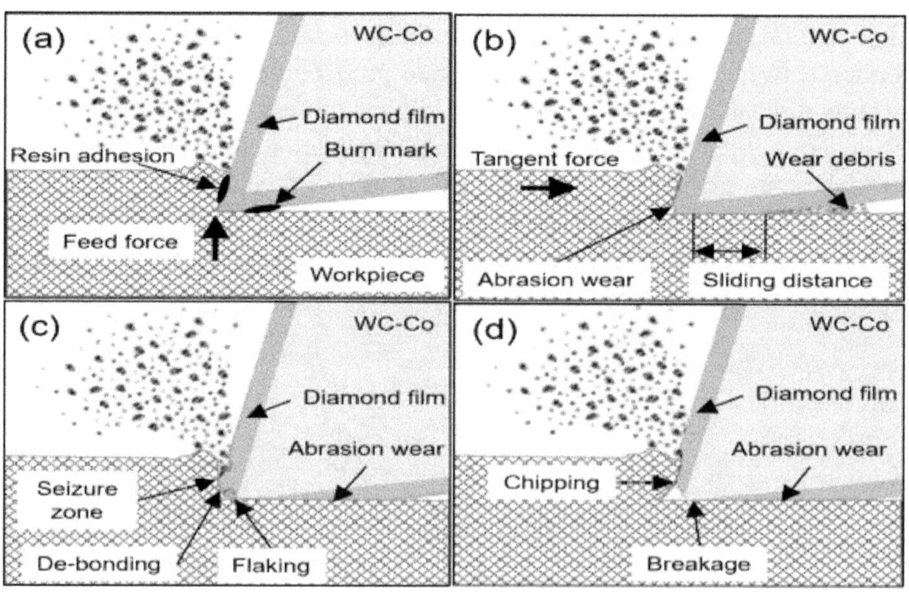

According to the cutting mechanism, Kuo et al. (2018) proposed the wear process of the CVD diamond tool during the drilling of CFRP, as shown in Figure 20. According to the tool wear evolution, the wear behavior could be classified into four stages. Figure 20(a) shows the continued adhesion of polymeric residue on the rake and flank faces of the cutting tool due to a temperature-induced by high cutting speed. The varied geometry at the cutting edge reduces the clearance angle thereby diverting the cutting actions into ploughing and rubbing actions. In Figure 20(b), abrasion wear on the flank face increases steadily due to the compression stress on the sliding zone being high and also due to the increase in friction heat and temperature. Scoring wear on the successive flank face may occur due to the removed abrasive mixtures pushing by the fibers bouncing back on the machined surface.

The wear process may involve graphitization in the diamond under a high cutting temperature, attrition in the columnar diamond grains, and abrasion wear which is produced by the mixture of the diamond debris and carbon fiber fragment. As shown in Figure 20(c), due to the increased intermittent loads on the seizure zone, diamond film failure in terms of flaking and delamination occurs along the cutting edge, which exposes the sharp edge of the tungsten carbide substrate which is simultaneously re-sharpened under the grinding effects of the carbon fiber abrasives. As seen in Figure 20(d), when the CVD diamond-coated layer had entirely flaked off, the exposed tungsten carbide edge was rapidly worn away; thereby greatly increasing the cutting force. This was due to the rounding of the cutting edge as well as a loss of clearance and rake angle during cutting.

Finally, it can be summarized, that tool wear in machining CFRP occurs primarily by abrasion and micro-chipping. Abrasive wear is caused by sliding of the tool surfaces under pressure against the abrasive fiber material embedded in the polymer matrix. Abrasive wear is manifested by clear rounding of the cutting edge and in the form of wear lands on the rake and the clearance surfaces of the cutting tool.

5. DIAMOND SYNTHESIS BY CVD

5.1 CVD Systems

The growth of diamond films from vapour phase on non-diamond substrates at practical rates was accomplished with the development of thermal- and plasma-enhanced CVD methods, in which a hydrocarbon gas mixed in low concentrations with hydrogen is energized thermally or in a plasma, prior to contact with a heated substrate. Between these methods, hot filament- assisted thermal CVD (HFCVD), microwave-plasma CVD, combustion flame-assisted CVD, and direct-current (DC) arc plasma jet CVD are predominant for commercial using (Gracio et al., 2010;

Haubner, 2013). The various CVD techniques, although different in their process details, have a several common features, the most important are the following:

Growth in the presence of atomic hydrogen: The deposition of polycrystalline diamond film in the presence of atomic hydrogen was based on phenomena that diamond is more stable towards atomic hydrogen influence than graphite.

Dissociation of carbon-containing source gases: In the various CVD methods, the carbon-containing compounds are dissociated by thermal, plasma or combustion processes to produce the reactant species responsible for the diamond nucleation and growth. The film growth rate depends on the ability of these reactant species to be transported to the substrate.

Growth at moderate substrate temperatures: In all CVD methods, diamond growth takes place at substrate temperatures between 773 K and 1473 K. Growth at temperatures above or below this range often leads to graphite deposit.

5.2 HFCVD

In this method, as shown in Figure 21, diamond particles are deposited on a heated substrate from a mixture of methane and hydrogen dissociated by hot refractory metal filament placed close to the substrate. The filament temperature may reach around 2473 K during this process. The main role of the hot-filament is to dissociate molecular hydrogen into atomic hydrogen. The dissociation mainly occurs near the filament due to the high gas temperature.

Besides atomic hydrogen creating, it was discovered that during diamond deposition the tungsten filament reacts with methane and undergoes carburization. This results in consumption of carbon from the methane, and thus a specific incubation time is needed for the nucleation of diamond films. Therefore, this process may affect the early stages of thin film growth. In addition, the resistance of the filament should be monitored and adjustments to the supplied voltage and current made in order to maintain the temperature of the filament constant.

Due to the temperature upper limit of the filament material, hot-filament processes operate at significantly lower gas temperatures than plasma processes, and consequently produce less atomic hydrogen. The low gas phase concentrations give relatively low growth rates compared to the plasma methods.

Figure 21. Schematic diagram of HFCVD method

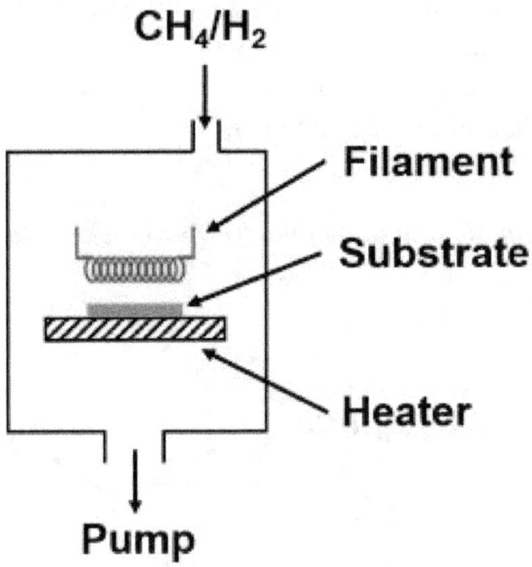

Despite these drawbacks, hot-filament assisted deposition has remained popular because of its low capital cost and simplicity. Also, hot-filament reactors are directly scalable to large sizes and can be used to coat complex shapes and internal surfaces.

5.3 Plasma-Enhanced CVD Methods

These methods of diamond growth involve various forms of plasma-assisted CVD using carbon-containing species mixed in low concentration with hydrogen. Plasma is generated either by various forms of electrical discharges or by induction heating. The role of the plasma is to generate atomic hydrogen and to produce proper carbon precursors for the growth of diamond. Atomic hydrogen is produced by electron impact dissociation of molecular hydrogen. Such atomic hydrogen produced in plasmas generally has high kinetic energy due to the difference between the hydrogen dissociation energy and the electron kinetic energy. This accounts reported for the significantly higher growth rates in plasmas, compared with HFCVD.

Among the plasma-enhanced methods, microwave-plasma-assisted CVD, as shown in Figure 22, has been used much more extensively than others for the growth of diamond films. This method of diamond growth has a number of distinct advantages over the other methods. Microwave deposition, being an electrode less process, avoids contamination of the films due to electrode erosion Furthermore, the microwave discharge at 2.45 GHz produces higher plasma density with energy electrons, compared with the RF discharges at 13.5 MHz, due to higher process frequency. This should result in higher concentrations of atomic hydrogen and hydrocarbon radicals. An additional advantage is that the plasma is confined in the center of the deposition chamber in the form of a ball and this prevents carbon deposition on the walls of the chamber.

Figure 22. Schematic diagram of microwave CVD method

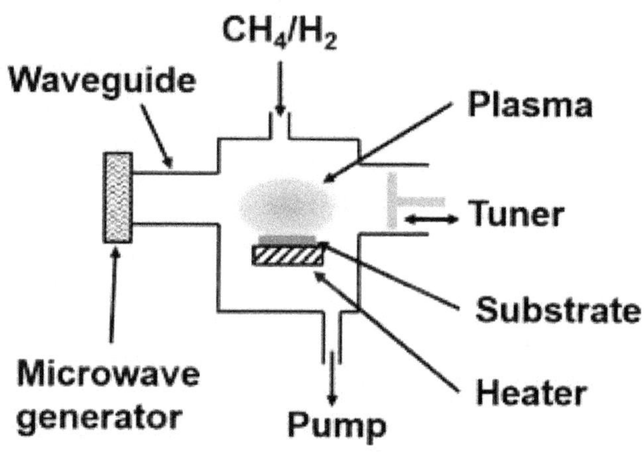

Other important plasma-based method is DC arc jet CVD, as shown in Figure 23. The DC arc jet is a general name for a high pressure direct-current plasma discharge. Electrical energy is converted to thermal and kinetic energy of a flowing gas mixture by an electric arc discharge.

Like other CVD methods, a major constituent of the gas mixture is hydrogen, while methane is introduced into the plasma jet to provide a source of carbon. The plasma temperature is 1000- 5000 K as average. The plasma jet containing these reactive species affects onto a cooled substrate surface for film growth. The electric discharge is sustained between a concentric cathode rod and a surrounding cylindrical anode, creating ionized arc column. Extremely high diamond growth rates (~1 mm/hour) can be achieved using this method (Asmussen & Reinhard, 2002). In addition, arc jets have a distinct advantage over other diamond CVD methods in that the deposition of diamond can proceed simultaneously with the deposition of other ceramics and metals by introducing powders of various types into the plasma stream. Such a functional gradient material can reduce the thermal stress between the diamond film and the substrate with, consequently, increasing the adhesion strength between them.

Figure 23. Schematic diagram of DC arc jet CVD method

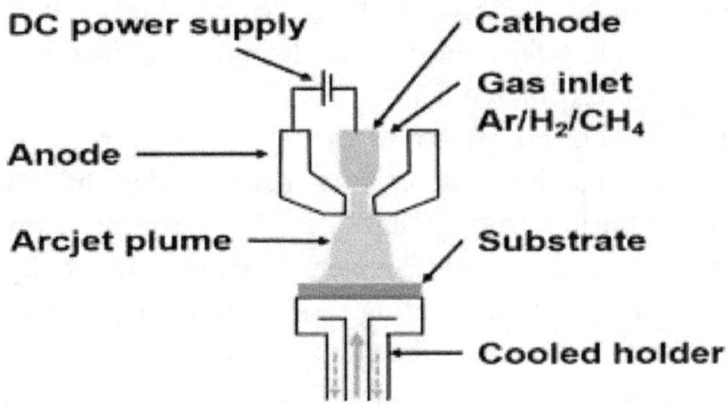

5.4 Combustion-Flame Assisted CVD

This method allows the growth of diamond at atmospheric pressures using combustion flames from an oxygen-acetylene brazing torch. Because of its simplicity and low cost of the experimental apparatus as well as the high growth rate, this method looks promised in diamond growth.

A schematic diagram of the method is shown in Figure 24. It consists of a brazing torch supplied with oxygen and acetylene (C_2H_2), and a water-cooled substrate. The substrate temperature is adjusted by varying distance of the substrate surface position related the torch. Normally the oxygen-acetylene torch is used with an excess of oxygen. The acetylene is completely burned to CO_2 and H_2O. Hydrogen addition to the oxygen-acetylene flame was found to reduce the amount of amorphous carbon in the diamond films (Asmussen & Reinhard, 2002).

Combustion flames operating in the fuel-rich mode have three distinct regions, as shown in Figure 24. The inner cone is the primary combustion zone followed by a diffused intermediate region and an outer zone. The temperature in the primary combustion zone can reach up to 3300 K. The combustion reaction at this region leads to the formation of CO and H_2 with a number of reactive intermediate species (H, OH, C_2, C_2H, etc.). In the fuel-rich mode, the un-burnt hydrocarbons as well as the combustion products form the diffuse intermediate and reaction free region, which is called feather. The outer zone, which is also known as a secondary combustion zone, consists of a flame caused by turbulent diffusion of oxygen from the surrounding atmosphere. In this region, the products of the combustion reaction are oxidized to CO_2 and H_2O. The substrate is usually placed in the feather region of the flame, where there is an abundance of atomic hydrogen and hydrocarbon radicals as required for diamond film grows.

All mentioned above CVD methods were evaluated in terms of diamond coating deposition on cutting tools for CFRP machining. Such criteria as coating growth rate and its uniformity, scaling feasibility for commercial application were compared. Some conclusions are considered below.

The deposition area of a microwave reactor is limited by the frequency of the plasma generator. The maximum deposition area in modern microwave systems is limited to about 150 mm in diameter (Herlinger, 2006). Some DC torches are capable to deposit areas up to 200 mm in diameter. Therefore, scalability of the plasma approaches is quite difficult engineering task for large deposition area compared with HFCVD. Indeed, the hot filament reactors are simple in their design. DC power supplies are used, control mainly comprises gas ratios and flow rates, vacuum level, and the amount of DC power in the filaments.

In growth rate point of view, HFCVD reactors typically grow diamond at 0.3 to 2.0 μm per hour, microwave systems from 1.0 to 5.0 μm per hour, and the DC arc jets have demonstrated growth rates exceeding 20 μm per hour (Herlinger, 2006). On the other hand, a film uniformity is an area where hot filament reactors have demonstrated their superiority over both microwave and DC arc jet approaches. HFCVD reactor has the advantage of uniform temperature across the entire deposition area. In contrast, microwave reactors and DC arc jets create a sphere or plume of energy that is hotter at the center compared its periphery. Such plasma temperature effects can be compensating by rotating the substrate in an attempt to normalize deposition temperatures.

This rotation adds complexity and cost, reduces reliability and often is a source of contamination. Another phenomenon observed is that substrates in microwave reactors and torch reactors tend to get hot spots on sharp corners of the tool substrates. The hot corners accelerate film growth on the ones, with the result film thickness non-uniformity. Hot filament reactors have demonstrated uniform temperature across large deposition areas and uniform coatings across an individual substrate. Since a temperature control is critical to diamond deposition, it is much easier to adjust the voltage on a hot wire than to try adjust the plasma sphere proximity to a substrate in ordinary microwave system.

Figure 24. Schematic diagram of combustion-flame assisted CVD method (Haubner, 2013)

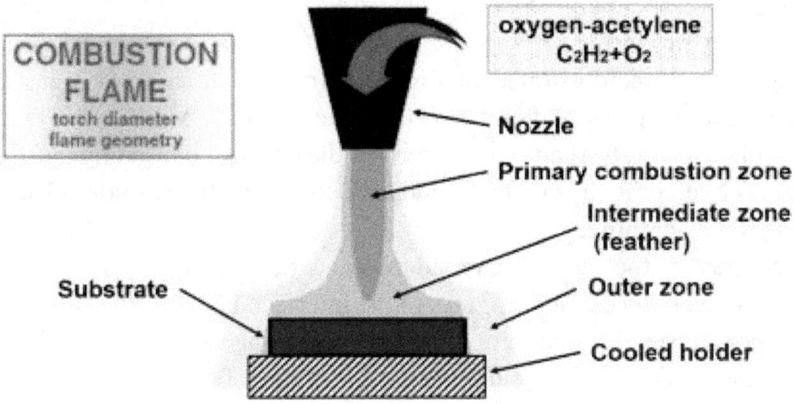

6.0 CVD DIAMOND COATINGS FOR CUTTING TOOL APPLICATION

As discussed in previous sections, the HFCVD process is the most suitable process technology for tool coating with CVD diamond thin films. In this technique, the required atomic hydrogen is produced with electrically heated glowing filaments made from refractory metals (Ta, W, or Re). An appropriate spatial filament arrangement provides distribution of the required species, even when processing substrates with complex geometry.

The basic principle of the HFCVD process, illustrated in Figure 25, can be briefly described as following. Atomic hydrogen is produced in the filament activation zone. Then atomic hydrogen diffuses to the tool substrate, recombines at the methane (CH_4) molecule and forms the methyl radical CH_3, which is the actual growth species for CVD diamond deposition. Further recombination of atomic hydrogen at the substrate surface etches graphitic carbon at the growth front. Growth of the diamond film initially starts with crystallization at nucleation sites, the finally closed film forms when small nuclei islands intergrowth. Crystals with beneficial orientation to the substrate gradually overgrow adjacent crystals. Conventional deposition of CVD diamond therefore forms characteristic columnar structures with rough surfaces.

Control of gas concentration ratio between H_2 and carbon-containing gas, such as CH_4, during the CVD process, determines the crystal shape and grain size of the deposited diamond coating (Sarangi et al., 2012). The coating morphologies play an essential role in the properties of the deposited film, and consequently, the cutting performance of the diamond-coated tools. According to the grain size, diamond films applied for cutting tools can be divided into a microcrystalline diamond (MCD) coating and a nanocrystalline diamond (NCD) coating, presented in Figure 26(a) and (b) respectively (Deuerler et al., 2002; Hu et al., 2007; Wang et al., 2017).

Beneficial properties of the MCD and NCD coatings can be combined in multilayer structure, illustrated in Figure 26(c). This concept increases fracture toughness and the growth of cracks is prevented. In pure MCD or NCD structures, as shown in Figure 26(d) and (e), a crack in the coating, caused by tensile residual stress, may overcome the crystal boundaries and reach the carbide substrate below it during the machining.

By contrast, in the case of diamond multilayer coatings, any crack is "absorbed" by the next layer and cannot progress to the interface between the substrate and the first diamond layer. Thus, failure of the entire coating does not happen. Formation of compressive stresses due to different coefficients of thermal expansion between substrate and coating, and between micro- and nanocrystalline diamonds, has the positive effect that vertical cracks change direction and rather grow parallel to the substrate at multilayer interfaces (Gåhlin et al., 1996).

Figure 25. Principle of the HFCVD process (Haubner, 2013)

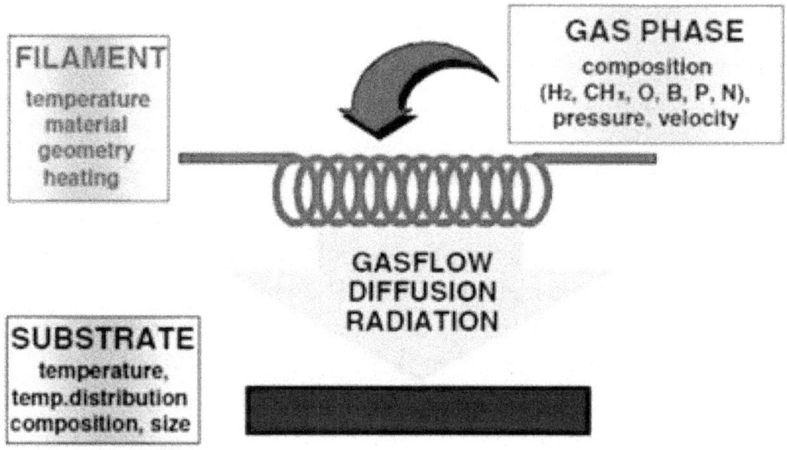

But it should be noted that such crack propagation stability of the multilayer diamond coating is not evidently confirmed by research society. On the other hand, application of dual- layer diamond films has widely described. Usually, the dual-layer film consists MCD layer, grown on a substrate directly, and NCD outlayer, grown in order to create smooth diamond film surface (Herlinger, 2006).

Figure 26. Kinds of HFCVD diamond coatings applied for cutting tools: (a) MCD, (b) NCD, (c) multilayer and schematic representation of crack propagation (red line marked) for (d) MCD, (e) NCD, (f) multilayer, respectively

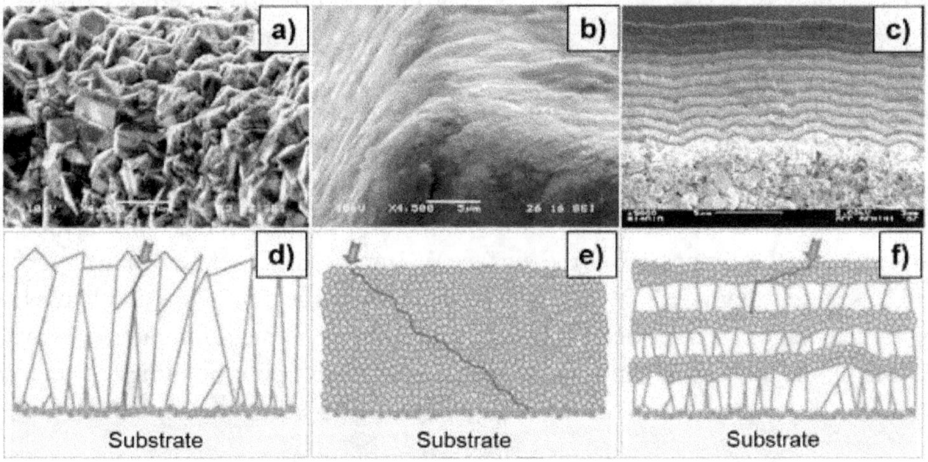

When diamond is deposited in cemented carbide materials such as WC-Co substrates, the cobalt will play a detrimental catalytic role in the diamond film adhesion. Any proposed solution must suppress the detrimental effect of cobalt and increase the mechanical interlocking between the film and the substrate in order to enhance the film interfacial toughness. The diamond-substrate interface requires the formation of strong interfacial chemical bonds between the diamond crystallites nucleated at the surface and the atoms of the substrate.

The most successful techniques, discussed in the corresponding papers, can be briefly divided on chemical etching procedures, creating intermediate layers, and selective thermal treatment (Haubner & Kalss, 2010). The chemical etching procedures are related to the cobalt removal in depths ranging in about 3 – 10 μm from the substrate surface by etching the cobalt out of the surface. Intermediate layers halt the cobalt effect on the surface by inter diffusion barrier effects. Also, the layers reduce the thermal stresses caused during the diamond growth. Thermal treatment of the tool substrates applied to decrease the cobalt concentration on the substrate surface and to increase the surface roughness.

6.2 Wet-Chemical Etching Method

The wet-chemical etching (WCE) in the surface of WC-Co substrates represent the most applied pretreatment method. This method has the purpose to produce a selective etching of the cobalt binder by using a two-step chemical process. First step is wet treatment in a Murakami solution with the aim of reconstruct and rough the surface by attacking the WC grains and exposing the Co binder. Then, a second wet etching in an acid solution to reduce the exposed cobalt at a certain depth determined by the etching time (Haubner et al., 1995; Polini, 2006). Other methods have been proposed as well. Surface pretreatment with Copper Sulfate ($CuSO_4$) solutions involves a combination of Co etching and Cu depositing by a cementation reaction. During diamond deposition, the Cu dissolves into the Co binder resulting in a homogeneous interface (Sommer et al., 2000).

Cobalt removal involves complicated processes that have to be matched carefully to each hardmetal type. Indeed, if the binder etch depth extends further into the substrate than the dimension of free-surface WC grains, the toughness of the interface will be severely degraded. By contrast, the removal of the binder to a depth less than the dimension of free-surface WC grains may allow chemical interaction between Co and diamond, during deposition process, with result poor bonding between the film and substrate.

6.3 Intermediate Layers

Interlayer materials, used to avoid the cobalt-diamond interdiffusion, are normally deposited by PVD methods. The interlayer material must remain stable during the diamond deposition, provides a diffusion barrier between carbon and cobalt, and has a low thermal expansion coefficient to minimize internal stresses. A wide range of interlayer materials (amorphous carbon, metallic Cr, Ti, Ta, and ceramic CrN, TiC, and TiN) can be successfully applied by PVD on WC-Co substrates (Endler et al., 1996; Haubner & Lux, 2006; Sarangi et al., 2008).

Although the application of interlayers becomes effective way of improving the adhesion of CVD diamond to hardmetals, this approach is rare in tool coating due to the high costs involved. Moreover, the interlayer leads to increase a cutting edge rounding, and consequently, thrust force and torque induced by a cutting process. Such phenomenon is undesired to CFRP machining point of view.

6.4 Thermal Treatment

Thermal treatments of the hardmetal substrates can be applied to decrease the Co- concentration on the hardmetal surface and to increase the surface roughness.

To avoid an additional process step, the thermal treatment can be applied prior to the diamond deposition in the CVD equipment. Using the hot-filament process Co evaporation, the tool substrates reach high temperatures (~1273 K) at low gas pressure. Ullram et al. reported (Ullram & Haubner, 2006) that the cobalt content on the substrate surface can be decreased from 6% to 1.5% within 3 hours and the surface roughness is similar to the WCE process results. The surface roughness is influenced by the temperature treatment due to the cobalt evaporation and WC crystal growth.

On the other hand, some tool substrate warp is expected due to the heat impact. It can be concluded, that thermal treatment is not suitable for tools with complicated shape.

6.5 Doping Into Diamond Coatings

It is well known that boron doping technology was firstly proposed for modifying the electrical properties of the diamond film, mainly changing the insulating diamond film to a conductive film.

Nowadays, a mechanical properties of doped diamond films and its industrial applications have significant interest by related research society. For instance, Wang et al. (2014) reported about improving in the solid particle erosion behavior of a boron-doped diamond (BDD) film compared with undoped ones. Recently, Gaydaychuk et al. (2020) investigated wear and friction of heavily BDD films against steel at elevated temperature, as well as influence of boron concentration on diamond film oxidation resistance. The obtained data indicated that the coefficient of friction and the oxidation resistance decrease with increasing of boron concentration. The boron doping level also affects the growth rate, grain size, and diamond purity of deposited BDD films. Regarding application doped films for cutting tools, Zhang et al. (2013) reported that the adoption of boron and silicon could enhance the adhesion between the diamond coating and WC-Co substrate.

Moreover, the covalent radius of the B atom (0.085 nm) is larger than the radius of the carbon atom (0.077 nm). Thus, the incorporation of B into the diamond structures may induce a certain amount of intrinsic tensile stress. Because the residual stress in diamond film, deposited on the WC-Co, is mainly compressive residual stress, reducing of the absolute value of the total residual stress can be expected due to implement boron doping during film deposition.

Therefore, in addition to the mentioned in the previous sections pretreatment techniques, boron-doping technology can be considered as attractive method for enhancing the interfacial adhesive quality between the diamond film and the substrate.

7. MATERIALS AND METHODS

The work material employed in this inquiry is cylindrical, 50 mm in diameter and 100 mm in length, reinforced PEEK composite with 30% fiber carbon (PEEK CF30). Dry turning tests were performed using TiN-coated cutting tools (WNMG 080408-TF) and a tool holder of the MWLNL 2020K08 type, on a GORATU G CRONO 4S CNC with a spindle power of 26.5 KW and a maximum spindle speed of 3,350 rpm (Mata et al., 2012).

Three component turning forces (feed force, Fa, radial force, and cutting force, (Fp) were collected using a Kistler piezoelectric dynamometer model 9121 that was attached to a data collecting board and load amplifier, as illustrated in Figure 27. Table 4 provides an overview of the work material's mechanical and thermal characteristics.

Figure 27. Details of the experimental set-up used for measurement of machining force component: (a) CNC turning, (b) dynamometer, (c) charge amplifier, (d) analogue output multifunction data acquisition devices BNC-2110, (e) PC with software data acquisition, (f) a measurement record with periodic pulsation of the cutting forces (Mata et al., 2012)

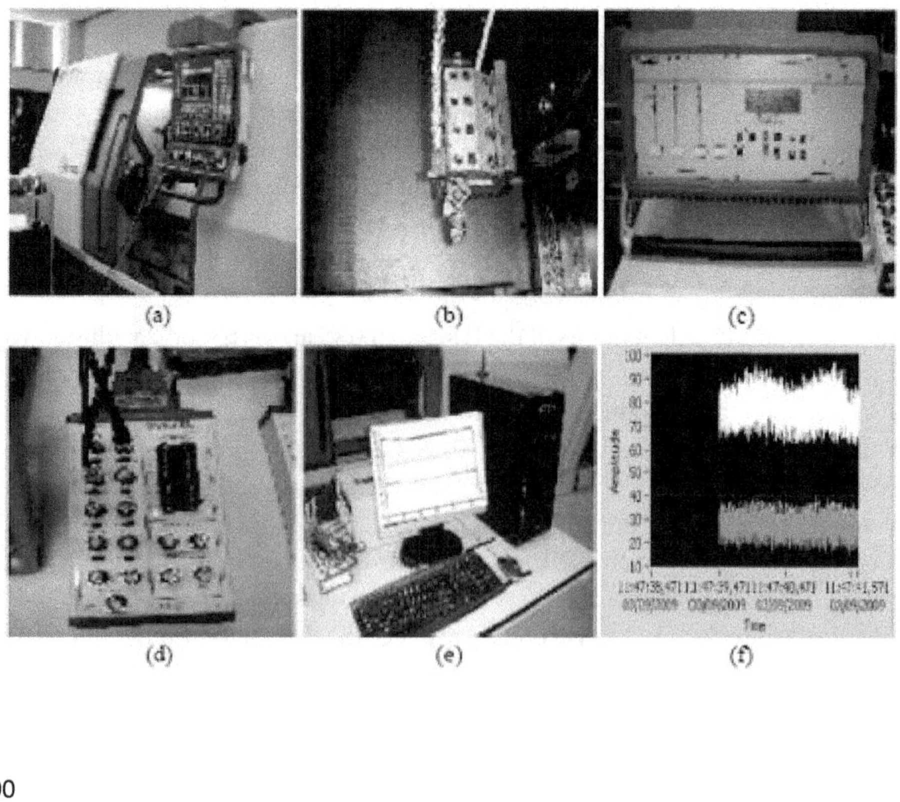

Table 4. Properties of PEEK CF30

Mechanical and thermal properties	*PEEK CF30*	*Unit*
Tensile modulus	7,700	MPa
Rockwell hardness	M102	-
Charpy impact resistance	35	KJ/m^2
Tensile strength	130	MPa
Melting temperature	340	°C
Density	1.41	g/cm^3
Coefficient of thermal expansion at (< 150°C)	25×10^{-6}	m/m/K
Coefficient of thermal expansion at (> 150°C)	55×10^{-6}	m/m/K

Because the machining processes are done in accordance with Taguchi's experimental design, fewer experiments are needed to get the same results. The current study will focus on three cutting parameters: depth of cut (d), feed rate (f), and cutting speed (v). For every factor, three level tests will be used because the variables under consideration are multi-level variables and the effects of those variables on the final product are nonlinear. Table 5 displays the machining parameters and their corresponding levels (Mata et al., 2012).

Table 5. Machining parameters and their levels

	Level	*Code*
Cutting speed (m/min)	300	1
	200	2
	100	3
Depth of cut (mm)	1.5	1
	0.75	2
	0.25	3
Feed rate (mm/rev)	0.20	1
	0.15	2
	0.10	3

An experimental layout plan based on full factorial design (FFD) of experiments has been selected. This consists of 27 experiments that are shown in Table 6.

Table 6. Experimental lay out and results in terms of cutting forces

Cutting speed v	Depth of cut d	Feed rate f	Radial force Fp (N)	Cutting force Fc (N)	Feed force Fa (N)
1	1	1	74.97	107.79	101.11
1	1	2	64.31	96.00	85.13
1	1	3	57.87	87.70	68.16
1	2	1	77.80	53.46	59.94
1	2	2	86.74	58.39	56.74
1	2	3	76.44	54.36	41.10
1	3	1	72.74	23.17	28.44
1	3	2	70.33	23.89	26.22
1	3	3	62.47	22.22	21.55
Cutting speed v	Depth of cut d	Feed rate f	Radial force Fp (N)	Cutting force Fc (N)	Feed force Fa (N)
2	1	1	111.18	133.03	107.79
2	1	2	101.55	125.55	93.42
2	1	3	89.62	116.70	73.67
2	2	1	111.39	76.28	68.68
2	2	2	100.93	70.93	57.69
2	2	3	86.58	62.82	44.72
2	3	1	85.42	28.75	33.57
2	3	2	78.08	26.65	29.37
2	3	3	67.52	23.65	23.61
3	1	1	121.14	145.49	118.21
3	1	2	112.73	136.84	101.71
3	1	3	98.94	123.86	78.75
3	2	1	98.43	66.91	54.20
3	2	2	102.63	67.71	56.97
3	2	3	102.39	67.94	54.59
3	3	1	91.84	28.43	36.19
3	3	2	86.60	27.19	31.93
3	3	3	76.86	25.19	26.93

8. CONCLUSION

Many solutions have been developed to increase the diamond adhesion when deposited on cobalt cemented carbide substrates, however, the majorities do not evaluate the practical conditions encountered when using commercial carbide tools existing in the market as the required substrate material. In most of these research studies, substrate samples correspond to WC-Co coupons with surface conditions that differ from the commercial tools. Moreover, the grinding effects in the surface/subsurface modifications of the carbide tool materials used as a substrate for further diamond deposition have not been evaluated enough. This analysis may exist as the internal know-how of CVD diamond tool makers, who normally optimize the process variables for particular customer applications.

It is possible to conclude that the application of diamond-coated one-shot drills shows advantages in machining CFRP laminates, such as hole quality through lower delamination and burr defects, compared with the conventional twist drills. Nevertheless, not enough data is available regarding the cutting performance of the mentioned above one-shot drills, especially in terms of required film thickness, morphology, and its influence on drilled hole quality. Also, adhesion enhancement of diamond coatings deposited on such tools is essential to reach optimal tool life in economic aspects.

Nowadays, CFRP based composites became popular for use in a wide range of industries because of their high specific strength and stiffness coupled with damping properties and almost zero thermal expansion coefficient. Aerospace industry is major of composite application field, since benefits from composites are especially important where weight control is critical. For instance, CFRP enables weight reduction in transportation applications such as automobiles and aerospace that leads to energy reductions through fuel savings. It is estimated that every 10% of vehicle mass savings results in a 6%–8% improvement in fuel economy. However, CFRP is one of the difficult-to-cut materials. The abrasive effect of hard carbon fibers in CFRP always leads to severe tool wear. Furthermore, when cutting tool conditions deteriorate due to wear and tear, the CFRP workpiece tends to show material-specific defects, such as delamination, burrs, and tearing of fibers.

REFERENCES

Abrão, A. M., Faria, P. E., & Davim, J. P. (2007). Drilling of fiber reinforced plastics: A review. *Journal of Materials Processing Technology*, *186*(1-3), 1–7. doi:10.1016/j.jmatprotec.2006.11.146

Arul, S., Vijayaraghavan, L., Malhotra, S. K., & Krishnamurthy, R. (2006). The effect of vibratory drilling on hole quality in polymeric composites. *International Journal of Machine Tools & Manufacture*, *46*(3-4), 252–259. doi:10.1016/j.ijmachtools.2005.05.023

Asmussen, J., & Reinhard, D. K. (2002). *Diamond Films Handbook*. Marcel Dekker Inc. doi:10.1201/9780203910603

Azmir, M. A., & Ahsan, A. K. (2008). Investigation on Glass/Epoxy Composite Surfaces Machined by Abrasive Water Jet Machining. *Journal of Materials Processing Technology*, *198*(1-3), 122–128. doi:10.1016/j.jmatprotec.2007.07.014

Bhatnagar, N., Ramakrishnan, N., Naik, N. K., & Komanduri, R. (1995). On the machining of fiber reinforced plastic (FRP) composite laminates. *International Journal of Machine Tools & Manufacture, 35*(5), 701–716. doi:10.1016/0890-6955(95)93039-9

Bombardier Aerospace. (2011). Drilling of composites and composite/metallic assemblies. *Process Specification.*

Bouzakis, K.D., Michailidis, N., & Skordaris, G. (2012). Cutting with coated tools: Coating technologies, characterization methods and performance optimization. *CIRP Annals - Manufacturing Technology, 61*, 703 – 723. doi:10.1016/j.cirp.2012.05.006

Brehl, D. E., & Dow, T. A. (2008). Review of vibration-assisted machining. *Precision Engineering, 32*(3), 153–172. doi:10.1016/j.precisioneng.2007.08.003

Brent, A. (2008). *Strong, Fundamentals of Composites Manufacturing: Materials, Methods, and Applications*. Society of Manufacturing Engineers.

Callister, W. D. Jr, & Rethwisch, D. G. (2008). *Fundamentals of materials science and engineering: an integrated approach (3rd* ed.). Wiley.

Chen, W. C. (1997). Some experimental investigations in the drilling of carbon fiber-reinforced plastic (CFRP) composite laminates. *International Journal of Machine Tools & Manufacture, 37*(8), 1097–1108. doi:10.1016/S0890-6955(96)00095-8

Cheng, C. F., Tsui, Y. C., & Clyne, T. W. (1998). Application of a three-dimensional heat flow model to treat laser drilling of carbon fibre composites. *Acta Materialia, 46*(12), 4273–4285. doi:10.1016/S1359-6454(98)00090-1

Chung, D. D. L. (1994). Carbon Fiber Composites. Butterworth-Heinemann. doi:10.1016/B978-0-08-050073-7.50009-9

Das, S., Warren, J., & West, D. (2016). Global Carbon Fiber Composites Supply Chain Competitiveness Analysis. Available from https://www.osti.gov/servlets/purl/1333049

Davim, J. P., & Mata, F. (2008). Chemical vapour deposition (CVD) diamond coated tools performance in machining of PEEK composites. *Materials & Design, 29*(8), 1568–1574. doi:10.1016/j.matdes.2007.11.002

Davim, J. P., Rubio, J. C., & Abrão, A. M. (2007). A novel approach based on digital image analysis to evaluate the delamination factor after drilling composite laminates. *Composites Science and Technology, 67*(9), 1939–1945. doi:10.1016/j.compscitech.2006.10.009

Deuerler, F., Lemmer, O., Frank, M., Pohl, M., & Heßing, C. (2002). Diamond films for wear protection of hardmetal tools. *International Journal of Refractory & Hard Metals*, *20*(2), 115–120. doi:10.1016/S0263-4368(02)00009-4

Durão, L. M. P., Gonçalves, D. J. S., Tavares, J. M. R. S., de Albuquerque, V. H. C., Aguiar Vieira, A., & Torres Marques, A. (2010). Drilling tool geometry evaluation for reinforced composite laminates. *Composite Structures*, *92*(7), 1545–1550. doi:10.1016/j.compstruct.2009.10.035

Endler, I., Leonhardt, A., & Born, R. (1996). Interlayers for diamond deposition on tool materials. *Diamond and Related Materials*, *5*(3-5), 299–303. doi:10.1016/0925-9635(95)00352-5

Faraz, A., Biermann, D., & Weinet, K. (2009). Cutting edge rounding: An innovative tool wear criterion in drilling CFRP composite laminates. *International Journal of Machine Tools & Manufacture*, *49*(15), 1185–1196. doi:10.1016/j.ijmachtools.2009.08.002

Feito, N., Álvarez, J. D., Cantero, J. L., & Miguélez, M. H. (2015). Influence of Special Tool Geometry in Drilling Woven CFRPs Materials. *Procedia Engineering*, *132*, 632–638. doi:10.1016/j.proeng.2015.12.541

Fernandes, M., & Cook, C. (2006). Drilling of carbon composites using a one shot drill bit. Part I: Five stage representation of drilling and affecting maximum force and torque. *International Journal of Machine Tools & Manufacture*, *46*(1), 70–75. doi:10.1016/j.ijmachtools.2005.03.015

Ferreira, J.R., Coppini, N.L., & Miranda, G.W.A. (1999). Machining optimization in carbon fibre reinforced composite materials. *Journal of Materials Processing Technology, 92 – 93*, 135 – 140. doi:10.1016/S0924-0136(99)00221-6

Gåhlin, R., Alahelisten, A., & Jacobson, S. (1996). The effects of compressive stresses on the abrasion of diamond coatings. *Wear, 196*(1 – 2), 226-233. doi:10.1016/0043-1648(95)06907-0

Gaitonde, V. N., Karnik, S. R., Rubio, J. C., Correia, A. E., Abrão, A. M., & Davim, J. P. (2008). Analysis of parametric influence on delamination in high-speed drilling of carbon fiber reinforced plastic composites. *Journal of Materials Processing Technology*, *203*(1-3), 431–438. doi:10.1016/j.jmatprotec.2007.10.050

Gay, D., Hoa, S., & Tsai, S. (2003). *Composite Materials – Design and Applications*. CRC Press.

Gaydaychuk, A., Linnik, S., & Zenkin, S. (2020). High temperature tribology of heavily boron doped diamond films against steel. *International Journal of Refractory & Hard Metals*, *88*, 105191. doi:10.1016/j.ijrmhm.2020.105191

Geier, N., Davim, J. P., & Szalay, T. (2019). Advanced cutting tools and technologies for drilling carbon fibre reinforced polymer (CFRP) composites: A review. *Composites. Part A, Applied Science and Manufacturing*, *125*, 105552. doi:10.1016/j.compositesa.2019.105552

Geng, D., Liu, Y., Shao, Z., Lu, Z., Cai, J., Li, X., Jiang, X., & Zhang, D. (2019). Delamination formation, evaluation and suppression during drilling of composite laminates: A review. *Composite Structures*, *216*, 168–186. doi:10.1016/j.compstruct.2019.02.099

Gracio, J. J., Fan, Q. H., & Madaleno, J. C. (2010). Diamond growth by chemical vapour deposition. *Journal of Physics. D, Applied Physics*, *43*(37), 374017. doi:10.1088/0022-3727/43/37/374017

Haubner, R. (2013). The history of hard CVD coatings for tool applications at the University of Technology Vienna. *International Journal of Refractory & Hard Metals*, *41*, 22–34. doi:10.1016/j.ijrmhm.2013.01.012

Haubner, R., & Kalss, W. (2010). Diamond deposition on hardmetal substrates – Comparison of substrate pre-treatments and industrial applications. *International Journal of Refractory & Hard Metals*, *28*(4), 475–483. doi:10.1016/j.ijrmhm.2010.03.004

Haubner, R., Kubelka, S., Lux, B., Griesser, M., & Grasserbauer, M. (1995). Murakami and H_2SO_4/H_2O_2 Pretreatment of WC- Co Hard Metal Substrates to Increase the Adhesion of CVD Diamond Coatings. *Journal de Physique. IV*, *5*(C5), 753–760. doi:10.1051/jphyscol:1995589

Haubner, R., & Lux, B. (2006). Diamond deposition on steel substrates using intermediate layers. *International Journal of Refractory & Hard Metals*, *24*(5), 380–386. doi:10.1016/j.ijrmhm.2005.11.008

Heath, P. J. (2001). Developments in applications of PCD tooling. *Journal of Materials Processing Technology*, *116*(1), 31–38. doi:10.1016/S0924-0136(01)00837-8

Heisel, U., & Pfeifroth, T. (2012). Influence of point angle on drill hole quality and machining forces when drilling CFRP. *Procedia CIRP*, *1*, 471–476. doi:10.1016/j.procir.2012.04.084

Herlinger, J. (2006). Sp3's experience using hot filament CVD reactors to grow diamond for an expanding set of applications. *Thin Solid Films*, *501*(1-2), 65–69. doi:10.1016/j.tsf.2005.07.108

HexcelP. T. (2013). Available from https://kevra.fi/wp- content/uploads/Prepreg_Technology_.pdf

Hu, J., Chou, Y. K., Thompson, R. G., Burgess, J., & Street, S. (2007). Characterizations of nanocrystalline diamond coating cutting tools. *Surface and Coatings Technology*, *202*(4-7), 1113–1117. doi:10.1016/j.surfcoat.2007.07.050

Iliescu, D., Gehin, D., Gutierrez, M. E., & Girot, F. (2010). Modeling and tool wear in drilling of CFRP. *International Journal of Machine Tools & Manufacture*, *50*(2), 204–213. doi:10.1016/j.ijmachtools.2009.10.004

Jia, Z., Fu, R., Niu, B., Qian, B., Bai, Y., & Wang, F. (2016). Novel drill structure for damage reduction in drilling CFRP composites. *International Journal of Machine Tools & Manufacture*, *110*, 55–65. doi:10.1016/j.ijmachtools.2016.08.006

John & Blackmon. (1990). Metal-Matrix Composites. In ASM Handbook Vol. 2: Properties and Selection: Nonferrous Alloys and Special-Purpose Materials. doi:10.31399/asm.hb.v02.a0001101

Johnson, T. (2018). History of Composites. ThoughtCo. Available from https://www.thoughtco.com/history-of-composites-820404

Kamo, M., Sato, Y., Matsumoto, S., & Setaka, N. (1983). Diamond synthesis from gas phase in microwave plasma. *Journal of Crystal Growth*, *62*(3), 642–644. doi:10.1016/0022-0248(83)90411-6

Karataş, M. A., & Gökkaya, H. (2018). A review on machinability of carbon fiber reinforced polymer (CFRP) and glass fiber reinforced polymer (GFRP) composite materials. *Defence Technology*, *14*(4), 318–326. doi:10.1016/j.dt.2018.02.001

König, W., & Graß, P. (1989). Quality Definition and Assessment in Drilling of Fibre Reinforced Thermosets. *CIRP Annals*, *38*(1), 119–124. doi:10.1016/S0007-8506(07)62665-1

König, W., Wulf, C., Graß, P., & Willerscheid, H. (1985). Machining of Fibre Reinforced Plastics. *CIRP Annals*, *34*(2), 537–548. doi:10.1016/S0007-8506(07)60186-3

Kuo, C., Soo, S., Aspinwall, D., Bradley, S., Thomas, W., M'Saoubi, R., Pearson, D., & Leahy, W. (2014). Tool wear and hole quality when single-shot drilling of metallic-composite stacks with diamond-coated tools. *Proceedings of the Institution of Mechanical Engineers. Part B, Journal of Engineering Manufacture*, *228*(10), 1314–1322. doi:10.1177/0954405413517388

Kuo, C., Wang, C., & Ko, S. (2018). Wear behavior of CVD diamond – coated tools in the drilling of woven CFRP composites. *Wear*, *398-399*, 1–12. doi:10.1016/j.wear.2017.11.015

Lei, X., Wang, L., Shen, B., Sun, F., & Zhang, Z. (2013). Comparison of chemical vapor deposition diamond-, diamond-like carbon- and TiAlN-coated microdrills in graphite machining. *Proceedings of the Institution of Mechanical Engineers. Part B, Journal of Engineering Manufacture*, *227*(9), 1299–1309. doi:10.1177/0954405413487898

Lifshitz, Y. (1999). Diamond – like carbon – present status. *Diamond and Related Materials*, *8*(8-9), 1659–1676. doi:10.1016/S0925-9635(99)00087-4

Linbo, Z., Lijiang, W., & Xin, W. (2003). Study on vibration drilling of fiber reinforced plastics with hybrid variation parameters method. *Composites. Part A, Applied Science and Manufacturing*, *34*(3), 237–244. doi:10.1016/S1359-835X(02)00207-5

Liu, D., Tang, Y., & Cong, W. L. (2012). A review of mechanical drilling for composite laminates. *Composite Structures*, *94*(4), 1265–1279. doi:10.1016/j.compstruct.2011.11.024

Makhdum, F., Phadnis, V. A., Roy, A., & Silberschmidt, V. V. (2014). Effect of ultrasonically- assisted drilling on carbon-fibre-reinforced plastics. *Journal of Sound and Vibration*, *333*(23), 5939–5952. doi:10.1016/j.jsv.2014.05.042

Mata, F., Hanafi, I., Beamud, E., Khamlichi, A., & Jabbouri, A. (2012). Modelling of machining force components during turning of PEEK CF30 by TiN coated cutting tools using artificial intelligence. *Int. J. Machining and Machinability of Materials*, *11*(3), 263. doi:10.1504/IJMMM.2012.046885

Matsumoto, S., Sato, Y., Kamo, M., & Setaka, N. (1982). Vapor Deposition of Diamond Particles from Methane. *Japanese Journal of Applied Physics*, *21*(4), 183–185. doi:10.1143/JJAP.21.L183

Mkaddem, A., Ben Soussia, A., & El Mansori, M. (2013). Wear resistance of CVD and PVD multilayer coatings when dry cutting fiber reinforced polymers (FRP). *Wear*, *302*(1-2), 946–954. doi:10.1016/j.wear.2013.03.017

Murphy, C., Byrne, G., & Gilchrist, M. D. (2002). The performance of coated tungsten carbide drills when machining carbon fibre-reinforced epoxy composite materials. *Proceedings of the Institution of Mechanical Engineers. Part B, Journal of Engineering Manufacture*, *216*(2), 143–152. doi:10.1243/0954405021519735

Ouyang, W., Jiao, J., Xu, Z., Xia, H., Ye, Y., Zou, Q., Tian, R., & Sheng, L. (2021). Experimental study on CFRP drilling with the picosecond laser "double rotation" cutting technique. *Optics & Laser Technology*, *142*, 107238. doi:10.1016/j.optlastec.2021.107238

Pan, C.T., & Hocheng, H. (1996). The anisotropic heat-affected zone in the laser grooving or fiber-reinforced composite material. *Journal of Materials Processing Technology*, *62*(1-3), 54 – 60. doi:10.1016/0924-0136(95)02192-2

Polini, R. (2006). Chemically vapour deposited diamond coatings on cemented tungsten carbides: Substrate pretreatments, adhesion and cutting performance. *Thin Solid Films*, *515*(1), 4–13. doi:10.1016/j.tsf.2005.12.042

Qiu, X., Li, P., & Niu, Q. (2019). New Compound Drill Bit for Damage Reduction in Drilling CFRP. *International Journal of Precision Engineering and Manufacturing – Green Technology*, *6*(1), 75 – 87. . doi:10.1007/s40684-019-00026-3

Sarangi, S. K., Chattopadhyay, A., & Chattopadhyay, A. K. (2008). Effect of pretreatment, seeding and interlayer on nucleation and growth of HFCVD diamond films on cemented carbide tools. *International Journal of Refractory & Hard Metals*, *26*(3), 220–231. doi:10.1016/j.ijrmhm.2007.05.002

Sarangi, S. K., Chattopadhyay, A., & Chattopadhyay, A. K. (2012). Influence of process parameters on growth of diamond crystal on cemented carbide substrates by HFCVD system. *International Journal of Refractory & Hard Metals*, *31*, 1–13. doi:10.1016/j.ijrmhm.2011.07.007

Shanmugam, D. K., Chen, F. L., Siores, E., & Brandt, M. (2002). Comparative study of jetting machining technologies over laser machining technology for cutting composite materials. *Composite Structures*, *57*(1-4), 289–296. doi:10.1016/S0263-8223(02)00096-X

Sheikh-Ahmed, J. Y. (2009). *Machining of polymer composites*. Springer. doi:10.1007/978-0-387-68619-6

Shyha, I. S., Aspinwall, D. K., Soo, S. L., & Bradley, S. (2009). Drill geometry and operating effects when cutting small diameter holes in CFRP. *International Journal of Machine Tools & Manufacture*, *49*(12-13), 1008–1014. doi:10.1016/j.ijmachtools.2009.05.009

Sommer, M., Haubner, R., & Lux, B. (2000). Diamond deposition on copper treated hardmetal substrates. *Diamond and Related Materials*, *9*(3-6), 351–357. doi:10.1016/S0925-9635(99)00250-2

Soutis, C. (2005). Fibre reinforced composites in aircraft construction. *Progress in Aerospace Sciences*, *41*(2), 143–151. doi:10.1016/j.paerosci.2005.02.004

Spitsyn, B., Bouilov, L., & Deryagin, B. (1988). Diamond and diamond-like films: Deposition from the vapour phase, structure and properties. *Progress in Crystal Growth and Characterization*, *17*(2), 79–170. doi:10.1016/0146-3535(88)90001-9

Su, F., Zheng, L., Sun, F., Wang, Z., Deng, Z., & Qiu, X. (2018). Novel drill bit based on the step-control scheme for reducing the CFRP delamination. *Journal of Materials Processing Technology*, *262*, 157–167. doi:10.1016/j.jmatprotec.2018.06.037

Teti, R. (2002). Machining of composite materials. *CIRP Annals*, *51*(2), 611–634. doi:10.1016/S0007-8506(07)61703-X

Trent, E., & Wright, P. (2001). *Metal cutting* (4th ed.). Butterworth – Heinemann.

Tsao, C. C. (2008). Investigation into the effects of drilling parameters on delamination by various step-core drills. *Journal of Materials Processing Technology*, *206*(1-3), 405–411. doi:10.1016/j.jmatprotec.2007.12.057

Tsao, C. C., & Chiu, Y. C. (2011). Evaluation of drilling parameters on thrust force in drilling carbon fiber reinforced plastic (CFRP) composite laminates using compound core-special drills. *International Journal of Machine Tools & Manufacture*, *51*(9), 740–744. doi:10.1016/j.ijmachtools.2011.05.004

Tsao, C. C., & Hocheng, H. (2003). The effect of chisel length and associated pilot hole on delamination when drilling composite materials. *International Journal of Machine Tools & Manufacture*, *43*(11), 1087–1092. doi:10.1016/S0890-6955(03)00127-5

Turki, Y., Habak, M., Velasco, R., Aboura, Z., Khellil, K., & Vantomme, P. (2014). Experimental investigation of drilling damage and stitching effects on the mechanical behavior of carbon/epoxy composites. *International Journal of Machine Tools & Manufacture*, *87*, 61–72. doi:10.1016/j.ijmachtools.2014.06.004

Ullram, S., & Haubner, R. (2006). Temperature pre-treatments of hardmetal substrates to reduce the cobalt content and improve diamond deposition. *Diamond and Related Materials*, *15*(4-8), 994–999. doi:10.1016/j.diamond.2005.12.055

UMECO. (2013). *An Introduction to Advanced Composites and Prepreg Technology*. Available from https://www.yumpu.com/en/document/read/9334619/introduction-to-advanced-composites-and-prepreg-technology

Vandevelde, T.C.S., Vandierendonck, K., & Van Stappen, M. (1999). Cutting applications of DLC, hard carbon and diamond films. *Surface and Coating Technology, 113*(1 – 2), 80 – 85. doi:10.1016/S0257-8972(98)00831-7

Wang, D. H., Ramulu, M., & Arola, D. (1995). Orthogonal cutting mechanisms of graphite/epoxy composite. Part I: Unidirectional laminate. *International Journal of Machine Tools & Manufacture, 35*(12), 1623–1638. doi:10.1016/0890-6955(95)00014-O

Wang, J. (1999). A machinability study of polymer matrix composites using abrasive waterjet cutting technology. *Journal of Materials Processing Technology, 94*(1), 30–35. doi:10.1016/S0924-0136(98)00443-9

Wang, X., Kwon, P. Y., Sturtevant, C., Kim, D. D.-W., & Lantrip, J. (2013). Tool wear of coated drills in drilling CFRP. *Journal of Manufacturing Processes, 15*(1), 127–135. doi:10.1016/j.jmapro.2012.09.019

Wang, X., Shen, X., Yang, G., & Sun, F. (2017). Evaluation of boron-doped-microcrystalline/nanocrystalline diamond composite coatings in drilling of CFRP. *Surface and Coatings Technology, 330*, 149–162. doi:10.1016/j.surfcoat.2017.10.002

Wang, X., Shen, X., Zeng, C., & Sun, F. (2018). Combined influences of tool shape and as- deposited diamond film on cutting performance of drills for CFRP machining. *Surface and Coatings Technology, 347*, 390–397. doi:10.1016/j.surfcoat.2018.05.024

Wang, X., Zhang, J., & Sun, F. (2014). Fracture and solid particle erosion of micro-crystalline, nano-crystalline and boron-doped diamond films. *International Journal of Refractory & Hard Metals, 45*, 31–40. doi:10.1016/j.ijrmhm.2014.02.005

Xu, J., An, Q., Cai, X., & Chen, M. (2013). Drilling machinability evaluation on new developed high- strength T800S/250F CFRP laminates. *International Journal of Precision Engineering and Manufacturing, 14*(10), 1687–1696. doi:10.1007/s12541-013-0252-2

Xu, J., An, Q., & Chen, M. (2014). A comparative evaluation of polycrystalline diamond drills in drilling high-strength T800S/250F CFRP. *Composite Structures, 117*, 71–82. doi:10.1016/j.compstruct.2014.06.034

Xu, J., Li, C., Mi, S., An, Q., & Chen, M. (2018). Study of drilling induced defects for CFRP composites using new criteria. *Composite Structures, 201*, 1076–1087. doi:10.1016/j.compstruct.2018.06.051

Zhang, H., Chen, W., Chen, D., & Zhang, L. C. (2001). Assessment of the Exit Defects in Carbon Fibre-Reinforced Plastic Plates Caused by Drilling. *Key Engineering Materials, 196*, 43–52. doi:10.4028/www.scientific.net/KEM.196.43

Zhang, J., Wang, X., Shen, B., & Sun, F. (2013). Effect of boron and silicon doping on improving the cutting performance of CVD diamond coated cutting tools in machining CFRP. *International Journal of Refractory & Hard Metals*, *41*, 285–292. doi:10.1016/j.ijrmhm.2013.04.017

Chapter 7
Understanding the Influence of Machining Parameters on Mechanical Properties of Polymer Matrix Composites

Mohd Faizul Hasan
RR Institute of Technology and Management, India

Gaurav Shukla
RR Institute of Modern Technology, India

Maheshwar Dayal Gupta
RR Institute of Modern Technology, India

Shubham Shukla
RR Institute of Modern Technology, India

ABSTRACT

Polymer matrix composites (PMCs) represent a class of advanced materials that have gained widespread attention and adoption across various industries due to their exceptional mechanical properties and versatility in applications. Understanding the full spectrum of mechanical properties and their interplay with other material characteristics is essential for optimizing the performance of PMCs in various real-world applications. This chapter offers a comprehensive exploration of the intricate relationship between machining parameters and the mechanical properties of PMCs. The scope of this chapter encompasses an in-depth exploration of various mechanical properties relevant to machining, including tensile strength, modulus, flexural strength, impact resistance, shear strength, fatigue resistance, and creep behavior. It explores the underlying mechanisms behind the alterations in mechanical properties. Furthermore, the chapter highlights strategies to mitigate the adverse effects of machining on the mechanical properties of PMCs.

DOI: 10.4018/978-1-6684-9927-6.ch007

1. INTRODUCTION

Polymer matrix composites (PMCs) exhibit a wide range of mechanical properties that are crucial for their performance in various applications. PMCs are engineered materials composed of two primary constituents: a polymer matrix and reinforcing fibers. The polymer matrix, typically composed of thermosetting or thermoplastic resins, serves as the continuous phase that binds and supports the reinforcing fibers (Alberto 2013). The reinforcing fibers, which can be unidirectional, woven, or chopped, provide strength, stiffness, and other tailored properties to the composite (Kumar,et al. 2021). PMCs have emerged as versatile materials with widespread applications across various industries including aerospace, automotive, marine, and construction. These composite materials consist of a polymer matrix reinforced with high-strength fibers such as carbon, glass, or aramid (Prasad et al. 2019). PMCs offer several advantages over traditional materials such as metals and plastics. Their high strength-to-weight ratio, excellent fatigue resistance, corrosion resistance, and tailorable properties make them ideal for applications where weight reduction, durability, and performance are critical. Moreover, PMCs can be designed to exhibit specific mechanical, thermal, and electrical properties, allowing for customization to meet the requirements of diverse applications (Kumar, Prasad, Patel, Kumain, et al. 2021).

The combination of the matrix and reinforcement imparts unique mechanical properties to PMCs, making them attractive for use in lightweight, high-performance structures and components. The mechanical properties of PMCs are influenced by factors such as the type of reinforcement fibers, matrix material, fiber volume fraction, and manufacturing process. Understanding these properties is essential for predicting the behavior of PMCs during machining processes. Machining processes play a crucial role in shaping PMCs into finished components with the desired dimensions, tolerances, and surface quality. However, machining PMCs presents unique challenges due to their heterogeneous and anisotropic nature, as well as the presence of reinforcing fibers (Fallahi, Taheri-Behrooz, and Asadi 2020). Understanding the different machining techniques and factors influencing machining performance is essential for achieving optimal results in PMC fabrication.

Machining is a crucial manufacturing process used to shape and finish components from raw materials. In the context of PMCs, machining involves the removal of material using cutting tools to achieve the desired shape, dimensions, and surface finish. The mechanical properties of PMCs play a pivotal role in determining their machinability and the performance of machining operations. Drilling, milling, and turning are among the most common machining techniques used for shaping PMCs (Varma et al. 2024). Each of these processes involves the removal of material through mechanical means, typically using cutting tools such as drills, end mills, and lathe tools. The selection of machining technique depends on factors such as the geometry of the component, material properties, and desired surface finish. For instance, the orientation, distribution, and type of reinforcing fibers affect parameters such as cutting forces, tool wear, chip formation, and surface finish. Understanding the mechanical properties and the effect of machining parameters on the mechanical properties of PMCs is essential for selecting appropriate machining techniques, tooling, cutting parameters, and process conditions to achieve efficient and high-quality machining outcomes (Francisco Mata Cabrera et al. 2010). As depicted in Fig. 1 (Ruszaj, Gawlik, and Skoczypiec 2016), the strength of PMCs are comparable with other available materials. In Fig. 1, the specific strength of PMCs is compared with different class of materials related to the temperature potential, not solely to composites. Also, this figure provides useful information on the potential temperature machining of PMCs in comparison with the other materials.

Figure 1. Specific strength and temperature potential of different composites (Ruszaj, Gawlik, and Skoczypiec 2016).

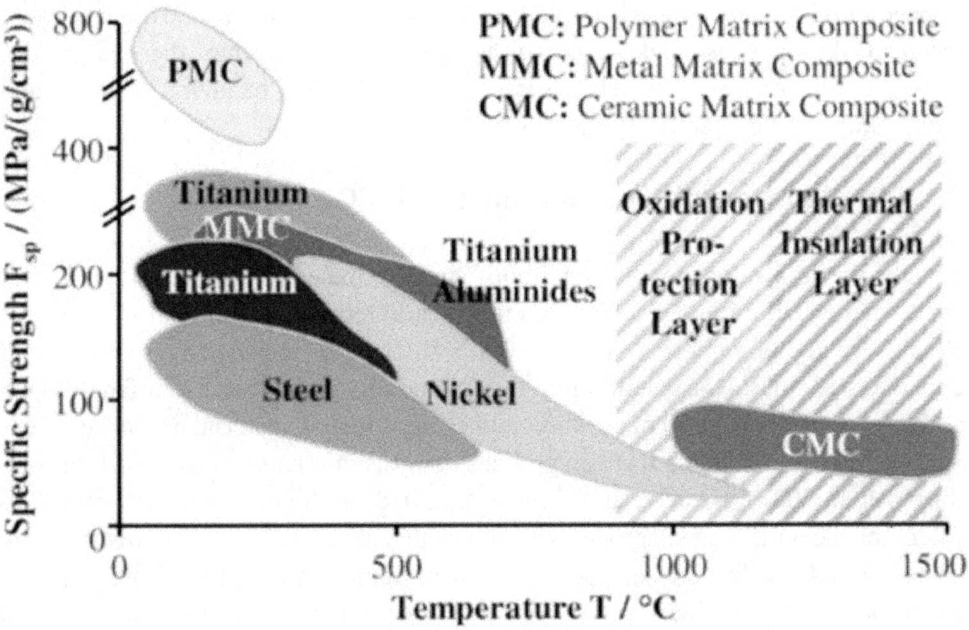

The aim of this chapter is to provide a comprehensive understanding of how machining performance influence the mechanical properties of PMCs. By examining the relationship between mechanical properties and machining behavior, this chapter seeks to elucidate the challenges, considerations, and strategies involved in machining

PMCs effectively. The scope of this chapter encompasses an in-depth exploration of various mechanical properties relevant to machining, including tensile strength, modulus, flexural strength, impact resistance, shear strength, fatigue resistance, and creep behavior.

2. MECHANICAL PROPERTIES OF PMCS

Mechanical properties encompass a broad range of characteristics that describe how materials behave under different types of mechanical loading. Mechanical properties are fundamental characteristics that govern the behavior of materials under mechanical loading conditions and are key determinants of their performance in real-world applications. In the context of PMCs, mechanical properties play a crucial role in dictating how these materials respond to external forces, withstand loads, and resist failure mechanisms (R. V. Patel, Yadav, and Winczek 2023). This section provides an in-depth exploration of the various mechanical properties exhibited by PMCs, including tensile strength and modulus, flexural strength and modulus, impact strength, shear strength, fatigue resistance, creep behavior, and other relevant mechanical properties. Understanding the significance of each of these properties is essential for comprehensively assessing the mechanical behavior of PMCs and their suitability for specific applications.

2.1 Tensile Strength and Modulus of PMCs

Tensile strength refers to the maximum stress a material can withstand before experiencing failure under tensile loading, while tensile modulus, also known as Young's modulus, represents the material's stiffness or resistance to deformation under tension. In PMCs, tensile strength and modulus are influenced by factors such as the type and orientation of reinforcement fibers, matrix material properties, fiber volume fraction, and manufacturing process parameters (Kumar et al. 2024; Prasad, Singh, et al. 2022). Kiran et al. (Kiran et al. 2018) highlighted that incorporating fillers enhances mechanical properties, with factors like filler dispersion, distribution, adhesion, and interface with the polymer matrix significantly influencing composite performance. Higher fiber content typically leads to increased tensile strength and modulus due to the reinforcing effect of the fibers, making PMCs particularly attractive for applications requiring high strength-to-weight ratios. For instance, Rajakumar et al. (Rajakumar et al. 2022) investigated the tensile properties of a high impact polyethylene composite reinforced with abaca fibre and the effects of fibre loading. The study found that with 55% fibre loading, tensile strength increased by 312% and Young's modulus increased by 545% compared to no fibre loading (Fig. 2). The alkaline treatment had a significant influence on the mechanical characteristics of the composite. Sharma and Lakkad (Sharma and Lakkad 2011) grew carbon nanotubes and nanofiberson carbon fiber substrates via thermal chemical vapor deposition. These coated fibers, used as reinforcement in epoxy matrices for unidirectional composites, exhibited uniform coverage. Tensile tests showed a significant 69% increase in strength compared to composites without nanotubes, highlighting the importance of vacuum conditions and catalyst precursors in modifying fiber properties and subsequently enhancing composite performance.

Figure 2. Tensile strength (MPa) vs abaca fibre loading (wt %) (Rajakumar et al. 2022)

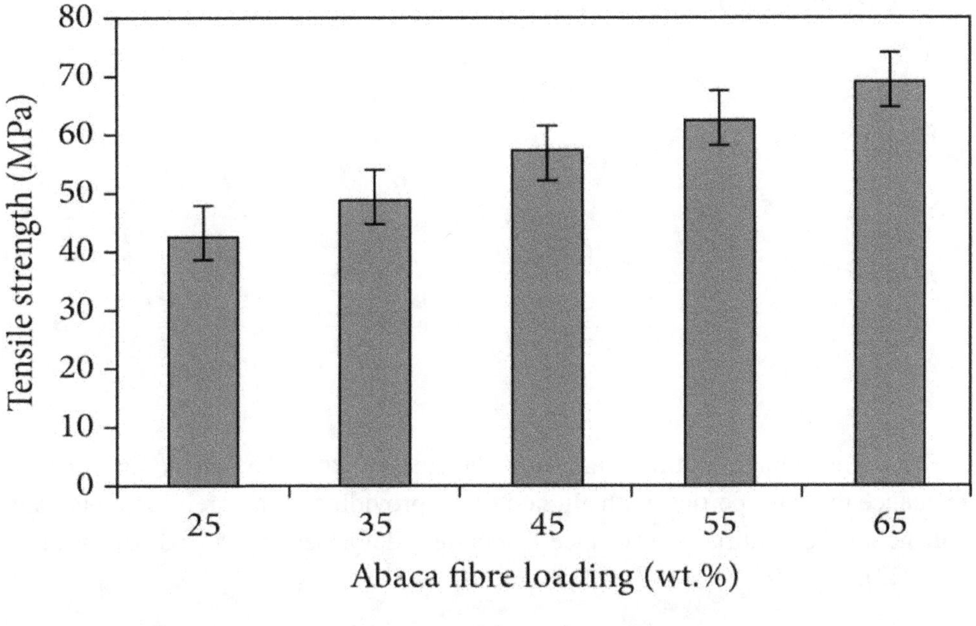

2.2 Flexural Strength and Modulus of PMCs

Flexural strength, measures a material's ability to resist bending or breaking under applied bending loads. Flexural modulus, similar to tensile modulus, quantifies the material's stiffness in bending. PMCs often exhibit high flexural strength and modulus compared to unreinforced polymers, owing to the reinforcing effect of the fibers (Hulugappa, Achutha, and Suresha 2016). Figure 3 (Amuthakkannan et al. 2013) presents the flexural properties of polyester composites reinforced with short basalt fibers of varying lengths and content. It was observed that flexural strength and modulus increased significantly with higher fiber content, peaking at 68 wt.%.

However, lower strength and modulus were noted in composites with lower fiber content. The flexural strength consistently increased with increasing fiber content. The highest flexural strength was achieved with a fiber length of 21 mm, while the highest modulus was attained with a length of 10 mm. Longer fibers exhibited greater resistance to bending loads, while shorter fibers demonstrated higher modulus. Furthermore, comparison between 4 mm and 10 mm lengths showed superior flexural modulus for the latter, with significant improvements in strength at higher fiber percentages. Similarly, comparison between 10 mm and 21 mm lengths favored the 10 mm length for flexural modulus.

Figure 3. Flexural strength and modulus of PMCs with variation of fiber content (Amuthakkannan et al. 2013)

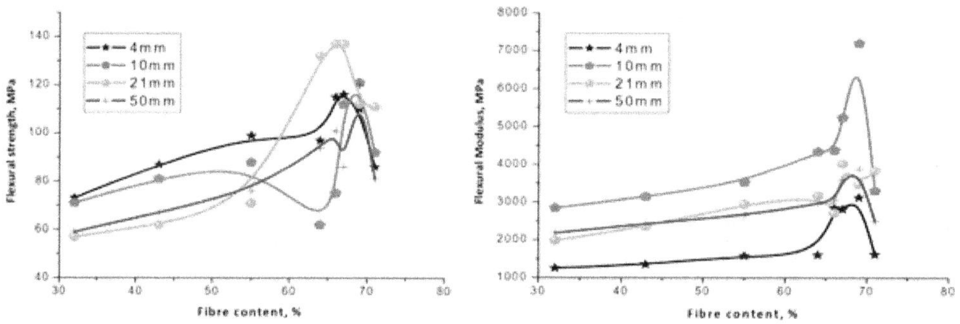

The arrangement and orientation of the fibers within the matrix significantly influence these properties, with aligned fibers providing greater reinforcement and enhancing the material's resistance to bending deformation (Prasad, Bairwan, et al. 2022).

2.3 Impact Strength

Impact strength, also referred to as toughness, measures a material's ability to absorb energy and resist fracture under sudden, high-energy loading conditions. In PMCs, impact strength is influenced by the toughness of the matrix material and the interfacial bonding between the fibers and the matrix (Negi et al. 2022; Prasad et al. 2020). Figure 4 depicts how both fiber content and length affect impact strength. Specifically, at a fixed length of 50 mm, impact strength increased steadily from 45 J/cm^2 to 55 J/cm^2 with higher fiber content, while other fiber lengths exhibited lower strength. This suggests that short basalt fibers may not effectively dissipate higher energy. The observed rise in impact strength is attributed to increased fiber content and compression pressure, which reduces voids in the composites, with fiber length also playing a contributing role. Raja et al. (Raja, Manisekar, and Manikandan 2013) showed how the Charpy impact energy changes with decreasing fly ash particle size. It was evident that as the size of the fly ash filler decreases, more energy is absorbed during specimen fracture. This increase in impact strength with smaller particle sizes confirms the presence of void spaces in larger particles. Consequently, stress propagation is more pronounced in composites filled with larger fly ash particles compared to those with smaller particles. Properly designed and engineered PMCs can exhibit superior impact resistance compared to traditional materials, making them suitable for applications subjected to dynamic loading conditions such as impact or crash events (Amuthakkannan et al. 2013).

Figure 4. Impact strength of PMCs with variation of fiber content (Amuthakkannan et al. 2013)

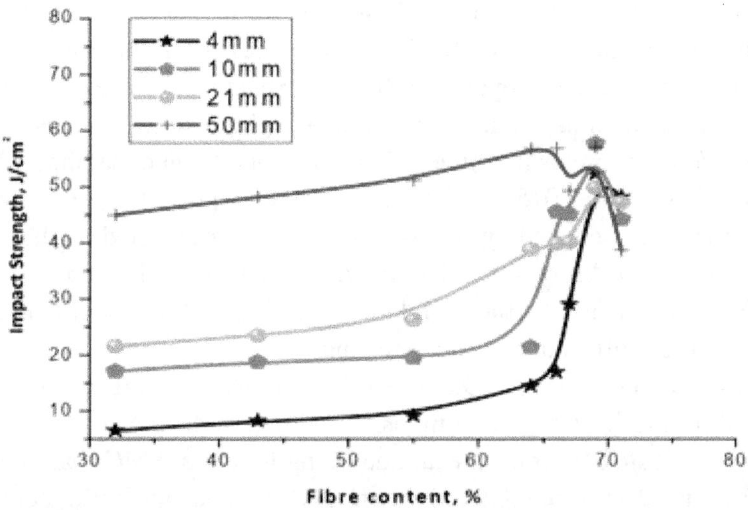

2.4 Shear Strength of PMCs

Shear strength is a measure of a material's resistance to shear stresses, which occur when adjacent layers of material slide past each other in response to applied forces. Zhi et al. (Zhi, Long, and Miao 2017) validated force-displacement curves and shear strength results obtained from finite element simulated model pullout tests against experimental data conducted under comparable conditions. Utilizing the von Mises and contact friction stress distributions from simulation models, they aimed to clarify the effects of experimental parameters on IFSS in microbond tests. The study highlighted that fiber diameter exerts the most significant influence on shear strength, with bead size and blade position also significantly affecting microbond test results. Patel et al. (J. S. Patel et al. 2016) investigated glass fiber structure and epoxy polymer effects on flexural strength, interlaminar shear stress, and energy absorption in glass fiber-reinforced polymer composites. In PMCs, shear strength is influenced by factors such as the shear strength of the matrix material, the interfacial bonding between the fibers and the matrix, and the orientation of the fibers relative to the direction of shear forces (Prasad, Kumain, et al. 2022). Hence, understanding shear behavior is essential for predicting the material's response to machining processes such as cutting and drilling, where shear forces are prevalent.

2.5 Fatigue Resistance of PMCs

Fatigue resistance is a critical mechanical property for materials subjected to cyclic loading conditions, such as those experienced in structural components under repetitive loading. PMCs can exhibit superior fatigue resistance compared to traditional materials due to the reinforcing effect of the fibers, which helps to distribute stresses and prevent crack propagation (Mohanavel et al. 2022; Prasad et al. 2023). To evaluate critical damage states and track changing damage processes in woven carbon fiber/epoxy composites under quasi-static and fatigue loading, (Montesano, Fawaz, and Bougherara 2015) used infrared thermography. Matrix fractures, mostly in weft yarns, were detected by quasi-static testing and verified by thermographic techniques and SEM observations. Thermography has been validated as a dependable fatigue damage indicator by the correlation between stiffness degradation profiles and temperature profiles during cycle testing.

Song and Otani (Song and Otani 1997) examined the fatigue process of satin woven carbon fabric laminate composites with two distinct matrix resins (CF/epoxy, CF/PEEK) to determine the fatigue strength (life) of PMCs based on material properties. They discovered that fatigue strength could be approximately estimated using a function of material properties. Furthermore, it was observed that the primary fatigue damage shifted from transverse crack formation in the transverse fiber bundle to fiber fracture in the longitudinal fiber bundle, depending on the value of κ. Hence, properly designed PMCs can withstand millions of load cycles without failure, making them ideal for applications requiring long-term durability and reliability.

2.6 Creep Behavior of PMCs

Creep behavior refers to the tendency of a material to deform over time under constant load or stress, particularly at elevated temperatures. PMCs with high-temperature matrix materials may exhibit viscoelastic behavior and creep under prolonged loading at elevated temperatures (Hassanzadeh-Aghdam et al. 2018; Kawai and Masuko 2004). Understanding creep behavior is essential for designing components that require long-term dimensional stability and structural integrity, particularly in high-temperature applications such as aerospace and automotive engine components (Kumar et al. 2022; Robinson, Binienda, and Ruggles 2003).

3. MACHINING PROCESSES IN PMCS

Machining processes play a pivotal role in the fabrication of PMCs, enabling the conversion of raw materials into finished components with precise dimensions, tolerances, and surface characteristics. Drilling, milling, turning, and grinding are among the most common machining techniques employed for shaping PMCs. Each of these processes involves the removal of material through mechanical means, typically using cutting tools such as drills, end mills, lathe tools, and grinding wheels.

3.1 Driling Process of PMCs

Drilling is a machining process used for creating holes in PMC components, ranging from simple clearance holes to complex features such as fastener holes and pockets. Drilling operations in PMCs require specialized drills with high-performance cutting geometries and coatings to withstand the abrasive nature of composite materials and minimize tool wear. Factors influencing drilling performance include fiber orientation, cutting speed, feed rate, tool geometry, and coolant application (Geier, Patra, et al. 2023). Khashaba (U. Khashaba 2013) investigated the influence of drilling parameters on machinability parameters when drilling polymeric composite materials. Parameters considered included feed, speed, drill wear, geometry, use of special tools, and composite material properties. Delamination, a common issue in drilling processes, often limits structural applications by weakening strength, compromising assembly tolerance, and potentially leading to long-term performance decline. They focused on identifying factors affecting delamination onset, mechanisms, measurement techniques, assessments, and delamination-free drilling methods in fiber-reinforced composites.

3.2. Milling Process of PMCs

Milling is a versatile machining process used for producing complex 3D shapes and features in PMC components. It involves the rotation of a multi-toothed cutting tool (end mill) to remove material from the workpiece surface. Milling operations in PMCs require careful consideration of cutting parameters such as spindle speed, feed rate, depth of cut, and tool path strategy to minimize tool wear, delamination, and surface defects (Geier, Xu, et al. 2023). High-speed machining techniques may be employed to achieve efficient material removal while reducing cutting forces and heat generation.

3.3. Turning Process of PMCs

Turning is a machining process used for producing cylindrical components by rotating a workpiece against a stationary cutting tool. In PMC turning operations, the selection of cutting tools, tool geometry, and cutting parameters is critical for achieving desired surface finish, dimensional accuracy, and minimizing tool wear (Yousefi and Zohoor 2019). Adaptive machining strategies such as variable cutting speed and feed rate may be employed to address variations in material properties and fiber orientation.

3.4 Grinding Process of PMCs

Grinding is a precision machining process used for achieving tight tolerances, surface finish, and dimensional accuracy in PMC components. It involves the use of abrasive particles bonded to a rotating wheel to remove material from the workpiece surface. Grinding operations in PMCs require specialized grinding wheels with controlled abrasive properties to minimize surface damage, fiber pullout, and heat-affected zones (Yousefi and Zohoor 2019). Factors influencing grinding performance include wheel speed, feed rate, wheel dressing, and coolant application.

3.5 Factors Influencing Machining Performance in PMCs

Several factors influence machining performance in PMCs, including material properties, tool geometry, cutting parameters, machining environment, and process conditions. One of the primary challenges in machining PMCs is the presence of reinforcing fibers, which can significantly affect material removal rates, tool wear, surface finish, and dimensional accuracy (Wan et al. 2019).

Fiber orientation plays a crucial role in determining machining performance, with aligned fibers providing enhanced reinforcement and higher mechanical properties in the direction of alignment. However, machining across the direction of fiber alignment may result in increased cutting forces, tool wear, and surface roughness due to the cutting action encountering more resistance from the reinforcing fibers. Strategies such as toolpath optimization, adaptive machining, and fiber orientation analysis can be employed to mitigate these challenges and achieve optimal machining performance (F.M. Cabrera et al. 2012; Manjunath Patel et al. 2020; Wan et al. 2019).

The *matrix material* also influences machining performance, with variations in matrix toughness, hardness, thermal conductivity, and thermal expansion affecting material removal rates, chip formation, tool wear, and surface finish. In addition, the presence of additives, fillers, or reinforcements in the matrix material may require adjustments to cutting parameters and tooling strategies to optimize machining performance (Jani et al. 2020; Wan et al. 2019).

Tool geometry and selection are critical factors in machining PMCs, with specialized cutting tools designed to withstand the abrasive nature of composite materials, minimize fiber pullout, and achieve desired surface finish and dimensional accuracy. Factors such as tool material, geometry, coating, and edge preparation influence cutting forces, chip formation, tool wear, and machining efficiency. High-performance cutting tools with diamond coatings, polycrystalline diamond (PCD) inserts, or carbide substrates may be employed to enhance machining performance and tool life in PMCs (F. M. Cabrera et al. 2011; Wan et al. 2019).

Cutting parameters such as cutting speed, feed rate, depth of cut, and coolant application significantly impact machining performance in PMCs. Optimal cutting conditions vary depending on material properties, tooling, workpiece geometry, and machining environment. Excessive cutting speeds or feed rates may result in overheating, tool wear, and surface damage, while inadequate coolant application may lead to poor chip evacuation, workpiece overheating, and dimensional inaccuracies. Balancing cutting parameters with material properties and process requirements is essential for achieving efficient and reliable machining performance in PMCs (Li et al. 2018; Wan et al. 2019).

Machining environment and process conditions, including temperature, humidity, vibration, and fixturing, also influence machining performance in PMCs. Variations in environmental conditions can affect material properties, tool behavior, chip formation, and surface finish. Stable machining setups, proper fixturing, and vibration damping techniques are essential for minimizing dimensional variations, surface defects, and tool chatter during machining operations (Wan et al. 2019).

4. EFFECT OF DRILLING PROCESS ON MECHANICAL PROPERTIES OF PMCS

The effect of drilling on the mechanical properties (tensile, flexural, impact, shear fatigue and creep behaviour) of PMCs is a critical aspect to consider in the machining of these materials. Drilling operations are commonly performed on PMCs to create holes for fasteners, joining components, or accommodating other mechanical features.

However, drilling introduces localized stress concentrations and material damage that can significantly impact the mechanical properties of the composite material (Merino-Pérez et al. 2016).

During drilling operations on PMCs, several damage mechanisms occur that can affect the tensile properties of the material. When the drill bit penetrates the composite material, it encounters the reinforcing fibers, which are typically stiffer and stronger than the matrix material (Rao et al. 2019). As a result, the fibers may undergo microcracking, debonding, or breakage due to the localized stresses generated by the drilling process. Additionally, the matrix material surrounding the drilled hole may experience cracking or chipping, leading to reduced load-bearing capacity and stiffness in the vicinity of the hole. Delamination, or separation of layers within the composite material, can occur at the hole's exit due to excessive cutting forces or improper tool geometry (Mata et al. 2012). Furthermore, residual stresses may develop around the drilled hole as a result of thermal and mechanical gradients induced during drilling, which can further degrade the material's tensile properties. Tao et al. (Tao et al. 2016) utilized an energy-based continuum damage mechanics model to explore the impact of drilling-induced delamination on the tensile strength of carbon fiber-reinforced polymer laminates, predicting residual strength. The FEM model results showed a significant agreement between numerical simulations and experimental results, suggesting that early outer plies failures were the primary cause of the strength loss.

Drilling-induced damage can have a significant impact on the tensile strength of PMCs. The presence of fiber damage, matrix cracking, and delamination reduces the load-bearing capacity of the material, leading to a decrease in tensile strength. Fiber breakage or debonding weakens the load transfer between the reinforcing fibers and the matrix, resulting in premature failure under tensile loading conditions. Additionally, residual stresses around the drilled hole can contribute to premature failure under tensile loading, particularly in regions where the material is already weakened by drilling-induced damage (Crivelli et al. 2015).

Hole-making is necessary in both synthetic and natural fiber composites for assembling different composite parts. However, drilling operations often cause damage such as delamination and matrix cracking around the hole, leading to long-term mechanical performance degradation. (Nasir, Azmi, and Khalil 2015) looked at how drilling circumstances affected residual tensile strength parametrically. The residual tensile strength in woven flax epoxy composites was analyzed using the Taguchi approach and statistical studies following drilling. Holes were made in composite specimens using a three-axis CNC machine according to two tiers of drilling parameters and drill bit geometries. Static loading tests were then carried out in accordance with ASTM Standard D3039. The analysis revealed that the feed rate is the most influential parameter affecting residual tensile strength.

In drilling glass fiber reinforced polymer matrix (GFRP) composites, common issues include burrs, splintering, microcracks, delamination, fiber pull-out, and thermal degradation of the polymer matrix. Cengiz (Cengiz et al. 2024) utilized molded-hole GFRP components as an alternative to drilling, yet research in this area is limited, particularly regarding the mechanical response of molded-hole GFRP. In this study, hand lay-up fabricated molded and drilled GFRP composites with epoxy matrix were analyzed using experimental and numerical techniques. Ansys Transient Structural Analysis was employed for stress and strain analysis. Although the fiber orientations were similar between the layer geometries [0°/90°] and 3D models, an additional layer was included in molded-hole composites to account for the fiber accumulation effect. Experimental results showed a 12% higher flexural strength in molded composites compared to drilled ones (Figure 5). Numerical analysis achieved a 98.7% proximity to experimental results.

Figure 5. Flexural stress for (a) net sections (NS) and (b) gross sections (GS) calculations

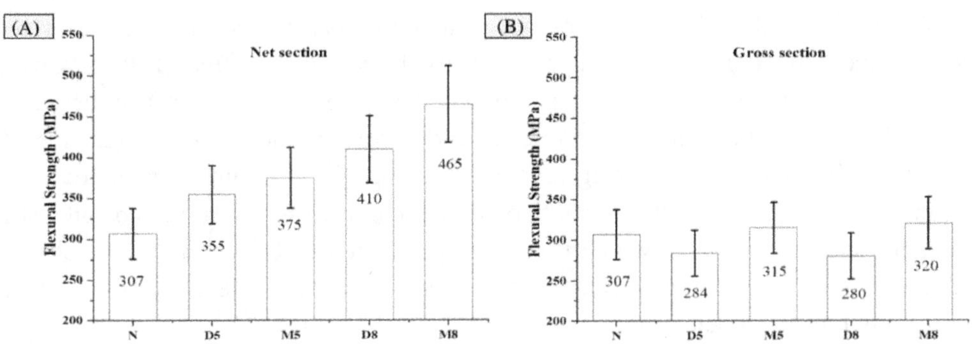

The drilling process in composite laminates is challenging to manage and often results in delamination, significantly impacting structural strength. Despite this, little attention has been given to the effect of drilling-induced damage on flexural strength, a key mechanical property of composite materials. (Heidary, Karimi, and Minak 2018) examined the thrust force, residual flexural strength, and delamination factor during the drilling process of woven epoxy composites reinforced with functionalized multi-walled carbon nanotubes and E-glass fibers.. Experiments were conducted considering process parameters such and the weight percentage of carbon nanotubes in nanocomposite laminates. Taguchi design of experiment was utilized for drilling experiments, followed by three-point bending tests to evaluate residual flexural strength.

Residual stresses induced around the drilled hole during the machining process can exacerbate the weakening effect, particularly in regions where the material is already compromised by drilling-induced damage (Rodriguez et al. 2023). Prakash et al. (Prakash and Dileep Aditya Dhar 2018) conducted experimental investigations to analyze the impact of machining parameters. Time and frequency domain analysis of vibration signals, measured using sound sensors, were employed to predict the effect of machining parameters on delamination and develop a tool replacement strategy.

Drilling polymer matrix composites is crucial for achieving precise structural integration, but it often leads to delamination, particularly at the hole exit plane, significantly impacting material strength. Limited research has explored the role of nano-fillers in reducing drilling-induced delamination in reinforced plastics. Pramod et al. (Pramod et al. 2022) investigated experimental design followed Taguchi's L16 approach, analyzing delamination factor and circularity ratio using image processing techniques in Matlab and ANOVA. Their results indicated reduced delamination factor in nanoclay and graphene reinforced polymer matrix toughened composites compared to base composites, associated with decreased thrust force and specific cutting energy, and increased circularity ratio (Wisnom 2012). Khasaba et al. (U. A. Khashaba et al. 2010) examined how machining parameters such as feed rate, speed, and drill diameter influence the thrust force and machinability of woven glass fiber-reinforced epoxy (GFRE) composites. The strong correlation between thrust force and machinability parameters underscores the need to minimize thrust force to enhance the load-bearing capacity of composite structures assembled with rivets or bolted joints. Photos for both drill diameters are shown in Figs 6 and 7 (U. A. Khashaba et al. 2010), which illustrate how feed affects peel-up and push-out delamination sizes, respectively. As can be expected, push-out delamination is more severe than peel-up delamination.

Figure 6. Images that show how feed affects push-out delamination for both drill diameters (U. A. Khashaba et al. 2010)

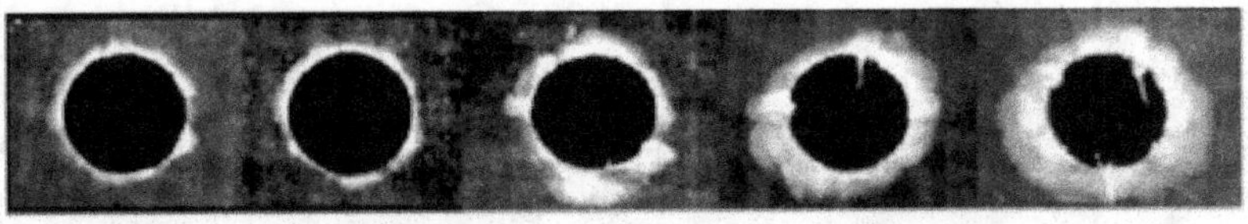

$f = 0.056$ $f = 0.112$ $f = 0.220$ $f = 0.315$ $f = 0.450$

(a) Push-out delamination, 8 mm diameter drill

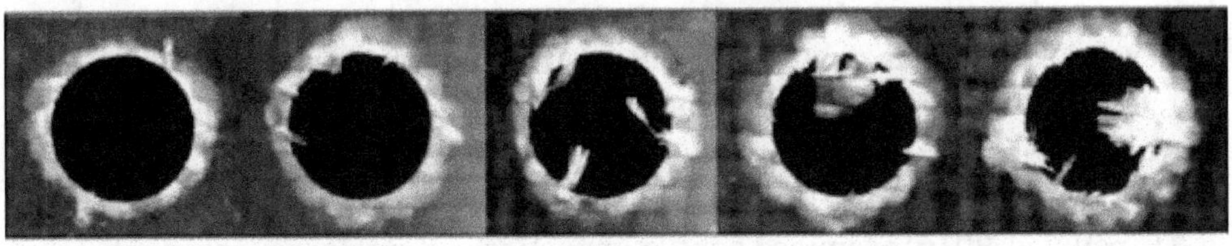

$f = 0.056$ $f = 0.112$ $f = 0.220$ $f = 0.315$ $f = 0.450$

(b) Push-out delamination, 13 mm diameter drill

Figure 7. Images showing how feed affects peel-up delamination at both drill diameters (U. A. Khashaba et al. 2010)

$f = 0.056$ $f = 0.112$ $f = 0.220$ $f = 0.315$ $f = 0.450$

(a) Peel-up delamination, 8 mm diameter drill

$f = 0.056$ $f = 0.112$ $f = 0.220$ $f = 0.315$ $f = 0.450$

(b) Peel-up delamination, 13 mm diameter drill

In addition to affecting shear strength, drilling-induced damage can also influence the interfacial bonding between the reinforcing fibers and the matrix material in PMCs.

Interfacial bonding plays a critical role in determining the material's mechanical properties, including shear strength, by facilitating load transfer between the fibers and the matrix (Fallahi, Kaynan, and Asadi 2023). Conventional drilling methods often struggle to produce high-quality holes efficiently. To address this, Arul et al. (Arul et al. 2006) developed a drilling technology minimizing defects and extending tool life. They introduced a new method involving low-frequency, high-amplitude vibration applied to the workpiece during drilling. Using a high-speed steel (HSS) drill, they conducted experiments comparing vibratory drilling with conventional drilling on glass fiber-reinforced plastics composites. They evaluated thrust force, flank wear, and delamination factor, while also monitoring acoustic emission during vibratory drilling. Results indicated that vibratory drilling shows promise by utilizing the regeneration effect to induce axial chatter, aiding chip breaking and reducing thrust force (Alam et al. 2019).

(Ismail et al. 2016) examined how delamination and surface roughness of carbon fiber reinforced polymer (CFRP) and hemp fiber reinforced polymer (HFRP) composite laminates were affected by drilling parameters (feed rate, cutting speed, and thrust force), drill diameters, and chip formation. HSS drills, or high-speed steel, were used in dry machining settings. The results show that, in contrast to cutting speed, increasing feed rate and thrust force increased delamination and surface roughness in both materials. Furthermore, when the material removal rate (MRR) increased, greater drill diameters and different kinds of chip production led to an increase in delamination and surface roughness. Feed rates of 0.05–0.10 mm/rev and cutting speeds of 30 m/min were found to be the ideal drilling parameters for reducing surface roughness and delamination factor for both sample types.

5. EFFECT OF MILLING PROCESS ON MECHANICAL PROPERTIES OF PMCS

The effect of milling on mechanical properties (tensile, flexural, impact, shear fatigue and creep behaviour) of PMCs is a critical consideration in machining operations involving these materials. Milling introduces mechanical forces and thermal effects that can significantly influence the mechanical behavior of PMCs (He et al. 2022). When a milling cutter engages with the composite material, it applies cutting forces and generates heat, leading to material removal and deformation. These mechanical and thermal effects can induce localized stress concentrations and initiate damage within the composite structure. Zhang et al. (G. Zhang et al. 2008) used a ball milling procedure at room temperature to fill polyetheretherketone (PEEK) with nano-SiO2 particles in a range of 0.5 to 4 vol.%. After that, they compressed the milled PEEK/nano-SiO2 mixes without doing any further melt mixing to create composites. The addition of nano-SiO2 considerably reduced the ductility of the material while increasing its stiffness. Fractography investigations revealed that the parabolic marks and dimples seen in the fracture initiation locations of nanocomposites with different nano-filler concentrations were generated by secondary cracks that originated from stress concentration at nanoparticles. Moreover, adding nano-SiO2 significantly improved PEEK's wear resistance; the best wear resistance was seen at a modest nano-SiO2 loading of 1 vol.%. The stress-strain curves of 0.5SP, 4SP, and pure PEEK are shown in Figure 8a. All of the materials exhibit initially linear elastic behavior during tensile testing. When nano-fillers are added, PEEK's Young's modulus increases and its elongation at break decreases. The mean Young's modulus, tensile strength, and elongation at break of PEEK reinforced with nano-SiO_2 are shown in Figures 8b-d in relation to filler content. As expected, adding nano-SiO_2 particles causes the fillers to become more rigid in the polymer matrix (Figure 8b). According to (G. Zhang et al. 2008), elastic deformation transfer at the polymer/filler interface is responsible for this increase in stiffness. The tensile modulus of PEEK increases when the nano-SiO_2 concentration increases from 0.5 to 2 vol.% because the interfacial area between the matrix and nanoparticles is larger. Nevertheless, there is a little decrease in tensile modulus when the nano-SiO_2 concentration is increased from 2 to 4 vol.%. The addition of nano-SiO_2 does not considerably change PEEK's tensile strength within the range under investigation (Figure 8c). However, it does lessen the matrix's ductility. When less than 1 vol.% of nano-SiO_2 is added to the matrix, the elongation at break decreases sharply, as seen in Figure 8d. Composite ductility somewhat decreases with further increases in nano-SiO_2 concentration.

Figure 8. PEEK composites' tensile qualities Pure PEEK, 0.5SP, and 4SP stress-strain curves, Young's modulus, tensile strength, and elongation at break (G. Zhang et al., 2008)

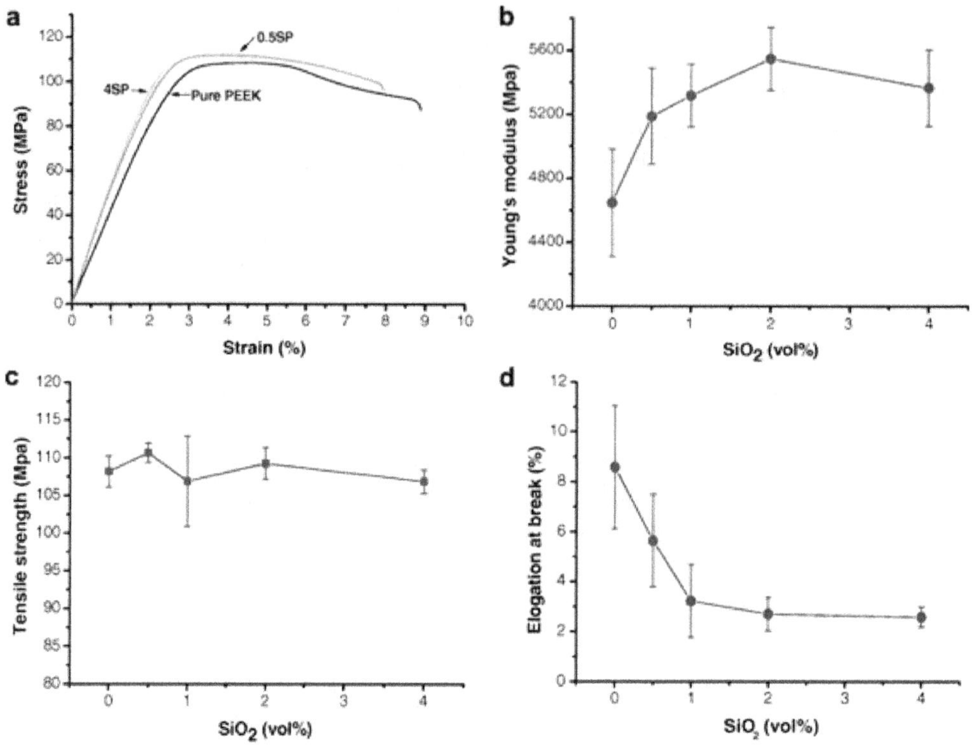

The surface quality resulting from milling operations can also influence the tensile strength of PMCs. Surface roughness, waviness, and defects introduced during milling can affect stress distribution and crack propagation behavior within the material, thereby influencing its tensile response. High surface roughness or irregularities can act as stress concentration sites, leading to localized deformation and premature failure under tensile loading conditions.

Because milling procedures on PMCs introduce several damage mechanisms such fiber breaking, matrix cracking, delamination, and residual stresses, the material's flexural strength may be affected. In this study, (Farooq et al. 2020) examined the factors that affect the flexural characteristics of UHMW-PE composites reinforced with nanoparticles. The flexural characteristics of the composite are impacted by the distribution of strengthening particles, which is determined by both macroscopic and microscopic examinations of the data. Combinations with low ν values and medium levels of other factors yield the best flexural characteristics. On the other hand, increased values of τ and ψ cause material deterioration. Furthermore, examining the impact of augmenting the quantity of passes considerably amplifies the composite material's flexural characteristics.

Salamani et al. (Slamani, Chafai, and Chatelain 2022) Salamani et al. provided the composite machining community with insights and cooperative solutions by identifying critical elements influencing the milled surface quality in flax fiber-reinforced composites. The study's main goal was to determine how cutting forces and surface roughness during the milling of flax/epoxy composites were affected by various cutting parameters, such as cutting speed, feed rate, and fiber orientation. A carbide end mill tool is used to perform slotting tests on flax fiber-reinforced polymer plates based on a complete factorial design of experiment. The results showed that the most important factor influencing cutting forces and roughness characteristics is feed rate. The torque versus time plot (Figure 9) can be divided into two regions, namely Region I and Region II (Naik et al. 2007). Region I depicts the material's behavior until it reaches peak torque, which corresponds to the interlaminar shear strength as indicated in Figure 9c. Region II illustrates the material's post-failure behavior. During the initial loading stage, indicated by the rise time, the strain rate varies.

Figure 9. Outcomes of the interlaminar shear test conducted on plain weave carbon/ epoxy composites: (a) time versus strain rate, (b) time versus strain, and (c) time versus stress (Naik et al., 2007)

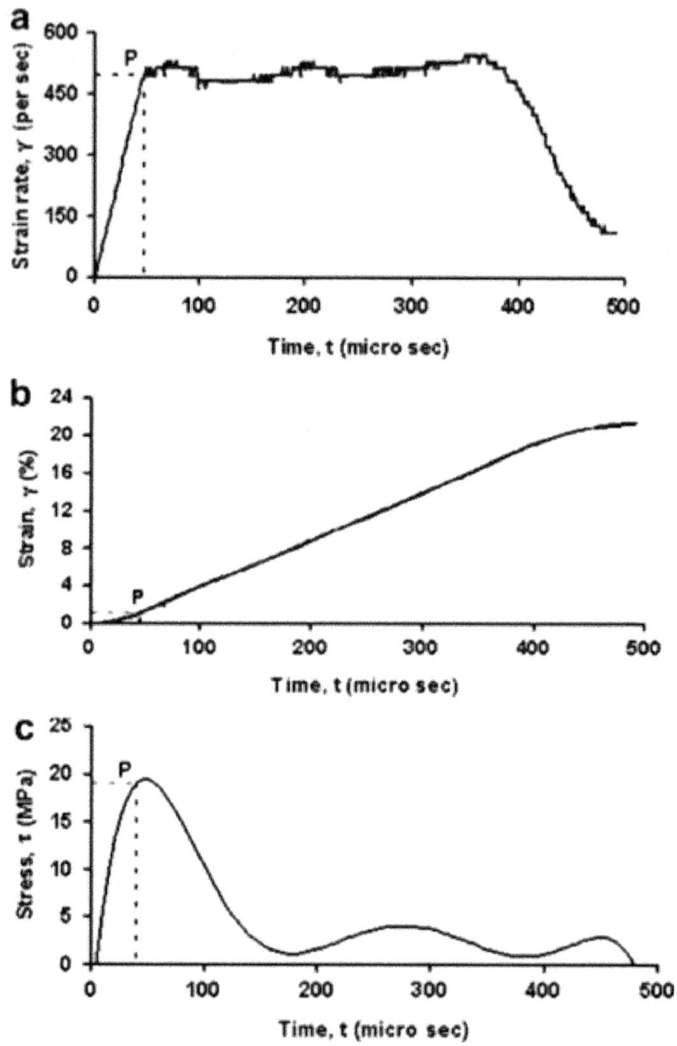

Prasshanth et al. (Prasanth, Ravishankar, and Hussain 2017) conducted milling experiments aimed at examining how machining conditions impact the cutting force, surface roughness, and damage factor of Glass Fiber Reinforced Polymer (GFRP) composites. Using a specially designed carbide end mill tool, experiments were conducted with a random set of milling process parameters. The findings indicated that spindle speed had the most significant influence on machined surface integrity, followed by the feed rate. The results were visually presented and analyzed through interaction plots and Scanning Electron Microscope (SEM) images. At exceeding spindle speeds, the machining force substantially increased, resulting in unintended friction-generated heat and subsequent damage to the machined slot (Figure 10). SEM images in Figures depict various surface failures experienced by some fibers, including crack formation, fiber shear failure, and fiber bending.

Figure 10. SEM image showing failure in composites at different milling conditions (Prasanth, Ravishankar, & Hussain, 2017)

In addition to affecting fatigue resistance, milling-induced damage can also influence the interfacial bonding between the reinforcing fibers and the matrix material in PMCs. Milling operations on PMCs can compromise the material's creep resistance due to the introduction of damage mechanisms such as fiber breakage, matrix cracking, delamination, and residual stresses. Poly(ether ketone) (PEK) matrix with graphite nanoplatelets (GNP) were fabricated using planetary ball milling followed by hot pressing for subsequent study of creep and creep recovery behaviour (Gaikwad et al. 2019). The research focused on investigating creep behavior within the linear viscoelastic region at varying temperatures ranging from 75 to

125°C. Results revealed that the inclusion of 0.5 wt.% GNP notably enhanced creep resistance. However, further GNP addition to the matrix resulted in a reduced rate of increase in creep resistance. Compared to pure PEK, all nanocomposites exhibited heightened glass transition temperature, dimensional stability, and storage modulus.

6. EFFECT OF TURNING PROCESS ON MECHANICAL PROPERTIES OF PMCS

The effect of turning on the mechanical properties (tensile, flexural, impact, shear fatigue and creep behaviour) of PMCs is an essential consideration in machining operations involving these materials. Turning introduces mechanical forces and thermal effects that can significantly influence the mechanical behavior of PMCs. Turning operations on PMCs can influence the material's tensile strength due to the introduction of various damage mechanisms such as fiber breakage, matrix cracking, delamination, and residual stresses. Fiber breakage may occur due to excessive cutting forces or improper tool geometry, leading to a loss of reinforcement and reduced tensile strength. Additionally, residual stresses induced during turning may further contribute to changes in tensile strength by promoting crack initiation and propagation, particularly in regions where the material is already compromised by turning-induced damage. The effects of changes in feed rate and cutting speed during turning on the fractal dimension (D), maximum peak to valley (Rt), and arithmetic mean roughness (Ra) of a glass fiber polymer composite were investigated by (Kechagias et al. 2009). A statistical analysis showed that the feed rate parameter mostly affected the fractal size, maximum peak to valley, and arithmetic mean roughness. Furthermore, interaction charts and assessment experiments indicated that regression modeling solely applied to the arithmetic mean roughness.

(Sivakiran et al. 2018) studied the machining of composites using a hybrid polymer matrix and chopped banana fiber (10 mm length) oriented randomly as reinforcement. There are two different hybrid polymers used: one that is composed of 15% cashew nutshell liquid (CNSL) and 85% isophthalic polyester, and another that is composed of 15% cardanol and 85% isophthalic polyester. Conventional machining methods are used, paying close attention to three parameters: depth of cut, feed, and speed. The response surface approach is utilized in the design of experiments. Surface roughness and material removal rate (MRR) were the main subjects of the inquiry. The effect of depth of cut on MRR for composites based on cardanol and CNSL, respectively, is shown in Figures 11(a) and 12(a). These numbers clearly show that MRR increases as cut depth increases. The effect of feed rate on MRR for composites based on cardanol and CNSL is shown in Figures 11(b) and 12(b). These numbers make it quite evident that a greater feed rate corresponds to a larger

MRR. Figures 11(c) and 12(c) illustrate how speed affects MRR. It is noticed that the MRR increases with speed, exhibiting a linear pattern with a positive slope.

Figure 11. Plots of MRR for cardonal sample: (a) MRR vs. depth of cut, (b) MRR vs. feed rate, (c) MRR vs. speed (Sivakiran et al., 2018)

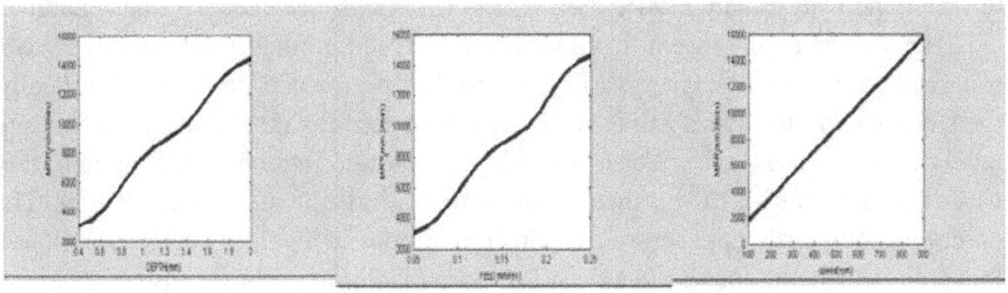

Figure 12. Plots of MRR for CNSL sample: (a) MRR vs. depth of cut, (b) MRR vs. feed rate, (c) MRR vs. speed (Sivakiran et al., 2018)

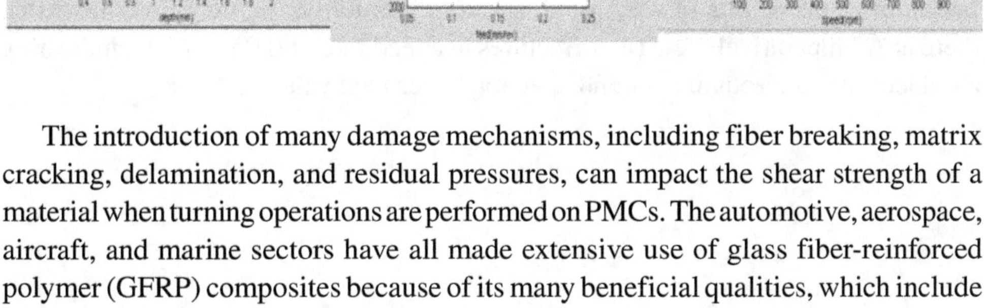

The introduction of many damage mechanisms, including fiber breaking, matrix cracking, delamination, and residual pressures, can impact the shear strength of a material when turning operations are performed on PMCs. The automotive, aerospace, aircraft, and marine sectors have all made extensive use of glass fiber-reinforced polymer (GFRP) composites because of its many beneficial qualities, which include stiffness, corrosion resistance, high strength-to-weight ratio, lightweight, and dimensional stability. Equbal et al. (Equbal et al. 2020) conducted study focusing on turning GFRP composites to ascertain the best cutting parameters for optimal outcomes. Using a full factorial design of experiments (FFDE), the study examined the effects of spindle rotational speed (N), feed rate (f), and depth of cut (d), as well as their interactions, on Material Removal Rate (MRR), Tool Wear Rate (TWR), and Surface Roughness (R_a). Analysis of variance (ANOVA) was employed to assess the statistical significance of these parameters and their interactions. Additionally, empirical models were developed to correlate output responses and cutting parameters. Furthermore, Artificial Neural Network (ANN) in combination with Genetic Algorithm (GA) was utilized for multi-response optimization, aiming to simultaneously enhance MRR, TWR, and Ra. Rajasekhran et al. (Rajasekaran et al. 2012) identified the machining parameter that predominantly influences surface roughness, as it directly impacts various material properties and strength. Additionally, the study sought to identify the set of machining settings that result in the desired level of surface roughness. For more efficient testing, Taguchi's orthogonal array was used, and analysis of variance was used to examine the machining settings. The outcomes showed that the ceramic cutting tool produced levels of surface roughness that were adequate, opening the door for more machining process improvement.

The SEM of machined CFRP material at different machining settings, such as changing feed conditions while maintaining a constant cutting speed and depth of cut, is shown in Figure 13 (Rajasekaran, Palanikumar, and Vinayagam 2012). Generally speaking, brittle materials deform when they rupture, however with CFRP composites, the deformation is thought to be the result of both shearing and rupturing together. The composite comprising the resin and the appropriately specified carbon fiber is shown in Figure 13(a). The state of the machined composite material shown in Figure 13(b) shows that more fiber has broken, which is consistent with a higher feed rate of 0.15 mm/rev during cutting, which supports greater surface roughness at such a high feed rate. When one looks at the chips made from machining composite material, they turn into tiny, fine powders when the feed is increased. This shows that the chip is under less strain, producing very fine powder and well-polished surfaces. In contrast, this results in very rough surfaces. Similarly, Figure 13(c) shows that there are comparatively few fiber fractures at a feed rate of 0.05 mm/rev, indicating a reduced surface roughness condition at a lower feed value.

Figure 13. SEM of turned CFRP composite (Rajasekaran, Palanikumar, & Vinayagam, 2012)

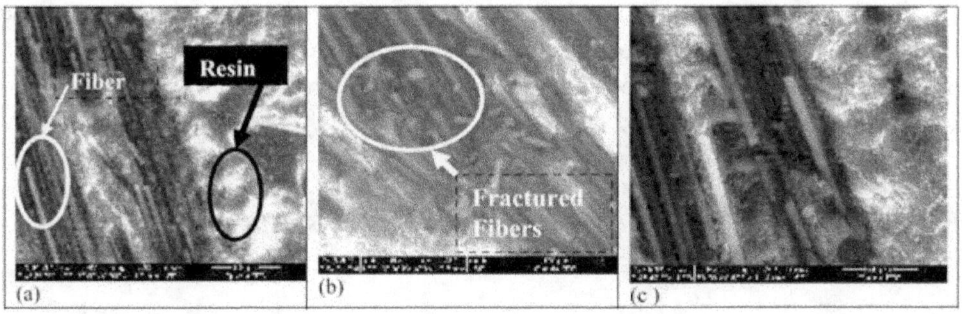

High surface roughness or irregularities can act as stress concentration sites, promoting localized creep deformation and accelerating material degradation. Surface defects such as scratches, gouges, or machining marks may serve as initiation points for creep-related damage mechanisms, leading to premature failure under prolonged loading conditions. Therefore, achieving the desired surface finish and minimizing surface defects are essential considerations in maintaining the creep resistance of PMCs after turning operations.

7. EFFECT OF GRINDING PROCESS ON MECHANICAL PROPERTIES OF PMCS

The effect of grinding on the mechanical properties (tensile, flexural, impact, shear fatigue and creep behaviour) of PMCs is an essential consideration in machining operations involving these materials. Grinding introduces mechanical forces and thermal effects that can significantly influence the mechanical behavior of PMCs (B. Zhang et al. 2023). Grinding operations on PMCs can influence the material's tensile strength due to the introduction of various damage mechanisms such as fiber breakage, matrix cracking, delamination, and residual stresses. When a grinding wheel engages with the composite material, it applies mechanical loads and generates heat, leading to material removal and deformation. These mechanical and thermal effects can induce localized stress concentrations and initiate damage within the composite structure. Fundamental mechanisms in CFRP machining were discovered by (Gao et al. 2021), especially when specific transversal grinding and lubrication conditions were met. The distinct geometries of grain and fiber interaction resulting from arbitrary fiber configurations and grain edge forms at various undeformed chip thicknesses were taken into consideration by developing four sub-models.

The tensile fracture force of a single fiber modeled as a bending beam fixed at both ends and constrained on an elastic foundation, extrusion and shearing force on the cut fiber section at the grinding groove and contact force between the grain tip and fibers were the main topics of these models. Additionally, the local contact stress of the elliptical region between the spherical grain edge and cylindrical fiber was also examined (LIU et al. 2023).

Figure 14. Modes of interaction between CFRPs and single grains at various ag values (Gao et al., 2021)

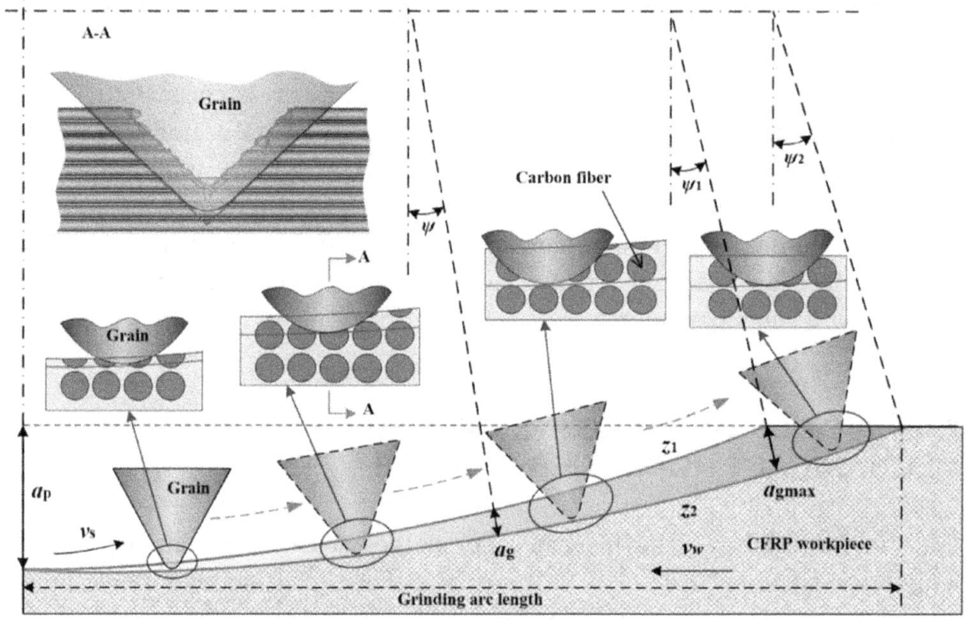

The flexural strength of PMCs can be affected by grinding processes because they introduce a number of damage mechanisms, including fiber breaking, matrix cracking, delamination, and residual stresses (Khan, Mithu, and Dhar 2009). Figure 15 (Gao et al. 2021) illustrates the input variables alongside the resultant shape of the grinding zone and undeformed chip thickness. The trend in undeformed chip thickness is similar until it reaches its maximum. Conversely, higher peripheral speed of the grinding wheel leads to a shorter and thinner grinding area, with thicker undeformed chips. Additionally, the rates of change in the grinding area's thickness and length drop as the grinding wheel's peripheral speed rises consistently, suggesting that the impact on the grinding area's size and shape is waning (Chen et al. 2022). The slope of the undeformed chip thickness curve during grinding varies as the grinding wheel's peripheral speed increases before the chip thickness reaches its maximum value. Nevertheless, this slope doesn't change after the chip thickness reaches its maximum. Higher workpiece feed speeds cause the grinding area's length and thickness to expand, but its form stays the same. As workpiece feed speed increases, the slope of the undeformed chip thickness curve decreases before reaching maximum chip thickness. This slope does not alter when the maximum chip thickness is reached.

Figure 15. Grinding parameters' impact on the grinding zone (Gao et al., 2021)

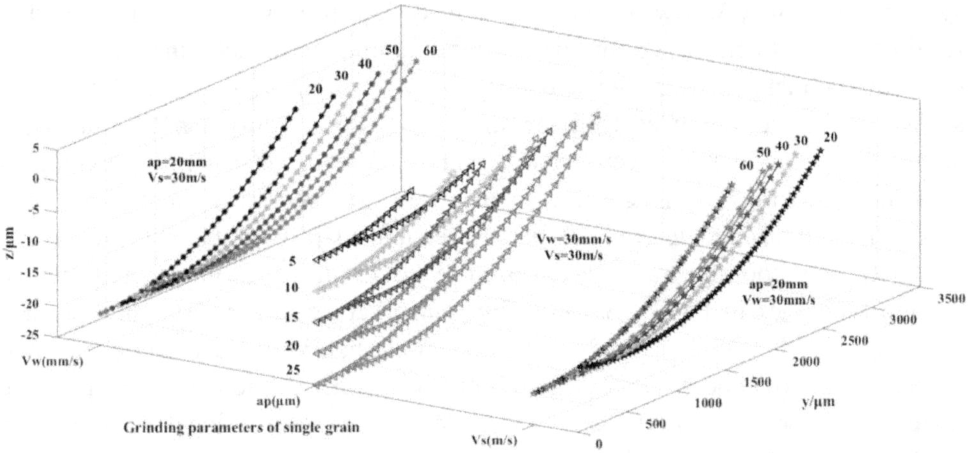

The temperature generated during grinding operations can significantly influence the creep behavior of PMCs. Elevated temperatures accelerate the rate of creep deformation in PMCs, particularly in thermoplastic matrix systems. The localized heating at the grinding interface can lead to softening of the matrix material and promote viscous flow, enhancing creep deformation.

8. STRATEGIES FOR OPTIMIZING MACHINING PERFORMANCE

Optimizing machining performance in PMCs requires a multidisciplinary approach encompassing material characterization, tooling selection, cutting parameter optimization, process monitoring, and adaptive machining strategies (Patole et al. 2021; Ullrich et al. 2024). *Material characterization* is essential for understanding the mechanical properties, microstructure, and behavior of PMCs under machining conditions. Analyzing fiber orientation, matrix composition, interfacial bonding, and material response to cutting forces provides insights into material behavior and guides machining process optimization. *Tooling selection* is critical for achieving optimal machining performance in PMCs, with specialized cutting tools designed to withstand the abrasive nature of composite materials, minimize tool wear, and achieve desired surface finish and dimensional accuracy. *Cutting parameter optimization* involves selecting appropriate cutting speeds, feed rates, depth of cut, and coolant application to achieve efficient material removal rates, minimize tool wear, and ensure dimensional accuracy and surface finish in PMCs. *Process monitoring and control systems* enable real-time monitoring of machining parameters, tool condition, and workpiece quality during machining operations. Sensors, gauges, and monitoring systems can detect tool wear, material defects, and process deviations, allowing for timely adjustments to cutting parameters, tooling, and machining strategies to maintain optimal performance and quality in PMCs. *Adaptive machining* strategies such as variable cutting speed, feed rate, and toolpath optimization can be employed to address variations in material properties, fiber orientation, and machining conditions encountered during PMC machining. Real-time feedback control systems, predictive modeling, and machine learning algorithms enable adaptive adjustments to cutting parameters and tooling strategies to optimize machining performance and achieve desired outcomes in PMCs. *Surface Finishing*: Performing post-turning surface finishing operations such as grinding or polishing can help improve surface quality and reduce surface roughness, thereby minimizing stress concentrations and enhancing the material's impact strength. *Post-Processing Treatments*: Performing post-processing treatments such as resin infusion, resin filling, or composite patching can help repair and reinforce the damaged areas around drilled holes, thereby improving the material's resistance to creep deformation. *Non-Destructive Testing*: Implementing non-destructive testing techniques such as ultrasonic testing or thermography can help detect and assess any damage or defects induced during machining operations without compromising the material's integrity. *Surface Finishing*: Performing post-machining surface finishing operations such as polishing or honing can help improve surface quality and reduce surface roughness, thereby enhancing the material's fatigue resistance. *Heat Treatment*: Implementing post-machining heat treatment processes such as stress relieving or annealing can help alleviate residual stresses and restore the material's mechanical properties.

9. SUMMARY

As PMCs find widespread use across industries for their lightweight, high strength-to-weight ratio, and corrosion resistance, understanding how machining affects their mechanical properties is paramount for optimizing manufacturing processes and ensuring component performance. Tensile strength and modulus, flexural strength and modulus, impact strength, shear strength, fatigue resistance, creep behavior, and other relevant mechanical properties play pivotal roles in shaping the material's response to external forces and its suitability for specific applications. By comprehensively understanding these mechanical properties and their interrelationships, engineers and researchers can effectively design, engineer, and manufacture PMCs with tailored performance characteristics optimized for a wide range of demanding applications. Through this comprehensive exploration of the mechanical properties of PMCs, this chapter aimed to provide readers with a deeper understanding of the underlying principles governing the behavior of PMCs under mechanical loading conditions. By elucidating the significance of each mechanical property and its influence on machining performance, this chapter seeks to facilitate informed decision-making and optimization of machining processes for PMCs in various industrial sectors.

REFERENCES

Alam, P., Mamalis, D., Robert, C., Floreani, C., & Ó Brádaigh, C. M. (2019). The Fatigue of Carbon Fibre Reinforced Plastics - A Review. *Composites. Part B, Engineering*, *166*, 555–579. doi:10.1016/j.compositesb.2019.02.016

Alberto, M. (2013). Introduction of Fibre-Reinforced Polymers – Polymers and Composites: Concepts, Properties and Processes. In Fiber Reinforced Polymers - The Technology Applied for Concrete Repair. InTech. doi:10.5772/54629

Amuthakkannan, P., Manikandan, V., Winowlin Jappes, J. T., & Uthayakumar, M. (2013). Effect of Fibre Length and Fibre Content on Mechanical Properties of Short Basalt Fibre Reinforced Polymer Matrix Composites. *Materials Physics and Mechanics*, *16*(2), 107–117.

Arul, S., Vijayaraghavan, L., Malhotra, S. K., & Krishnamurthy, R. (2006). The Effect of Vibratory Drilling on Hole Quality in Polymeric Composites. *International Journal of Machine Tools & Manufacture*, *46*(3–4), 252–259. doi:10.1016/j.ijmachtools.2005.05.023

Cabrera, F. M., Beamud, E., Hanafi, I., Khamlichi, A., & Jabbouri, A. (2011). Fuzzy Logic-Based Modeling of Surface Roughness Parameters for CNC Turning of PEEK CF30 by TiN-Coated Cutting Tools. *Journal of Thermoplastic Composite Materials*, *24*(3), 399–413. doi:10.1177/0892705710391562

Cabrera, F. M., Fuentes, D., Hanafi, I., Khamlichi, A., & Jabbouri, A. (2012). Multi-Criteria Optimization Using Taguchi and Grey Relational Analysis in CNC Turning of PEEK CF30. *Journal of Thermoplastic Composite Materials*, *25*(1), 101–114. doi:10.1177/0892705711405248

Cabrera, F. M., Hanafi, I., Khamlichi, A., Jabbouri, A., & Bezzazi, M. (2010). Sur l'usinabilité Des Composites à Matrices Polymères Renforcée Par Des Fibres. *Mechanics & Industry*, *11*(2), 93–103. doi:10.1051/meca/2010021

Cengiz, A., Öztürk, M. F., Sabah, A., & Avcu, E. (2024). Experimental and Numerical Analysis of Drilled- and Molded-hole Glass Fiber Reinforced Polymer Matrix (<scp>GFRP</Scp>) Composites. *Polymer Composites*, pc.28305. Advance online publication. doi:10.1002/pc.28305

Chen, C., Chen, B., Wu, C., Gu, X., Liu, X., & Guo, F. (2022). Theoretical Analysis of Grinding Wheel Deflection Angle on Peripheral Grinding Parameters and Grinding Force. *Metals*, *12*(7), 1209. doi:10.3390/met12071209

Crivelli, D., Guagliano, M., Eaton, M., Pearson, M., Al-Jumaili, S., Holford, K., & Pullin, R. (2015). Localisation and Identification of Fatigue Matrix Cracking and Delamination in a Carbon Fibre Panel by Acoustic Emission. *Composites. Part B, Engineering*, *74*, 1–12. doi:10.1016/j.compositesb.2014.12.032

Equbal, A., Shamim, M., Badruddin, I. A., Equbal, M. I., Sood, A. K., Nik Ghazali, N. N., & Khan, Z. A. (2020). Application of the Combined ANN and GA for Multi-Response Optimization of Cutting Parameters for the Turning of Glass Fiber-Reinforced Polymer Composites. *Mathematics*, *8*(6), 947. doi:10.3390/math8060947

Fallahi, H., Kaynan, O., & Asadi, A. (2023). Insights into the Effect of Fiber–Matrix Interphase Physiochemical- Mechanical Properties on Delamination Resistance and Fracture Toughness of Hybrid Composites. *Composites. Part A, Applied Science and Manufacturing*, *166*, 107390. doi:10.1016/j.compositesa.2022.107390

Fallahi, H., Taheri-Behrooz, F., & Asadi, A. (2020). Nonlinear Mechanical Response of Polymer Matrix Composites: A Review. *Polymer Reviews (Philadelphia, Pa.)*, *60*(1), 42–85. doi:10.1080/15583724.2019.1656236

Farooq, U., Khan, I., Asif, M., Hira, F., Hussain, G., Ullah, A., Ahmad, S. M., Umar, M., & Alhassan, M. A. (2020). Investigation on the Effects of the Processing Parameters and the Number of Passes on the Flexural Properties of Polymer Nanocomposite Fabricated through FSP Method. *Materials Research Express*, *7*(5), 055310. doi:10.1088/2053-1591/ab9524

Gaikwad, S. D., Dravid, S. V., Sonawane, B. U., Abraham, M., & Goyal, R. K. (2019). Dynamic Mechanical and Creep Properties of Poly(Ether Ketone)-Graphite Nanoplatelet (GNP) Nanocomposites. *Polymer-Plastics Technology and Materials*, *58*(15), 1691–1701. doi:10.1080/25740881.2018.1563141

Gao, T., Li, C., Yang, M., Zhang, Y., Jia, D., Ding, W., Debnath, S., Yu, T., Said, Z., & Wang, J. (2021). Mechanics Analysis and Predictive Force Models for the Single-Diamond Grain Grinding of Carbon Fiber Reinforced Polymers Using CNT Nano-Lubricant. *Journal of Materials Processing Technology*, *290*, 116976. doi:10.1016/j.jmatprotec.2020.116976

Geier, N., Patra, K., Anand, R. S., Ashworth, S., Balázs, B. Z., Lukács, T., Magyar, G., Tamás-Bényei, P., Xu, J., & Davim, J. P. (2023). A Critical Review on Mechanical Micro-Drilling of Glass and Carbon Fibre Reinforced Polymer (GFRP and CFRP) Composites. *Composites. Part B, Engineering*, *254*, 110589. doi:10.1016/j.compositesb.2023.110589

Geier, N., Xu, J., Poór, D. I., Dege, J. H., & Davim, J. P. (2023). A Review on Advanced Cutting Tools and Technologies for Edge Trimming of Carbon Fibre Reinforced Polymer (CFRP) Composites. *Composites. Part B, Engineering*, *266*, 111037. doi:10.1016/j.compositesb.2023.111037

Hassanzadeh-Aghdam, M. K., Ansari, R., Mahmoodi, M. J., & Darvizeh, A. (2018). Effect of Nanoparticle Aggregation on the Creep Behavior of Polymer Nanocomposites. *Composites Science and Technology*, *162*, 93–100. doi:10.1016/j.compscitech.2018.04.025

He, Y., Li, L., Wan, M., & Xue, H. (2022). Three-Dimensional Finite Element Simulations of Milling Carbon/Epoxy Composites. *Composite Structures*, *282*, 115037. doi:10.1016/j.compstruct.2021.115037

Heidary, H., Karimi, N. Z., & Minak, G. (2018). Investigation on Delamination and Flexural Properties in Drilling of Carbon Nanotube/Polymer Composites. *Composite Structures*, *201*, 112–120. doi:10.1016/j.compstruct.2018.06.041

Hulugappa, B., Achutha, M. V., & Suresha, B. (2016). Effect of Fillers on Mechanical Properties and Fracture Toughness of Glass Fabric Reinforced Epoxy Composites. *Journal of Minerals & Materials Characterization & Engineering*, *04*(01), 1–14. doi:10.4236/jmmce.2016.41001

Iñigo, R. (2023). Improving Surface Integrity When Drilling CFRPs and Ti-6Al-4V Using Sustainable Lubricated Liquid Carbon Dioxide. *Chinese Journal of Aeronautics*, *36*(7), 129–146. doi:10.1016/j.cja.2022.09.004

Ismail, Dhakal, Popov, & Beaugrand. (2016). Comprehensive Study on Machinability of Sustainable and Conventional Fibre Reinforced Polymer Composites. *Engineering Science and Technology, an International Journal,* 19(4), 2043–52. . doi:10.1016/j.jestch.2016.07.010

Jani, S. P., Senthil Kumar, A., Adam Khan, M., & Uthayakumar, M. (2020). Surface Roughness and Morphology Studies on Machining Hybrid Composite Material Using Abrasive Water Jet Cutting Process. doi:10.1007/978-3-030-43232-4_6

Kawai, M., & Masuko, Y. (2004). Creep Behavior of Unidirectional and Angle-Ply T800H/3631 Laminates at High Temperature and Simulations Using a Phenomenological Viscoplasticity Model. *Composites Science and Technology, 64*(15), 2373–2384. doi:10.1016/j.compscitech.2004.04.004

Kechagias, J., Petropoulos, G., Iakovakis, V., & Maropoulos, S. (2009). An Investigation of Surface Texture Parameters during Turning of a Reinforced Polymer Composite Using Design of Experiments and Analysis. *International Journal of Experimental Design and Process Optimisation, 1*(2/3), 164. doi:10.1504/IJEDPO.2009.030317

Khan, M. M. A., Mithu, M. A. H., & Dhar, N. R. (2009). Effects of Minimum Quantity Lubrication on Turning AISI 9310 Alloy Steel Using Vegetable Oil-Based Cutting Fluid. *Journal of Materials Processing Technology, 209*(15–16), 5573–5583. doi:10.1016/j.jmatprotec.2009.05.014

Khashaba, U. A. (2013). Drilling of Polymer Matrix Composites: A Review. *Journal of Composite Materials, 47*(15), 1817–1832. doi:10.1177/0021998312451609

Khashaba, U. A., El-Sonbaty, I. A., Selmy, A. I., & Megahed, A. A. (2010). Machinability Analysis in Drilling Woven GFR/Epoxy Composites: Part I – Effect of Machining Parameters. *Composites. Part A, Applied Science and Manufacturing, 41*(3), 391–400. doi:10.1016/j.compositesa.2009.11.006

Kiran, M. D., Govindaraju, H. K., Jayaraju, T., & Kumar, N. (2018). Review-Effect of Fillers on Mechanical Properties of Polymer Matrix Composites. *Materials Today: Proceedings, 5*(10), 22421–22424. doi:10.1016/j.matpr.2018.06.611

Kumar, S., Gariya, N., Shaikh, A., Ahmad, F., & Yadav, A. (2024). Physico-Mechanical and Thermal Properties of Fly Ash and Benzoylated Himalayan Agave Fiber Reinforced Polyester Composite. *Biomass Conversion and Biorefinery*. Advance online publication. doi:10.1007/s13399-024-05518-0

Kumar, S., Prasad, L., Bijlwan, P. P., & Yadav, A. (2022). Thermogravimetric Analysis of Lignocellulosic Leaf-Based Fiber-Reinforced Thermosets Polymer Composites: An Overview. *Biomass Conversion and Biorefinery*. Advance online publication. doi:10.1007/s13399-022-03332-0

Kumar, S., Prasad, L., Patel, V. K., Kumain, A., & Yadav, A. (2021). Experimental and Numerical Study on Physico-mechanical Properties and Taguchi's Designed Abrasive Wear Behavior of Hemp/Nettle-polyester Hybrid Composite. *Polymer Composites*, *42*(12), 6912–6927. doi:10.1002/pc.26350

Kumar, S., Prasad, L., Patel, V. K., Kumar, V., Kumar, A., Yadav, A., & Winczek, J. (2021). Physical and Mechanical Properties of Natural Leaf Fiber-Reinforced Epoxy Polyester Composites. *Polymers*, *13*(9), 1369. doi:10.3390/polym13091369 PMID:33922169

Li, X., Liu, F., Gong, N., Yang, C., & Wang, B. (2018). Surface Topography Induced High Injection Joining Strength of Polymer-Metal Composite and Fracture Mechanism. *Composite Structures*, *184*, 545–553. doi:10.1016/j.compstruct.2017.10.020

Manjunath Patel, Jagadish, Kumar, & Naidu. (2020). Optimization of Abrasive Water Jet Machining for Green Composites Using Multi-Variant Hybrid Techniques. doi:10.1007/978-3-030-19638-7_6

Mata, F., Hanafi, I., Beamud, E., Khamlichi, A., & Jabbouri, A. (2012). Modelling of Machining Force Components during Turning of PEEK CF30 by TiN Coated Cutting Tools Using Artificial Intelligence. *International Journal of Machining and Machinability of Materials*, *11*(3), 263. doi:10.1504/IJMMM.2012.046885

Merino-Pérez, J. L., Royer, R., Merson, E., Lockwood, A., Ayvar-Soberanis, S., & Marshall, M. B. (2016). Influence of Workpiece Constituents and Cutting Speed on the Cutting Forces Developed in the Conventional Drilling of CFRP Composites. *Composite Structures*, *140*, 621–629. doi:10.1016/j.compstruct.2016.01.008

Mingzheng, L. I. U. (2023). Analysis of Grinding Mechanics and Improved Grinding Force Model Based on Randomized Grain Geometric Characteristics. *Chinese Journal of Aeronautics*, *36*(7), 160–193. doi:10.1016/j.cja.2022.11.005

Mohanavel, V., Raja, T., Yadav, A., Ravichandran, M., & Winczek, J. (2022). Evaluation of Mechanical and Thermal Properties of Jute and Ramie Reinforced Epoxy-Based Hybrid Composites. *Journal of Natural Fibers*, *19*(14), 8022–8032. doi:10.1080/15440478.2021.1958432

Montesano, J., Fawaz, Z., & Bougherara, H. (2015). Non-Destructive Assessment of the Fatigue Strength and Damage Progression of Satin Woven Fiber Reinforced Polymer Matrix Composites. *Composites. Part B, Engineering*, *71*, 122–130. doi:10.1016/j.compositesb.2014.11.005

Naik, N. K., Asmelash, A., Kavala, V. R., & Ch, V. (2007). Interlaminar Shear Properties of Polymer Matrix Composites: Strain Rate Effect. *Mechanics of Materials*, *39*(12), 1043–1052. doi:10.1016/j.mechmat.2007.05.003

Nasir, A. A., Azmi, A. I., & Khalil, A. N. M. (2015). Parametric Study on the Residual Tensile Strength of Flax Natural Fibre Composites after Drilling Operation. *Procedia Manufacturing*, *2*, 97–101. doi:10.1016/j.promfg.2015.07.017

Negi, R. S., Prasad, L., Yadav, A., & Winczek, J. (2022). Physical and Mechanical Properties of Pinecone Scale Fiber/Vigna Mungo Powder Reinforced Polypropylene Based Hybrid Composites. *Journal of Natural Fibers*, *19*(15), 11458–11468. doi: 10.1080/15440478.2022.2025983

Patel, J. S., Boddu, V. M., Brenner, M. W., & Kumar, A. (2016). Effect of Fabric Structure and Polymer Matrix on Flexural Strength, Interlaminar Shear Stress, and Energy Dissipation of Glass Fiber-Reinforced Polymer Composites. *Textile Research Journal*, *86*(2), 127–137. doi:10.1177/0040517515586165

Patel, R. V., Yadav, A., & Winczek, J. (2023). Physical, Mechanical, and Thermal Properties of Natural Fiber-Reinforced Epoxy Composites for Construction and Automotive Applications. *Applied Sciences (Basel, Switzerland)*, *13*(8), 5126. doi:10.3390/app13085126

Patole, P. B., Pol, G. J., Desai, A. A., & Kamble, S. B. (2021). Analysis of Surface Roughness and Cutting Force under MQL Turning Using Nano Fluids. *Materials Today: Proceedings*, *45*, 5684–5688. doi:10.1016/j.matpr.2021.02.501

Prakash, M., & Dileep Aditya Dhar, P. V. S. (2018). Investigation on the Effect of Drilling Parameters on the Tool Wear and Delamination of Glass Fibre-Reinforced Polymer Composite Using Vibration Signal Analysis. *Journal of Composite Materials*, *52*(12), 1641–1648. doi:10.1177/0021998317728109

Pramod, R., Basavarajappa, S., Veeresh Kumar, G. B., & Chavali, M. (2022). Drilling Induced Delamination Assessment of Nanoparticles Reinforced Polymer Matrix Composites. *Proceedings of the Institution of Mechanical Engineers. Part C, Journal of Mechanical Engineering Science*, *236*(6), 2931–2948. doi:10.1177/09544062211030967

Prasad, L., Bairwan, R., Yadav, A., Kumar, A., Kumar, V., & Winczek, J. (2022). Evaluation of Physical, Mechanical, and Wear Properties of Jatropha Shell Powder Reinforced Epoxy Glass Fiber Composites. *Journal of Natural Fibers*, *19*(15), 12195–12207. doi:10.1080/15440478.2022.2054892

Prasad, L., Kapri, P., Patel, R. V., Yadav, A., & Winczek, J. (2023). Physical and Mechanical Behavior of Ramie and Glass Fiber Reinforced Epoxy Resin-Based Hybrid Composites. *Journal of Natural Fibers*, *20*(2), 2234080. Advance online publication. doi:10.1080/15440478.2023.2234080

Prasad, L., Kumain, A., Patel, R. V., Yadav, A., & Winczek, J. (2022). Physical and Mechanical Behavior of Hemp and Nettle Fiber-Reinforced Polyester Resin-Based Hybrid Composites. *Journal of Natural Fibers*, *19*(7), 2632–2647. doi:10.1080/15440478.2020.1821284

Prasad, L., Kumar, S., Patel, R. V., Yadav, A., Kumar, V., & Winczek, J. (2020). Physical and Mechanical Behaviour of Sugarcane Bagasse Fibre-Reinforced Epoxy Bio-Composites. *Materials (Basel)*, *13*(23), 5387. doi:10.3390/ma13235387 PMID:33260868

Prasad, L., Singh, G., Yadav, A., Kumar, V., & Kumar, A. (2019). Properties of Functionally Gradient Composites Reinforced with Waste Natural Fillers. *Acta Periodica Technologica*, (50), 250–259. doi:10.2298/APT1950250P

Prasad, L., Singh, V., Patel, R. V., Yadav, A., Kumar, V., & Winczek, J. (2022). Physical and Mechanical Properties of Rambans (Agave) Fiber Reinforced with Polyester Composite Materials. *Journal of Natural Fibers*, *19*(13), 6104–6118. doi:10.1080/15440478.2021.1904481

Prasanth, I. S. N. V. R., Ravishankar, D. V., & Manzoor Hussain, M. (2017). Analysis of Milling Process Parameters and Their Influence on Glass Fiber Reinforced Polymer Composites. *International Journal of Engineering*, *30*(7), 1074–1080. doi:10.5829/ije.2017.30.07a.17

Raja, Satheesh, Manisekar, & Manikandan. (2013). Effect of Fly Ash Filler Size on Mechanical Properties of Polymer Matrix Composites. *International Journal of Mining, Metallurgy & Mechanical Engineering*, *1*(1), 34–37.

Rajakumar, I., Raguraman, D., Isaac, J. S., Suthan, R., Bhattacharya, S., Seikh, A. H., Khan, S. M. A., & Raghavan, I. K. (2022). Mechanical Properties of Polymer Composites Reinforced with Alkaline-Treated Natural Fibre. *Advances in Polymer Technology*, *2022*, 1–7. doi:10.1155/2022/1458547

Rajasekaran, T., Palanikumar, K., & Vinayagam, B. K. (2012). Turning CFRP Composites with Ceramic Tool for Surface Roughness Analysis. *Procedia Engineering, 38*, 2922–2929. doi:10.1016/j.proeng.2012.06.341

Rao, Y. S., Mohan, N. S., Shetty, N., & Shivamurthy, B. (2019). Drilling and Structural Property Study of Multi-Layered Fiber and Fabric Reinforced Polymer Composite - a Review. *Materials and Manufacturing Processes, 34*(14), 1549–1579. doi:10.1080/10426914.2019.1686522

Robinson, D. N., Binienda, W. K., & Ruggles, M. B. (2003). Creep of Polymer Matrix Composites. I: Norton/Bailey Creep Law for Transverse Isotropy. *Journal of Engineering Mechanics, 129*(3), 310–317. doi:10.1061/(ASCE)0733-9399(2003)129:3(310)

Ruszaj, A., Gawlik, J., & Skoczypiec, S. (2016). Electrochemical Machining – Special Equipment and Applications in Aircraft Industry. *Management and Production Engineering Review, 7*(2), 34–41. doi:10.1515/mper-2016-0015

Sharma, S. P., & Lakkad, S. C. (2011). Effect of CNTs Growth on Carbon Fibers on the Tensile Strength of CNTs Grown Carbon Fiber-Reinforced Polymer Matrix Composites. *Composites. Part A, Applied Science and Manufacturing, 42*(1), 8–15. doi:10.1016/j.compositesa.2010.09.008

Sivakiran, G., Gangwal, Y., Venkatachalam, G., Pandivelan, C., & Ayyappan, S. (2018). Investigations on Machining of Banana Fibre Reinforced Hybrid Polymer Matrix Composite Materials. *Materials Today: Proceedings, 5*(2), 7908–7914. doi:10.1016/j.matpr.2017.11.472

Slamani, M., Chafai, H., & Chatelain, J. F. (2022). Effect of Milling Parameters on the Surface Quality of a Flax Fiber-Reinforced Polymer Composite. *Proceedings of the Institution of Mechanical Engineers, Part E: Journal of Process Mechanical Engineering*. 10.1177/09544089221126087

Song, D.-Y., & Otani, N. (1997). Fatigue Life Prediction of Cross-Ply Composite Laminates. *Materials Science and Engineering A, 238*(2), 329–335. doi:10.1016/S0921-5093(97)00427-9

Tao, C., Qiu, J., Yao, W., & Ji, H. (2016). The Effect of Drilling-Induced Delamination on Tensile Strength and Prediction of Residual Strength of Carbon Fiber-Reinforced Polymer Laminate. *Journal of Composite Materials, 50*(24), 3373–3384. doi:10.1177/0021998315620292

Ullrich, K., von Elling, M., Gutzeit, K., Dix, M., Weigold, M., Aurich, J. C., Wertheim, R., Jawahir, I. S., & Ghadbeigi, H. (2024). AI-Based Optimisation of Total Machining Performance: A Review. *CIRP Journal of Manufacturing Science and Technology*, *50*, 40–54. doi:10.1016/j.cirpj.2024.01.012

Varma, M., Chandran, S., Vijay Kumar, V., Suyambulingam, I., & Siengchin, S. (2024). A Comprehensive Review on the Machining and Joining Characteristics of Natural Fiber-reinforced Polymeric Composites. *Polymer Composites*, *45*(6), 4850–4875. Advance online publication. doi:10.1002/pc.28148

Wan, M., Li, S.-E., Yuan, H., & Zhang, W.-H. (2019). Cutting Force Modelling in Machining of Fiber-Reinforced Polymer Matrix Composites (PMCs): A Review. *Composites. Part A, Applied Science and Manufacturing*, *117*, 34–55. doi:10.1016/j.compositesa.2018.11.003

Wisnom, M. R. (1965). The Role of Delamination in Failure of Fibre-Reinforced Composites. *Philosophical Transactions of the Royal Society of London. Series A, Mathematical and Physical Sciences*, *370*(1965), 1850–1870. doi:10.1098/rsta.2011.0441 PMID:22431760

Yousefi, S., & Zohoor, M. (2019). Effect of Cutting Parameters on the Dimensional Accuracy and Surface Finish in the Hard Turning of MDN250 Steel with Cubic Boron Nitride Tool, for Developing a Knowledged Base Expert System. *International Journal of Mechanical and Materials Engineering*, *14*(1), 1. doi:10.1186/s40712-018-0097-7

Zhang, B., Lu, S., Rabiey, M., Axinte, D., & Bleicher, F. (2023). Grinding of Composite Materials. *CIRP Annals*, *72*(2), 645–671. doi:10.1016/j.cirp.2023.05.001

Zhang, G., Schlarb, A. K., Tria, S., & Elkedim, O. (2008). Tensile and Tribological Behaviors of PEEK/Nano-SiO2 Composites Compounded Using a Ball Milling Technique. *Composites Science and Technology*, *68*(15–16), 3073–3080. doi:10.1016/j.compscitech.2008.06.027

Zhi, C., Long, H., & Miao, M. (2017). Influence of Microbond Test Parameters on Interfacial Shear Strength of Fiber Reinforced Polymer-Matrix Composites. *Composites. Part A, Applied Science and Manufacturing*, *100*, 55–63. doi:10.1016/j.compositesa.2017.05.004

Compilation of References

Abdallah, R., Soo, S. L., & Hood, R. (2018). A feasibility study on wire electrical discharge machining of carbon fibre reinforced plastic composites. *Procedia CIRP*, *77*, 195–198. doi:10.1016/j.procir.2018.08.284

Abd-Elwahed, M. S. (2022). Drilling process of GFRP composites: Modeling and optimization using hybrid ANN. *Sustainability (Basel)*, *14*(11), 6599. doi:10.3390/su14116599

Abhishek, K., Kumar, V. R., Datta, S., & Mahapatra, S. S. (2017). Application of JAYA algorithm for the optimization of machining performance characteristics during the turning of CFRP (epoxy) composites : Comparison with TLBO, GA, and ICA. *Engineering with Computers*, *33*(3), 457–475. doi:10.1007/s00366-016-0484-8

Abouelatta, O. B., & Mádl, J. (2001). Surface roughness prediction based on cutting parameters and tool vibrations in turning operations. *Journal of Materials Processing Technology*, *118*(1), 269–277. doi:10.1016/S0924-0136(01)00959-1

Abrão, A. M., Faria, P. E., Rubio, J. C. C., Reis, P., & Davim, J. P. (2007). Drilling of fiber reinforced plastics : A review. In Journal of Materials Processing Technology (Vol. 186, Numéros 1-3, p. 1-7). Elsevier. doi:10.1016/j.jmatprotec.2006.11.146

Ahmadifar, M., Benfriha, K., Shirinbayan, M., Fitoussi, J., & Tcharkhtchi, A. (2022). Mechanical behavior of polymer-based composites using fused filament fabrication under monotonic and fatigue loadings. *Polymers & Polymer Composites*, *30*. doi:10.1177/09673911221082480

Alam, P., Mamalis, D., Robert, C., Floreani, C., & Ó Brádaigh, C. M. (2019). The Fatigue of Carbon Fibre Reinforced Plastics - A Review. *Composites. Part B, Engineering*, *166*, 555–579. doi:10.1016/j.compositesb.2019.02.016

Alarifi, I. M. (2023). A Review on Factors Affecting Machinability and Properties of Fiber-Reinforced Polymer Composites. *Journal of Natural Fibers*, *20*(1), 2154304. doi:10.1080/15440478.2022.2154304

Alberto, M. (2013). Introduction of Fibre-Reinforced Polymers – Polymers and Composites: Concepts, Properties and Processes. In Fiber Reinforced Polymers - The Technology Applied for Concrete Repair. InTech. doi:10.5772/54629

Al-Turaihi, A. S., Hunain, M. B., Hamzah, A. F., & Fadhel, E. Z. (n.d.). *Experimental and Numerical Investigation of Fatigue Behavior of Chopped GFRP Composite Rod under Rotating Bending Load*. Academic Press.

Amin, M., Yuan, S., Khan, M. Z., Zhang, C., & Israr, A. (2017). Parametric analysis for slot milling of carbon fiber reinforced polymers based on ultrasonic machining. *Proceedings of 2017 14th International Bhurban Conference on Applied Sciences and Technology, IBCAST 2017*, 91-99. 10.1109/IBCAST.2017.7868040

Amuthakkannan, P., Manikandan, V., Winowlin Jappes, J. T., & Uthayakumar, M. (2013). Effect of Fibre Length and Fibre Content on Mechanical Properties of Short Basalt Fibre Reinforced Polymer Matrix Composites. *Materials Physics and Mechanics*, *16*(2), 107–117.

An, S.-O., Lee, E.-S., & Noh, S.-L. (1997). A study on the cutting characteristics of glass fiber reinforced plastics with respect to tool materials and geometries. *Journal of Materials Processing Technology*, *68*(1), 60–67. doi:10.1016/S0924-0136(96)02534-4

Arola, D., & Ramulu, M. (1997). Orthogonal cutting of fiber-reinforced composites : A finite element analysis. *International Journal of Mechanical Sciences*, *39*(5), 597–613. doi:10.1016/S0020-7403(96)00061-6

Arola, D., Ramulu, M., & Wang, D. H. (1996). Chip formation in orthogonal trimming of graphite/epoxy composite. *Composites. Part A, Applied Science and Manufacturing*, *27A*(2), 121–133. doi:10.1016/1359-835X(95)00013-R

Arul, S., Vijayaraghavan, L., Malhotra, S. K., & Krishnamurthy, R. (2006). The effect of vibratory drilling on hole quality in polymeric composites. *International Journal of Machine Tools & Manufacture*, *46*(3-4), 252–259. doi:10.1016/j.ijmachtools.2005.05.023

Arumugam, P. U., Malshe, A. P., & Batzer, S. A. (2006). Dry machining of aluminum–silicon alloy using polished CVD diamond-coated cutting tools inserts. *Surface and Coatings Technology*, *200*(11), 3399–3403. doi:10.1016/j.surfcoat.2005.08.127

Arumugam, V., Sidharth, A. A. P., & Santulli, C. (2014). Failure modes characterization of impacted carbon fibre reinforced plastics laminates under compression loading using acoustic emission. *Journal of Composite Materials*, *48*(28), 3457–3468. doi:10.1177/0021998313509504

Asmussen, J., & Reinhard, D. K. (2002). *Diamond Films Handbook*. Marcel Dekker Inc. doi:10.1201/9780203910603

Azmi, A. I., Lin, R. J. T., & Bhattacharyya, D. (2012). Experimental study of machinability of GFRP composites by end milling. *Materials and Manufacturing Processes*, *27*(10), 1045–1050. doi:10.1080/10426914.2012.677917

Azmir, M. A., & Ahsan, A. K. (2008). Investigation on Glass/Epoxy Composite Surfaces Machined by Abrasive Water Jet Machining. *Journal of Materials Processing Technology*, *198*(1-3), 122–128. doi:10.1016/j.jmatprotec.2007.07.014

Bai, Q. S., Yao, Y. X., Bex, P., & Zhang, G. (2004). Study on wear mechanisms and grain effects of PCD tool in machining laminated flooring. *International Journal of Refractory & Hard Metals*, *22*(2), 111–115. doi:10.1016/j.ijrmhm.2004.01.007

Bakis, C. E., Bank, L. C., Brown, V. L., Cosenza, E., Davalos, J. F., Lesko, J. J., Machida, A., Rizkalla, S. H., & Triantafillou, T. C. (2002). Fiber-Reinforced Polymer Composites for Construction—State-of-the-Art Review. *Journal of Composites for Construction*, *6*(2), 73–87. doi:10.1061/(ASCE)1090-0268(2002)6:2(73)

Bayraktar, S., & Turgut, Y. (2016). Investigation of the cutting forces and surface roughness in milling carbon-fiber-reinforced polymer composite material. *Materiali in Tehnologije*, *50*(4), 591–600. doi:10.17222/mit.2015.199

Belmonte, M., Oliveira, F. J., Lanna, M. A., Silva, C. R. M., Corat, E. J., & Silva, R. F. (2004). Turning of CFRC Composites Using Si3N4 and Thin CVD Diamond Coated Si3N4 Tools. *Materials Science Forum*, *455-456*, 609–613.. doi:10.4028/www.scientific.net/MSF.455-456.609

Benardos, P. G., & Vosniakos, G.-C. (2003). Predicting surface roughness in machining : A review. *International Journal of Machine Tools & Manufacture*, *43*(8), 833–844. doi:10.1016/S0890-6955(03)00059-2

Bhatnagar, N., Ramakrishnan, N., Naik, N. K., & Komanduri, R. (1995). On the machining of fiber reinforced plastic (FRP) composite laminates. *International Journal of Machine Tools & Manufacture*, *35*(5), 701–716. doi:10.1016/0890-6955(95)93039-9

Bi, G., Wang, F., Fu, R., & Chen, P. (2022). Wear characteristics of multi-tooth milling cutter in milling CFRP and its impact on machining performance. *Journal of Manufacturing Processes*, *81*, 580–593. doi:10.1016/j.jmapro.2022.07.008

Bílek, O., Žaludek, M., & Čop, J. (2016). Cutting tool performance in end milling of glass fiber-reinforced polymer composites. *Manufacturing Technology*, *16*(1), 12–16. doi:10.21062/ujep/x.2016/a/1213-2489/MT/16/1/12

Bledzki, A. K., & Gassan, J. (1999). Composites reinforced with cellulose based fibres. *Progress in Polymer Science*, *24*(2), 221–274. doi:10.1016/S0079-6700(98)00018-5

Boey, J. Y., Lee, C. K., & Tay, G. S. (2022). Factors Affecting Mechanical Properties of Reinforced Bioplastics : A Review. *Polymers*, *14*(18), 18. Advance online publication. doi:10.3390/polym14183737 PMID:36145883

Bombardier Aerospace. (2011). Drilling of composites and composite/metallic assemblies. *Process Specification.*

Bonifacio, M. E. R., & Diniz, A. E. (1994). Correlating tool wear, tool life, surface roughness and tool vibration in finish turning with coated carbide tools. *Wear*, *173*(1), 137–144. doi:10.1016/0043-1648(94)90266-6

Boothroyd, G., & Knight, W. (1989). *Fundamentals of Machining and Machine Tools* (2nd ed.). Marcel Dekker.

Bouzakis, K.D., Michailidis, N., & Skordaris, G. (2012). Cutting with coated tools: Coating technologies, characterization methods and performance optimization. *CIRP Annals - Manufacturing Technology, 61*, 703 – 723. doi:10.1016/j.cirp.2012.05.006

Brehl, D. E., & Dow, T. A. (2008). Review of vibration-assisted machining. *Precision Engineering, 32*(3), 153–172. doi:10.1016/j.precisioneng.2007.08.003

Breidenstein, B., Denkena, B., Bergmann, B., Wolters, P., & Picker, T. (2022). Turning Copper and Aluminum Alloys with Natural Rocks as Cutting Tools. *Materials (Basel), 15*(6), 6. Advance online publication. doi:10.3390/ma15062187 PMID:35329638

Brent, A. (2008). *Strong, Fundamentals of Composites Manufacturing: Materials, Methods, and Applications*. Society of Manufacturing Engineers.

Brezocnik, M., Kovacic, M., & Ficko, M. (2004). Prediction of surface roughness with genetic programming. *Journal of Materials Processing Technology, 157-158*, 28–36. doi:10.1016/j.jmatprotec.2004.09.004

Cabral, G., Reis, P., Polini, R., Titus, E., Ali, N., Davim, J. P., & Grácio, J. (2006). Cutting performance of time-modulated chemical vapour deposited diamond coated tool inserts during machining graphite. *Diamond and Related Materials, 15*(10), 1753–1758. doi:10.1016/j.diamond.2006.03.007

Cabrera, F. M., Beamud, E., Hanafi, I., Khamlichi, A., & Jabbouri, A. (2011). Fuzzy Logic-Based Modeling of Surface Roughness Parameters for CNC Turning of PEEK CF30 by TiN-Coated Cutting Tools. *Journal of Thermoplastic Composite Materials, 24*(3), 399–413. doi:10.1177/0892705710391562

Cabrera, F. M., Fuentes, D., Hanafi, I., Khamlichi, A., & Jabbouri, A. (2012). Multi-Criteria Optimization Using Taguchi and Grey Relational Analysis in CNC Turning of PEEK CF30. *Journal of Thermoplastic Composite Materials, 25*(1), 101–114. doi:10.1177/0892705711405248

Cabrera, F. M., Hanafi, I., Khamlichi, A., Jabbouri, A., & Bezzazi, M. (2010). Sur l'usinabilité des composites à matrices polymères renforcée par des fibres. *Mechanics & Industry, 11*(2), 2. Advance online publication. doi:10.1051/meca/2010021

Caggiano, A. (2018a). Machining of Fibre Reinforced Plastic Composite Materials. *Materials (Basel), 11*(3), 3. Advance online publication. doi:10.3390/ma11030442 PMID:29562635

Caggiano, A., Nele, L., & Teti, R. (2019). Drilling of Fiber-Reinforced Composite Materials for Aeronautical Assembly Processes. In *Characterizations of Some Composite Materials*. IntechOpen., doi:10.5772/intechopen.80466

Cai, C., An, Q., Ming, W., & Chen, M. (2021). Modelling of machined surface topography and anisotropic texture direction considering stochastic tool grinding error and wear in peripheral milling. *Journal of Materials Processing Technology, 292*, 117065. doi:10.1016/j.jmatprotec.2021.117065

Callister, W. D. Jr, & Rethwisch, D. G. (2008). *Fundamentals of materials science and engineering: an integrated approach (3rd* ed.). Wiley.

Caprino, G., Santo, L., & Nele, L. (1998). Interpretation of size effect in orthogonal machining of composite materials. Part I : Unidirectional glass-fibre-reinforced plastics. *Composites. Part A, Applied Science and Manufacturing*, *29*(8), 887–892. doi:10.1016/S1359-835X(98)00028-1

Caprino, G., & Tagliaferri, V. (1995). Damage development in drilling glass fibre reinforced plastics. *International Journal of Machine Tools & Manufacture*, *35*(6), 817–829. doi:10.1016/0890-6955(94)00055-O

Çelik, Y. H., & Türkan, C. (2020). Investigation of mechanical characteristics of GFRP composites produced from chopped glass fiber and application of taguchi methods to turning operations. *SN Applied Sciences*, *2*(5), 849. doi:10.1007/s42452-020-2684-5

Cengiz, A., Öztürk, M. F., Sabah, A., & Avcu, E. (2024). Experimental and Numerical Analysis of Drilled- and Molded-hole Glass Fiber Reinforced Polymer Matrix (<scp>GFRP</Scp>) Composites. *Polymer Composites*, pc.28305. Advance online publication. doi:10.1002/pc.28305

Cenna, A. A., & Mathew, P. (1997). Evaluation of cut quality of fibre-reinforced plastics—A review. *International Journal of Machine Tools & Manufacture*, *37*(6), 723–736. doi:10.1016/S0890-6955(96)00085-5

Chairi, M., El Bahaoui, J., Hanafi, I., Mata, F., & Di Bella, G. (2023). Composite Materials : A Review of Polymer and Metal Matrix Composites, Their Mechanical Characterization, and Mechanical Properties. In Next Generation Fiber-Reinforced Composites—New Insights. IntechOpen. doi:10.5772/intechopen.106624

Chambers, A., & Bishop, G. (1995). The drilling of carbon fiber polymer matrix composites. *Tenth International Conference on Composite Materials. III. Processing and Manufacturing*, 565-572.

Chang, C. S. (2006). Turning of glass-fiber reinforced plastics materials with chamfered main cutting edge carbide tools. *Journal of Materials Processing Technology*, *180*(1-3), 117–129. doi:10.1016/j.jmatprotec.2006.05.011

Chatelain, J.-F., & Zaghbani, I. (2012). Effect of tool geometry special features on cutting forces of multilayered CFRP laminates. *Journal of Composite Materials*, *85*, 90.

Chaubey, S. K., & Gupta, K. (2022). A Review on Fabrication of Cylindrical and Rotating Parts by Thermoelectric-Erosion Based Turning Processes. *Metals*, *12*(7), 7. Advance online publication. doi:10.3390/met12071227

Chawla, K. K. (2012). Composite materials : Science and engineering, third edition. In Composite Materials: Science and Engineering, Third Edition (Numéro 1994). doi:10.1007/978-0-387-74365-3

Chegdani, F., & Mansori, M. E. (2018). Mechanics of material removal when cutting natural fiber reinforced thermoplastic composites. *Polymer Testing*, *67*, 275–283. doi:10.1016/j.polymertesting.2018.03.016

Chegdani, F., Mezghani, S., El Mansori, M., & Mkaddem, A. (2015). Fiber type effect on tribological behavior when cutting natural fiber reinforced plastics. *Wear*, *332-333*, 772–779. doi:10.1016/j.wear.2014.12.039

Chen, C., Chen, B., Wu, C., Gu, X., Liu, X., & Guo, F. (2022). Theoretical Analysis of Grinding Wheel Deflection Angle on Peripheral Grinding Parameters and Grinding Force. *Metals*, *12*(7), 1209. doi:10.3390/met12071209

Cheng, C. F., Tsui, Y. C., & Clyne, T. W. (1998). Application of a three-dimensional heat flow model to treat laser drilling of carbon fibre composites. *Acta Materialia*, *46*(12), 4273–4285. doi:10.1016/S1359-6454(98)00090-1

Chen, W.-C. (1997). Some experimental investigations in the drilling of carbon fiber-reinforced plastic (CFRP) composite laminates. *International Journal of Machine Tools & Manufacture*, *37*(8), 1097–1108. doi:10.1016/S0890-6955(96)00095-8

Chohan, J. S., Boparai, K. S., Singh, R., & Hashmi, M. S. J. (2022). Manufacturing techniques and applications of polymer matrix composites : A brief review. *Advances in Materials and Processing Technologies*, *8*(1), 884–894. doi:10.1080/2374068X.2020.1835012

Choi, I., Lee, S.-J., Shin, D., & Suh, J. (2021). Green Picosecond Laser Machining of Thermoset and Thermoplastic Carbon Fiber Reinforced Polymers. *Micromachines*, *12*(2), 2. Advance online publication. doi:10.3390/mi12020205 PMID:33671185

Chou, Y. K., & Liu, J. (2005). CVD diamond tool performance in metal matrix composite machining. *Surface and Coatings Technology*, *200*(5), 1872–1878. doi:10.1016/j.surfcoat.2005.08.094

Christman, T., Needleman, A., & Suresh, S. (1989). An experimental and numerical study of deformation in metal-ceramic composites. *Acta Metallurgica*, *37*(11), 3029–3050. doi:10.1016/0001-6160(89)90339-8

Chung, D. D. L. (1994). Carbon Fiber Composites. Butterworth-Heinemann. doi:10.1016/B978-0-08-050073-7.50009-9

Cococcetta, N. M., Pearl, D., Jahan, M. P., & Ma, J. (2020). Investigating surface finish, burr formation, and tool wear during machining of 3D printed carbon fiber reinforced polymer composite. *Journal of Manufacturing Processes*, *56*, 1304–1316. doi:10.1016/j.jmapro.2020.04.025

Crivelli, D., Guagliano, M., Eaton, M., Pearson, M., Al-Jumaili, S., Holford, K., & Pullin, R. (2015). Localisation and Identification of Fatigue Matrix Cracking and Delamination in a Carbon Fibre Panel by Acoustic Emission. *Composites. Part B, Engineering*, *74*, 1–12. doi:10.1016/j.compositesb.2014.12.032

Daniel, I. M., & Ishai, O. (2006). *Engineering Mechanics of Composite Materials* (2nd ed.). Oxford University Press.

Das, S., Warren, J., & West, D. (2016). Global Carbon Fiber Composites Supply Chain Competitiveness Analysis. Available from https://www.osti.gov/servlets/purl/1333049

Davim, J. P. (2002). Diamond tool performance in machining metal–matrix composites. *Journal of Materials Processing Technology*, *128*(1), 100–105. doi:10.1016/S0924-0136(02)00431-4

Davim, J. P., & Mata, F. (2005). A new machinability index in turning fiber reinforced plastics. *Journal of Materials Processing Technology*, *170*(1-2), 436–440. doi:10.1016/j.jmatprotec.2005.05.047

Davim, J. P., & Mata, F. (2007). New machinability study of glass fiber reinforced plastics using polycrystalline diamond and cemented carbide (K15) tools. *Materials & Design*, *28*(3), 1050–1054. doi:10.1016/j.matdes.2005.09.019

Davim, J. P., & Mata, F. (2008). Chemical vapour deposition (CVD) diamond coated tools performance in machining of PEEK composites. *Materials & Design*, *29*(8), 1568–1574. doi:10.1016/j.matdes.2007.11.002

Davim, J. P., & Reis, P. (2003a). Drilling carbon fiber reinforced plastics manufactured by autoclave—Experimental and statistical study. *Materials & Design*, *24*(5), 315–324. doi:10.1016/S0261-3069(03)00062-1

Davim, J. P., & Reis, P. (2003b). Study of delamination in drilling carbon fiber reinforced plastics (CFRP) using design experiments. *Composite Structures*, *59*(4), 481–487. doi:10.1016/S0263-8223(02)00257-X

Davim, J. P., & Reis, P. (2005). Damage and dimensional precision on milling carbon fiber-reinforced plastics using design experiments. *Journal of Materials Processing Technology*, *160*(2), 160–167. doi:10.1016/j.jmatprotec.2004.06.003

Davim, J. P., Reis, P., & António, C. C. (2004). A study on milling of glass fiber reinforced plastics manufactured by hand-lay up using statistical analysis (ANOVA). *Composite Structures*, *64*(3-4), 493–500. doi:10.1016/j.compstruct.2003.09.054

Davim, J. P., Reis, P., & Antonio, C. C. (2004). Experimental study of drilling glass fiber reinforced plastics (GFRP) manufactured by hand lay-up. *Composites Science and Technology*, *64*(2), 289–297. doi:10.1016/S0266-3538(03)00253-7

Davim, J. P., Rubio, J. C., & Abrão, A. M. (2007). A novel approach based on digital image analysis to evaluate the delamination factor after drilling composite laminates. *Composites Science and Technology*, *67*(9), 1939–1945. doi:10.1016/j.compscitech.2006.10.009

De Lacalle López, L. N., & Lamikiz, A. (2010). Milling of carbon fiber reinforced plastics. *Advanced Materials Research*, *83-86*, 49–55. . doi:10.4028/www.scientific.net/AMR.83-86.49

De Melo, E. G., Klein, T. B., Reinkober, S., Gomes, J. D. O., & Uhlmann, E. (2020). Pocket milling of composite fibre-reinforced polymer using industrial robot. *Procedia CIRP*, *85*, 180–185. doi:10.1016/j.procir.2019.09.006

De Oliveira-Leite, W., Campos-Rubio, J. C., Mata-Cabrera, F., Tejero-Manzanares, J., & Hanafi, I. (2014). Utilization of artificial neural networks to predict the influence of milling type on the quality product. *Dyna (Spain), 89*(4), 457-466. *Scopus*. Advance online publication. doi:10.6036/6981

Deuerler, F., Lemmer, O., Frank, M., Pohl, M., & Heßing, C. (2002). Diamond films for wear protection of hardmetal tools. *International Journal of Refractory & Hard Metals*, *20*(2), 115–120. doi:10.1016/S0263-4368(02)00009-4

Dhand, V., Mittal, G., Rhee, K. Y., Park, S.-J., & Hui, D. (2015). A short review on basalt fiber reinforced polymer composites. *Composites. Part B, Engineering*, *73*, 166–180. doi:10.1016/j.compositesb.2014.12.011

Dolinšek, S., & Kopač, J. (1999). Acoustic emission signals for tool wear identification. *Wear*, *225-229*, 295–303. doi:10.1016/S0043-1648(98)00363-9

Durão, L. M. P., Gonçalves, D. J. S., Tavares, J. M. R. S., de Albuquerque, V. H. C., Aguiar Vieira, A., & Torres Marques, A. (2010). Drilling tool geometry evaluation for reinforced composite laminates. *Composite Structures*, *92*(7), 1545–1550. doi:10.1016/j.compstruct.2009.10.035

Dutta, H., Debnath, K., & Sarma, D. K. (2020). A Study of Wire Electrical Discharge Machining of Carbon Fibre Reinforced Plastic. In M. S. Shunmugam & M. Kanthababu (Eds.), *Advances in Unconventional Machining and Composites* (pp. 451–460). Springer Singapore. doi:10.1007/978-981-32-9471-4_36

Elfarhani, M., Guesmi, F., Mkaddem, A., Ghazali, S., Rubaiee, S., & Jarraya, A. (2022). Thermal Aspects in Edge Trimming of Bio-Filled GFRP: Influence of Fiber Orientation and Silica Sand Filler in Heat Generation. *Materials (Basel)*, *15*(14), 4792. Advance online publication. doi:10.3390/ma15144792 PMID:35888259

El-Hofy, M. H., & El-Hofy, H. (2019). Laser beam machining of carbon fiber reinforced composites : A review. *International Journal of Advanced Manufacturing Technology*, *101*(9-12), 2965–2975. doi:10.1007/s00170-018-2978-6

El-Sonbaty, I., Khashaba, U. A., & Machaly, T. (2004). Factors affecting the machinability of GFR/epoxy composites. *Composite Structures*, *63*(3), 329–338. doi:10.1016/S0263-8223(03)00181-8

Endler, I., Leonhardt, A., & Born, R. (1996). Interlayers for diamond deposition on tool materials. *Diamond and Related Materials*, *5*(3-5), 299–303. doi:10.1016/0925-9635(95)00352-5

Equbal, A., Shamim, M., Badruddin, I. A., Equbal, I., Sood, A. K., Ghazali, N. N. N., & Khan, Z. A. (2020). Application of the Combined ANN and GA for Multi-Response Optimization of Cutting Parameters for the Turning of Glass Fiber-Reinforced Polymer Composites. *Mathematics*, *8*(6), 947. doi:10.3390/math8060947

Erden, S., & Ho, K. (2017). Fiber reinforced composites. In *Fiber Technology for Fiber-Reinforced Composites* (pp. 51–79). Woodhead Publishing. doi:10.1016/B978-0-08-101871-2.00003-5

Eriksen, E. (1999). Influence from production parameters on the surface roughness of a machined short fibre reinforced thermoplastic. *International Journal of Machine Tools & Manufacture*, *39*(10), 1611–1618. doi:10.1016/S0890-6955(99)00017-6

Erturk, A. T., Vatansever, F., Yarar, E., Guven, E. A., & Sinmazcelik, T. (2021). Effects of cutting temperature and process optimization in drilling of GFRP composites. *Journal of Composite Materials*, *55*(2), 235–249. doi:10.1177/0021998320947143

Eshelby, J. D., & Peierls, R. E. (1997). The determination of the elastic field of an ellipsoidal inclusion, and related problems. *Proceedings of the Royal Society of London. Series A, Mathematical and Physical Sciences*, *241*(1226), 376–396. doi:10.1098/rspa.1957.0133

Everstine, G. C., & Rogers, T. G. (1971). A Theory of Machining of Fiber-Reinforced Materials. *Journal of Composite Materials*, *5*(1), 94–106. doi:10.1177/002199837100500109

Fallahi, H., Kaynan, O., & Asadi, A. (2023). Insights into the Effect of Fiber–Matrix Interphase Physiochemical- Mechanical Properties on Delamination Resistance and Fracture Toughness of Hybrid Composites. *Composites. Part A, Applied Science and Manufacturing*, *166*, 107390. doi:10.1016/j.compositesa.2022.107390

Fallahi, H., Taheri-Behrooz, F., & Asadi, A. (2020). Nonlinear Mechanical Response of Polymer Matrix Composites: A Review. *Polymer Reviews (Philadelphia, Pa.)*, *60*(1), 42–85. doi:10.1080/15583724.2019.1656236

Faraz, A., Biermann, D., & Weinet, K. (2009). Cutting edge rounding: An innovative tool wear criterion in drilling CFRP composite laminates. *International Journal of Machine Tools & Manufacture*, *49*(15), 1185–1196. doi:10.1016/j.ijmachtools.2009.08.002

Farooq, U., Khan, I., Asif, M., Hira, F., Hussain, G., Ullah, A., Ahmad, S. M., Umar, M., & Alhassan, M. A. (2020). Investigation on the Effects of the Processing Parameters and the Number of Passes on the Flexural Properties of Polymer Nanocomposite Fabricated through FSP Method. *Materials Research Express*, *7*(5), 055310. doi:10.1088/2053-1591/ab9524

Faruk, O., Bledzki, A. K., Fink, H.-P., & Sain, M. (2012). Biocomposites reinforced with natural fibers : 2000–2010. *Progress in Polymer Science*, *37*(11), 1552–1596. doi:10.1016/j.progpolymsci.2012.04.003

Faure, C., Hänni, W., Schmutz, C. J., & Gervanoni, M. (1999). Diamond-coated tools. *Diamond and Related Materials*, *8*(2-5), 830–833. doi:10.1016/S0925-9635(98)00363-X

Feito, N., Álvarez, J. D., Cantero, J. L., & Miguélez, M. H. (2015). Influence of Special Tool Geometry in Drilling Woven CFRPs Materials. *Procedia Engineering*, *132*, 632–638. doi:10.1016/j.proeng.2015.12.541

Fernandes, M., & Cook, C. (2006). Drilling of carbon composites using a one shot drill bit. Part I: Five stage representation of drilling and affecting maximum force and torque. *International Journal of Machine Tools & Manufacture*, *46*(1), 70–75. doi:10.1016/j.ijmachtools.2005.03.015

Fernando, R., Gamage, J., & Karunathilake, H. (2022). Sustainable machining : Environmental performance analysis of turning. *International Journal of Sustainable Engineering*, *15*(1), 15–34. doi:10.1080/19397038.2021.1995524

Ferreira, J. R., Coppini, N. L., & Levy Neto, F. (2001). Characteristics of carbon–carbon composite turning. *Journal of Materials Processing Technology*, *109*(1), 65–71. doi:10.1016/S0924-0136(00)00776-7

Ferreira, J. R., Coppini, N. L., & Miranda, G. W. A. (1999). Machining optimisation in carbon fibre reinforced composite materials. *Journal of Materials Processing Technology*, *92-93*, 135–140. doi:10.1016/S0924-0136(99)00221-6

Ferriera, J.R., Coppini, N.L., & Miranda, G.W.A. (1999). *Machining optimization in carbon fibre reinforced composites materials.* Journal of Materials Processing Technology, 92, 135–140.

Frketic, J., Dickens, T., & Ramakrishnan, S. (2017). Automated manufacturing and processing of fiber-reinforced polymer (FRP) composites : An additive review of contemporary and modern techniques for advanced materials manufacturing. *Additive Manufacturing*, *14*, 69–86. doi:10.1016/j.addma.2017.01.003

Fuentes Molina, N., Fragozo Brito, Y., & Polo Benavides, J. M. (2021). Recycling of Residual Polymers Reinforced with Natural Fibers as a Sustainable Alternative : A Review. *Polymers*, *13*(21), 21. Advance online publication. doi:10.3390/polym13213612 PMID:34771169

Fukui, H., Okida, J., Omori, N., Moriguchi, H., & Tsuda, K. (2004). Cutting performance of DLC coated tools in dry machining aluminum alloys. *Surface and Coatings Technology*, *187*(1), 70–76. doi:10.1016/j.surfcoat.2004.01.014

Gåhlin, R., Alahelisten, A., & Jacobson, S. (1996). The effects of compressive stresses on the abrasion of diamond coatings. *Wear*, *196*(1 – 2), 226-233. doi:10.1016/0043-1648(95)06907-0

Gaikwad, S. D., Dravid, S. V., Sonawane, B. U., Abraham, M., & Goyal, R. K. (2019). Dynamic Mechanical and Creep Properties of Poly(Ether Ketone)-Graphite Nanoplatelet (GNP) Nanocomposites. *Polymer-Plastics Technology and Materials*, *58*(15), 1691–1701. doi:10.1080/25740881.2018.1563141

Gaitonde, V. N., Karnik, S. R., Rubio, J. C., Correia, A. E., Abrão, A. M., & Davim, J. P. (2008). Analysis of parametric influence on delamination in high-speed drilling of carbon fiber reinforced plastic composites. *Journal of Materials Processing Technology*, *203*(1-3), 431–438. doi:10.1016/j.jmatprotec.2007.10.050

Gao, T., Zhang, Y., Li, C., Wang, Y., Chen, Y., An, Q., Zhang, S., Li, H. N., Cao, H., Ali, H. M., Zhou, Z., & Sharma, S. (2022). Fiber-reinforced composites in milling and grinding : Machining bottlenecks and advanced strategies. In Frontiers of Mechanical Engineering (Vol. 17, Numéro 2, p. 1-35). Springer. doi:10.1007/s11465-022-0680-8

Gao, T., Li, C., Yang, M., Zhang, Y., Jia, D., Ding, W., Debnath, S., Yu, T., Said, Z., & Wang, J. (2021). Mechanics Analysis and Predictive Force Models for the Single-Diamond Grain Grinding of Carbon Fiber Reinforced Polymers Using CNT Nano-Lubricant. *Journal of Materials Processing Technology*, *290*, 116976. doi:10.1016/j.jmatprotec.2020.116976

Gay, D., Hoa, S., & Tsai, S. (2003). *Composite Materials – Design and Applications*. CRC Press.

Gaydaychuk, A., Linnik, S., & Zenkin, S. (2020). High temperature tribology of heavily boron doped diamond films against steel. *International Journal of Refractory & Hard Metals*, *88*, 105191. doi:10.1016/j.ijrmhm.2020.105191

Geier, N. (2020). Influence of fibre orientation on cutting force in up and down milling of UD-CFRP composites. *International Journal of Advanced Manufacturing Technology*, *111*(3-4), 881–893. doi:10.1007/s00170-020-06163-3

Geier, N., Davim, J. P., & Szalay, T. (2019). Advanced cutting tools and technologies for drilling carbon fibre reinforced polymer (CFRP) composites: A review. *Composites. Part A, Applied Science and Manufacturing*, *125*, 105552. doi:10.1016/j.compositesa.2019.105552

Geier, N., Patra, K., Anand, R. S., Ashworth, S., Balázs, B. Z., Lukács, T., Magyar, G., Tamás-Bényei, P., Xu, J., & Davim, J. P. (2023). A Critical Review on Mechanical Micro-Drilling of Glass and Carbon Fibre Reinforced Polymer (GFRP and CFRP) Composites. *Composites. Part B, Engineering*, *254*, 110589. doi:10.1016/j.compositesb.2023.110589

Geier, N., Xu, J., Poór, D. I., Dege, J. H., & Davim, J. P. (2023). A Review on Advanced Cutting Tools and Technologies for Edge Trimming of Carbon Fibre Reinforced Polymer (CFRP) Composites. *Composites. Part B, Engineering*, *266*, 111037. doi:10.1016/j.compositesb.2023.111037

Ge, J., Catalanotti, G., Falzon, B. G., McClelland, J., Higgins, C., Jin, Y., & Sun, D. (2022). Towards understanding the hole making performance and chip formation mechanism of thermoplastic carbon fibre/polyetherketoneketone composite. *Composites. Part B, Engineering*, *234*, 109752. doi:10.1016/j.compositesb.2022.109752

Geng, D., Liu, Y., Shao, Z., Lu, Z., Cai, J., Li, X., Jiang, X., & Zhang, D. (2019). Delamination formation, evaluation and suppression during drilling of composite laminates: A review. *Composite Structures*, *216*, 168–186. doi:10.1016/j.compstruct.2019.02.099

Ghidossi, P., El Mansori, M., & Pierron, F. (2006). Influence of specimen preparation by machining on the failure of polymer matrix off-axis tensile coupons. *Composites Science and Technology*, *66*(11-12), 1857–1872. doi:10.1016/j.compscitech.2005.10.009

Girot, F., López de Lacalle, L. N., Lamikiz, A., Iliescu, D., & Gutiérrez, M. E. (2009). Machinabilty aspects of polymer matrix composites. In J. P. Davim (Éd.), Machining Composites Materials. ISTE Publishing Knowledge and John Wiley & Sons Inc.

Gracio, J. J., Fan, Q. H., & Madaleno, J. C. (2010). Diamond growth by chemical vapour deposition. *Journal of Physics. D, Applied Physics*, *43*(37), 374017. doi:10.1088/0022-3727/43/37/374017

Groover, M. P. (2012). Fundamentals of Modern Manufacturing : Materials, Processes, and Systems 5th Edition (5th Edition Binder Ready Version). Wiley.

Groover, M. P. (2020). *Fundamentals of Modern Manufacturing : Materials, Processes, and Systems*. John Wiley & Sons.

Haddad M, et al. (2014) *Study of the surface defects and dust generated during trimming of CFRP: influence of tool geometry, machining parameters and cutting speed range. Compos* A Appl Sci Manuf., 66, 142–54.

Haddad, M., Zitoune, R., Bougherara, H., Eyma, F., & Castanié, B. (2014). Study of trimming damages of CFRP structures in function of the machining processes and their impact on the mechanical behavior. *Composites. Part B, Engineering, 57*, 136–143. doi:10.1016/j.compositesb.2013.09.051

Hanafi, I., Cabrera, F. M., Bahaoui, J. E., Rubio, J. C. C., & Leite, W. de O. (2023). 3D-Printed Polymer Composite : A Bibliometric Approach – What, Where, and Who Is Doing Research on This Topic. In Development, Properties, and Industrial Applications of 3D Printed Polymer Composites (pp. 9-30). IGI Global. doi:10.4018/978-1-6684-6009-2.ch002

Hanafi, I., Abdellatif, K., Cabrera, F. M., & Manzanares, J. T. (2013). Application of Desirability Function Based on Neural Network for Optimizing the Process Parameters in Turning of PEEK CF30 Composites. *Materials Science Forum, 766*, 1–20. . doi:10.4028/www.scientific.net/MSF.766.1

Hanafi, I., Cabrera, F. M., Khamlichi, A., Garrido, I., & Manzanares, J. T. (2013). Artificial neural networks back propagation algorithm for cutting force components predictions. *Mechanics & Industry, 14*(6), 6. Advance online publication. doi:10.1051/meca/2013084

Hanafi, I., Khamlichi, A., Cabrera, F. M., Almansa, E., & Jabbouri, A. (2012). Optimization of cutting conditions for sustainable machining of PEEK-CF30 using TiN tools. *Journal of Cleaner Production, 33*, 1–9. doi:10.1016/j.jclepro.2012.05.005

Hanafi, I., Khamlichi, A., Cabrera, F. M., & Nuñez López, P. J. (2012). Prediction of surface roughness in turning of PEEK cf30 by using an artificial neural network. *Journal of the Chinese Institute of Industrial Engineers, 29*(5), 337–347. doi:10.1080/10170669.2012.702690

Hanafi, I., Khamlichi, A., Cabrera, F. M., Nuñez López, P. J., & Jabbouri, A. (2012). Fuzzy rule based predictive model for cutting force in turning of reinforced PEEK composite. *Measurement, 45*(6), 1424–1435. doi:10.1016/j.measurement.2012.03.022

Hassanzadeh-Aghdam, M. K., Ansari, R., Mahmoodi, M. J., & Darvizeh, A. (2018). Effect of Nanoparticle Aggregation on the Creep Behavior of Polymer Nanocomposites. *Composites Science and Technology, 162*, 93–100. doi:10.1016/j.compscitech.2018.04.025

Haubner, R. (2013). The history of hard CVD coatings for tool applications at the University of Technology Vienna. *International Journal of Refractory & Hard Metals, 41*, 22–34. doi:10.1016/j.ijrmhm.2013.01.012

Haubner, R., & Kalss, W. (2010). Diamond deposition on hardmetal substrates – Comparison of substrate pre-treatments and industrial applications. *International Journal of Refractory & Hard Metals, 28*(4), 475–483. doi:10.1016/j.ijrmhm.2010.03.004

Haubner, R., Kubelka, S., Lux, B., Griesser, M., & Grasserbauer, M. (1995). Murakami and H_2SO_4/H_2O_2 Pretreatment of WC-Co Hard Metal Substrates to Increase the Adhesion of CVD Diamond Coatings. *Journal de Physique. IV*, *5*(C5), 753–760. doi:10.1051/jphyscol:1995589

Haubner, R., & Lux, B. (2006). Diamond deposition on steel substrates using intermediate layers. *International Journal of Refractory & Hard Metals*, *24*(5), 380–386. doi:10.1016/j.ijrmhm.2005.11.008

Hays, C. C., Kim, C. P., & Johnson, W. L. (2000). Microstructure Controlled Shear Band Pattern Formation and Enhanced Plasticity of Bulk Metallic Glasses Containing in situ Formed Ductile Phase Dendrite Dispersions. *Physical Review Letters*, *84*(13), 2901–2904. doi:10.1103/PhysRevLett.84.2901 PMID:11018971

Heath, P. J. (2001). Developments in applications of PCD tooling. *Journal of Materials Processing Technology*, *116*(1), 31–38. doi:10.1016/S0924-0136(01)00837-8

Heidary, H., Karimi, N. Z., & Minak, G. (2018). Investigation on Delamination and Flexural Properties in Drilling of Carbon Nanotube/Polymer Composites. *Composite Structures*, *201*, 112–120. doi:10.1016/j.compstruct.2018.06.041

Heisel, U., & Pfeifroth, T. (2012). Influence of point angle on drill hole quality and machining forces when drilling CFRP. *Procedia CIRP*, *1*, 471–476. doi:10.1016/j.procir.2012.04.084

Helmi, Y., & El-Hofy, H. (2020). Design for Machining. In Non-Traditional and Advanced Machining Technologies (2nd Editio, p. 62). CRC Press. doi:10.21062/ujep/x.2009/a/1213-2489/MT/9/1/81

Henry Ononiwu, N., Akinlabi, E. T., & Ozoegwu, C. G. (2021). Optimization techniques applied to machinability studies for turning aluminium metal matrix composites : A literature review. *Materials Today: Proceedings*, *44*, 1124–1129. doi:10.1016/j.matpr.2020.11.228

Herlinger, J. (2006). Sp3's experience using hot filament CVD reactors to grow diamond for an expanding set of applications. *Thin Solid Films*, *501*(1-2), 65–69. doi:10.1016/j.tsf.2005.07.108

HexcelP. T. (2013). Available from https://kevra.fi/wp- content/uploads/Prepreg_Technology_.pdf

He, Y., Li, L., Wan, M., & Xue, H. (2022). Three-Dimensional Finite Element Simulations of Milling Carbon/Epoxy Composites. *Composite Structures*, *282*, 115037. doi:10.1016/j.compstruct.2021.115037

Hintze, W., Hartmann, D., & Schütte, C. (2013). *Method for Milling Long Fiber Reinforced Composite Plastic* (Patent N° US20130020735A1).

Hintze, W., Cordes, M., & Koerkel, G. (2015). Influence of weave structure on delamination when milling CFRP. *Journal of Materials Processing Technology*, *216*, 199–205. doi:10.1016/j.jmatprotec.2014.09.004

Hocheng, H., Puw, H. Y., & Yao, K. C. (1992).*Experimental aspects of drilling of some fiber-reinforced plastics. Proceedings of the Machining of Composites Materials Symposium, ASM Materials Week,* 127–138.

Hocheng, H., Puw, H. Y., & Huang, Y. (1993). Preliminary study on milling of unidirectional carbon fibre-reinforced plastics. *Composites Manufacturing*, *4*(2), 103–108. doi:10.1016/0956-7143(93)90077-L

Hocheng, H., & Tsao, C. C. (2003). Comprehensive analysis of delamination in drilling of composite materials with various drill bits. *Journal of Materials Processing Technology*, *140*(1), 335–339. doi:10.1016/S0924-0136(03)00749-0

Hocheng, H., & Tsao, C. C. (2005). The path towards delamination-free drilling of composite materials. *Journal of Materials Processing Technology*, *167*(2), 251–264. doi:10.1016/j.jmatprotec.2005.06.039

Hocheng, H., & Tsao, C. C. (2006). Effects of special drill bits on drilling-induced delamination of composite materials. *International Journal of Machine Tools & Manufacture*, *46*(12), 1403–1416. doi:10.1016/j.ijmachtools.2005.10.004

Hu, J., Chou, Y. K., Thompson, R. G., Burgess, J., & Street, S. (2007). Characterizations of nanocrystalline diamond coating cutting tools. *Surface and Coatings Technology*, *202*(4-7), 1113–1117. doi:10.1016/j.surfcoat.2007.07.050

Hulugappa, B., Achutha, M. V., & Suresha, B. (2016). Effect of Fillers on Mechanical Properties and Fracture Toughness of Glass Fabric Reinforced Epoxy Composites. *Journal of Minerals & Materials Characterization & Engineering*, *04*(01), 1–14. doi:10.4236/jmmce.2016.41001

Hutyrova, Z., Scucka, J., Hloch, S., Hlavacek, P., & Zelenak, M. (2016). Turning of wood plastic composites by water jet and abrasive water jet. *International Journal of Advanced Manufacturing Technology*, *84*(5-8), 1615–1623. doi:10.1007/s00170-015-7831-6

Ibrahim, I. A., Mohamed, F. A., & Lavernia, E. J. (1991). Particulate reinforced metal matrix composites—A review. *Journal of Materials Science*, *26*(5), 1137–1156. doi:10.1007/BF00544448

Iliescu, D., Gehin, D., Gutierrez, M. E., & Girot, F. (2010). Modeling and tool wear in drilling of CFRP. *International Journal of Machine Tools & Manufacture*, *50*(2), 204–213. doi:10.1016/j.ijmachtools.2009.10.004

Iñigo, R. (2023). Improving Surface Integrity When Drilling CFRPs and Ti-6Al-4V Using Sustainable Lubricated Liquid Carbon Dioxide. *Chinese Journal of Aeronautics*, *36*(7), 129–146. doi:10.1016/j.cja.2022.09.004

Inoue, H., Aoyama, E., Hirogaki, T., Ogawa, K., Matushita, H., Kitahara, Y., & Katayama, T. (1997). Influence of tool wear on internal damage in small diameter drilling in GFRP. *Composite Structures*, *39*(1–2), 55–62. doi:10.1016/S0263-8223(97)00068-8

Inoue, T., Hagino, M., Matsui, M., & Gu, L. (2009). Cutting characteristics of CFRP materials with end milling. *Key Engineering Materials, 407-408*, 710–713. . doi:10.4028/www.scientific.net/KEM.407-408.710

Ishida, T., Noma, K., Kakinuma, Y., Aoyama, T., Hamada, S., Ogawa, H., & Higaino, T. (2014). New production technologies in aerospace industry—5th machining innovations conference (MIC 2014) helical milling of carbon fiber reinforced plastics using ultrasonic vibration and liquid nitrogen. *Procedia CIRP*, *24*(C), 13–18. doi:10.1016/j.procir.2014.07.139

Islam Shyha, D. H. (2021). Advances in Machining of Composite Materials. In I. Shyha & D. Huo (Eds.), *Springer International Publishing*. Springer International Publishing. doi:10.1007/978-3-030-71438-3

Ismail, Dhakal, Popov, & Beaugrand. (2016). Comprehensive Study on Machinability of Sustainable and Conventional Fibre Reinforced Polymer Composites. *Engineering Science and Technology, an International Journal,* 19(4), 2043–52. . doi:10.1016/j.jestch.2016.07.010

Jahanmir, S. (n.d.). Machining of ceramics and composites. https://cir.nii.ac.jp/crid/1130282269159410816

Jain, A., Kumar, C. S., & Shrivastava, Y. (2022) *Machining of composite materials using different conventional and unconventional machining processes: a short review.* Adv Mech Energy Technol Select Proc ICMET, 75–83.

Jani, S. P., Senthil Kumar, A., Adam Khan, M., & Uthayakumar, M. (2020). Surface Roughness and Morphology Studies on Machining Hybrid Composite Material Using Abrasive Water Jet Cutting Process. doi:10.1007/978-3-030-43232-4_6

Jäschke, P., Wippo, V., Bluemel, S., Staehr, R., & Dittmar, H. (2017). Laser Machining of Carbon Fiber-Reinforced Plastic Composites. In Advances in Laser Materials Processing : Technology, Research and Applications (p. 121-152). Woodhead Publishing. doi:10.1016/B978-0-08-101252-9.00006-6

Jawahir, I. S., & Balaji, A. K. (2018). *Machining*. Routledge Handbooks Online. doi:10.1201/9781351045636-140000423

Jessy, K., Dinakaran, D., & Seshagiri Rao, V. (2015). Influence of different cooling methods on drill temperature in drilling GFRP. *International Journal of Advanced Manufacturing Technology*, *76*(1), 609–621. doi:10.1007/s00170-014-6280-y

Jha, K., Kataria, R., Verma, J., Pradhan, S., Jha, K., Kataria, R., Verma, J., & Pradhan, S. (2019). Potential biodegradable matrices and fiber treatment for green composites : A review. *AIMS Materials Science*, *6*(1), 119–138. doi:10.3934/matersci.2019.1.119

Jialiang, Y., Zhigang, Z., & Hongzhuan, Z. (2019). Polymer Matrix Composites. In Highway Engineering Composite Material and Its Application (Vol. 70, Numéro 7, p. 1282-1283). Springer Singapore. doi:10.1007/s11837-018-2917-x

Jia, Z., Fu, R., Niu, B., Qian, B., Bai, Y., & Wang, F. (2016). Novel drill structure for damage reduction in drilling CFRP composites. *International Journal of Machine Tools & Manufacture*, *110*, 55–65. doi:10.1016/j.ijmachtools.2016.08.006

Jia, Z., Fu, R., Wang, F., Qian, B., & He, C. (2018). Temperature effects in end milling carbon fiber reinforced polymer composites. *Polymer Composites*, *39*(2), 437–447. doi:10.1002/pc.23954

John & Blackmon. (1990). Metal-Matrix Composites. In ASM Handbook Vol. 2: Properties and Selection: Nonferrous Alloys and Special-Purpose Materials. doi:10.31399/asm.hb.v02.a0001101

John, R., Lin, R., Jayaraman, K., & Bhattacharyya, D. (2023). Investigation on microstructure characteristics of tool wear and machined surface mechanisms while milling : Kenaf vs glass fiber-reinforced composites. *Journal of Materials Research and Technology*, *23*, 4716–4733. doi:10.1016/j.jmrt.2023.02.115

Johnson, T. (2018). History of Composites. ThoughtCo. Available from https://www.thoughtco.com/history-of-composites-820404

John, V. (1992). Material Removal Processes. In *Introduction to Engineering Materials* (pp. 360–380). Palgrave Macmillan. doi:10.1007/978-1-349-21976-6_25

Kalla, D., Sheikh-Ahmad, J., & Twomey, J. (2010). Prediction of cutting forces in helical end milling fiber reinforced polymers. *International Journal of Machine Tools & Manufacture*, *50*(10), 882–891. doi:10.1016/j.ijmachtools.2010.06.005

Kalpakjian, S., & Schmid, S. R. (2002). *Manufactura, ingeniería y tecnología*. Pearson Educación.

Kamada, Y., & Yan, J. (2021). Machinability Investigation for Cellulose Nanofiber-Reinforced Polymer Composite by Ultraprecision Diamond Turning. *International Journal of Automotive Technology*, *15*(4), 475–482. doi:10.20965/ijat.2021.p0475

Kamo, M., Sato, Y., Matsumoto, S., & Setaka, N. (1983). Diamond synthesis from gas phase in microwave plasma. *Journal of Crystal Growth*, *62*(3), 642–644. doi:10.1016/0022-0248(83)90411-6

Kandavalli, S. R., Khan, A. M., Iqbal, A., Jamil, M., Abbas, S., Laghari, R. A., & Cheok, Q. (2023). Application of sophisticated sensors to advance the monitoring of machining processes: Analysis and holistic review. *International Journal of Advanced Manufacturing Technology*, *125*(3-4), 989–1014. Advance online publication. doi:10.1007/s00170-022-10771-6

Kaneeda, T. (1989). CFRP cutting mechanism. *Proc. of 17th North American Manufacturing Research Conf.*, 220.

Karataş, M. A., & Gökkaya, H. (2018). A review on machinability of carbon fiber reinforced polymer (CFRP) and glass fiber reinforced polymer (GFRP) composite materials. *Defence Technology*, *14*(4), 318–326. doi:10.1016/j.dt.2018.02.001

Kawai, M., & Masuko, Y. (2004). Creep Behavior of Unidirectional and Angle-Ply T800H/3631 Laminates at High Temperature and Simulations Using a Phenomenological Viscoplasticity Model. *Composites Science and Technology*, *64*(15), 2373–2384. doi:10.1016/j.compscitech.2004.04.004

Kaw, K. (2006). *Mechanics of Composite Materials* (2nd ed.). Taylor & Francis.

Kechagias, J., Petropoulos, G., Iakovakis, V., & Maropoulos, S. (2009). An Investigation of Surface Texture Parameters during Turning of a Reinforced Polymer Composite Using Design of Experiments and Analysis. *International Journal of Experimental Design and Process Optimisation*, *1*(2/3), 164. doi:10.1504/IJEDPO.2009.030317

Kessler, M. R. (2012). Polymer matrix composites : A perspective for a special issue of polymer reviews. In Polymer Reviews (Vol. 52, Numéros 3-4, p. 229-233). Taylor & Francis Group. doi :10.1080/15583724.2012.708004

Khan, M. M. A., Mithu, M. A. H., & Dhar, N. R. (2009). Effects of Minimum Quantity Lubrication on Turning AISI 9310 Alloy Steel Using Vegetable Oil-Based Cutting Fluid. *Journal of Materials Processing Technology*, *209*(15–16), 5573–5583. doi:10.1016/j.jmatprotec.2009.05.014

Khashaba, U. A. (2013). Drilling of Polymer Matrix Composites: A Review. *Journal of Composite Materials*, *47*(15), 1817–1832. doi:10.1177/0021998312451609

Khashaba, U. A., El-Sonbaty, I. A., Selmy, A. I., & Megahed, A. A. (2010). Machinability Analysis in Drilling Woven GFR/Epoxy Composites: Part I – Effect of Machining Parameters. *Composites. Part A, Applied Science and Manufacturing*, *41*(3), 391–400. doi:10.1016/j.compositesa.2009.11.006

Kim, D., Kim, Y. H., Gururaja, S., & Ramulu, M. (2010). Processing and fiber content effects on the machinability of compression moulded random direction short GFRP composites. *International Journal of Automotive Technology*, *11*(6), 849–855. doi:10.1007/s12239-010-0101-6

Kiran, M. D., Govindaraju, H. K., Jayaraju, T., & Kumar, N. (2018). Review-Effect of Fillers on Mechanical Properties of Polymer Matrix Composites. *Materials Today: Proceedings*, *5*(10), 22421–22424. doi:10.1016/j.matpr.2018.06.611

Kishore, M., Amrita, M., & Kamesh, B. (2020). Experimental investigation of milling on basalt-jute hybrid composites with graphene as nanofiller. *Materials Today: Proceedings*, *43*, 726–730. doi:10.1016/j.matpr.2020.12.847

Kitajima, T., Kusuyama, J., Yui, A., Fujii, K., & Ito, Y. (2014). Development of PCD milling tool for carbon-fiber-reinforced plastics. *Advanced Materials Research*, *1017*, 411–414. . doi:10.4028/www.scientific.net/AMR.1017.411

Knápek, T., Dvořáčková, Š., & Knap, A. (2022). Wear Study of Coated Mills during Circumferential Milling of Carbon Fiber-Reinforced Composites and Their Influence on the Sustainable Quality of the Machined Surface. *Coatings*, *12*(10), 1379. doi:10.3390/coatings12101379

König, W., & Graß, P. (1989). Quality Definition and Assessment in Drilling of Fibre Reinforced Thermosets. *CIRP Annals*, *38*(1), 119–124. doi:10.1016/S0007-8506(07)62665-1

Konig, W., Wulf, Ch., Grab, P., & Willerscheid, H. (1985). Machining of fibre reinforced plastics. *CIRP Annals*, *34*(2), 537–547. doi:10.1016/S0007-8506(07)60186-3

Kopač, J., & Bahor, M. (1999). Interaction of the technological history of a workpiece material and the machining parameters on the desired quality of the surface roughness of a product. *Journal of Materials Processing Technology*, *92-93*, 381–387. doi:10.1016/S0924-0136(99)00239-3

Köpf, A., Feistritzer, S., & Udier, K. (2006). Diamond coated cutting tools for machining of non-ferrous metals and fibre reinforced polymers. *International Journal of Refractory & Hard Metals*, *24*(5), 354–359. doi:10.1016/j.ijrmhm.2005.11.013

Koplev, A. (1980). Cutting of CFRP with Single Edge Tools : 3rd International Conference of Advances in Composite Materials. *Advanced Composite Materials*, *2*, 1597–1605.

Koplev, A., Lystrup, A., & Vorm, T. (1983). The cutting process, chips, and cutting forces in machining CFRP. *Composites*, *14*(4), 371–376. doi:10.1016/0010-4361(83)90157-X

Krishnamurthy, R., Santhanakrishnan, G., & Malhotra, S. K. (1992).*Machining of polymeric composites.* Proceedings of the Machining of Composites Materials Symposium, ASM Materials Week, 139–148.

Kumar, A. L., & Prakash, M. (2021). The effect of fiber orientation on mechanical properties and machinability of GFRP composites by end milling using cutting force analysis. *Polymers & Polymer Composites*, *29*(9), S178–S187. doi:10.1177/0967391121991289

Kumar, A., Sharma, K., & Dixit, A. R. (2020). Carbon nanotube- and graphene-reinforced multiphase polymeric composites : Review on their properties and applications. *Journal of Materials Science*, *55*(7), 2682–2724. doi:10.1007/s10853-019-04196-y

Kumar, S., Gariya, N., Shaikh, A., Ahmad, F., & Yadav, A. (2024). Physico-Mechanical and Thermal Properties of Fly Ash and Benzoylated Himalayan Agave Fiber Reinforced Polyester Composite. *Biomass Conversion and Biorefinery*. Advance online publication. doi:10.1007/s13399-024-05518-0

Kumar, S., Prasad, L., Bijlwan, P. P., & Yadav, A. (2022). Thermogravimetric Analysis of Lignocellulosic Leaf-Based Fiber-Reinforced Thermosets Polymer Composites: An Overview. *Biomass Conversion and Biorefinery*. Advance online publication. doi:10.1007/s13399-022-03332-0

Kumar, S., Prasad, L., Patel, V. K., Kumain, A., & Yadav, A. (2021). Experimental and Numerical Study on Physico-mechanical Properties and Taguchi's Designed Abrasive Wear Behavior of Hemp/Nettle-polyester Hybrid Composite. *Polymer Composites*, *42*(12), 6912–6927. doi:10.1002/pc.26350

Kumar, S., Prasad, L., Patel, V. K., Kumar, V., Kumar, A., Yadav, A., & Winczek, J. (2021). Physical and Mechanical Properties of Natural Leaf Fiber-Reinforced Epoxy Polyester Composites. *Polymers*, *13*(9), 1369. doi:10.3390/polym13091369 PMID:33922169

Kuo, C., Soo, S., Aspinwall, D., Bradley, S., Thomas, W., M'Saoubi, R., Pearson, D., & Leahy, W. (2014). Tool wear and hole quality when single-shot drilling of metallic-composite stacks with diamond-coated tools. *Proceedings of the Institution of Mechanical Engineers. Part B, Journal of Engineering Manufacture*, *228*(10), 1314–1322. doi:10.1177/0954405413517388

Kuo, C., Wang, C., & Ko, S. (2018). Wear behavior of CVD diamond – coated tools in the drilling of woven CFRP composites. *Wear*, *398-399*, 1–12. doi:10.1016/j.wear.2017.11.015

Lee, B. Y., Tarng, Y. S., & Lii, H. R. (2000). An investigation of modeling of the machining database in turning operations. *Journal of Materials Processing Technology*, *105*(1), 1–6. doi:10.1016/S0924-0136(00)00535-5

Leite, W. O., Rubio, J. C. C., Tejero, J., Mata, F., & Hanafi, I. (2015). A comparison of milling cutting path strategies for aluminium 6262-T6 alloys fabrication. In Manufacturing Engineering : New Research (p. 77-90). Scopus.

Lei, X., Wang, L., Shen, B., Sun, F., & Zhang, Z. (2013). Comparison of chemical vapor deposition diamond-, diamond-like carbon- and TiAlN-coated microdrills in graphite machining. *Proceedings of the Institution of Mechanical Engineers. Part B, Journal of Engineering Manufacture*, *227*(9), 1299–1309. doi:10.1177/0954405413487898

Lifshitz, Y. (1999). Diamond – like carbon – present status. *Diamond and Related Materials*, *8*(8-9), 1659–1676. doi:10.1016/S0925-9635(99)00087-4

Linbo, Z., Lijiang, W., & Xin, W. (2003). Study on vibration drilling of fiber reinforced plastics with hybrid variation parameters method. *Composites. Part A, Applied Science and Manufacturing*, *34*(3), 237–244. doi:10.1016/S1359-835X(02)00207-5

Lin, S.-C., & Shen, J.-M. (1999). Drilling unidirectional glass fiber-reinforced composite materials at high speed. *Journal of Composite Materials*, *33*(9), 827–851. doi:10.1177/002199839903300903

Lin, S., & Chen, I. (1996). Drilling carbon fiber-reinforced composite material at high speed. *Wear*, *194*(1–2), 156–162. doi:10.1016/0043-1648(95)06831-7

Lin, Y.-C., Wu, K.-D., Shih, W.-C., Hsu, P.-K., & Hung, J.-P. (2020). Prediction of Surface Roughness Based on Cutting Parameters and Machining Vibration in End Milling Using Regression Method and Artificial Neural Network. *Applied Sciences (Basel, Switzerland)*, *10*(11), 11. Advance online publication. doi:10.3390/app10113941

Liu, J. D., Yang, X. F., Xu, Z. L., Li, Y. S., Yan, G. H., & Huang, L. K. (2017). Experimental study on the tool wear in milling carbon fiber reinforced plastics. *Materials Science Forum, 893 MSF*, 57-61. https://doi.org/ doi:10.4028/www.scientific.net/MSF.893.57

Liu, C., Ren, J., Shi, K., & Zhang, Y. (2023). Investigation of fracture mechanism evolution model for UD-CFRP and MD-CFRP during the milling process. *Composite Structures*, *306*, 116585. doi:10.1016/j.compstruct.2022.116585

Liu, D., Tang, Y., & Cong, W. L. (2012). A review of mechanical drilling for composite laminates. *Composite Structures*, *94*(4), 1265–1279. doi:10.1016/j.compstruct.2011.11.024

Liu, D., Wang, H., Xu, H., & Zhang, C. (2011). Study of milling force in high speed milling Carbon Fibre Reinforced Plastics(CFRP) using diamond tool. *Applied Mechanics and Materials*, *50-51*, 368–371. . doi:10.4028/www.scientific.net/AMM.50-51.368 PMID:31267820

Liu, J., Chen, G., Ren, C., Qin, X., Zou, Y., & Ge, J. (2020). Effects of axial and longitudinal-torsional vibration on fiber removal in ultrasonic vibration helical milling of CFRP composites. *Journal of Manufacturing Processes*, *58*, 868–883. doi:10.1016/j.jmapro.2020.08.071

Liu, L., Qi, C., Wu, F., Zhang, X., & Zhu, X. (2018). Analysis of thrust force and delamination in drilling GFRP composites with candle stick drills. *International Journal of Advanced Manufacturing Technology*, *95*(5), 2585–2600. doi:10.1007/s00170-017-1369-8

Li, W., Zhu, H., Wang, J., Ali, Y. M., & Huang, C. (2013). An investigation into the radial-mode abrasive waterjet turning process on high tensile steels. *International Journal of Mechanical Sciences*, *77*, 365–376. doi:10.1016/j.ijmecsci.2013.05.005

Li, X., & Du, R. (2002). Analysis and compensation of workpiece errors in turning. *International Journal of Production Research*, *40*(7), 1647–1667. doi:10.1080/00207540110100685

Li, X., Liu, F., Gong, N., Yang, C., & Wang, B. (2018). Surface Topography Induced High Injection Joining Strength of Polymer-Metal Composite and Fracture Mechanism. *Composite Structures*, *184*, 545–553. doi:10.1016/j.compstruct.2017.10.020

López De Lacalle, N., Lamikiz, A., Campa, F. J., Valdivielso, A. F., & Etxeberria, I. (2009). Design and test of a multitooth tool for CFRP milling. *Journal of Composite Materials*, *43*(26), 3275–3290. doi:10.1177/0021998309345354

Lu, Z., & Yoneyama, T. (1999). Micro cutting in the micro lathe turning system. *International Journal of Machine Tools & Manufacture*, *39*(7), 1171–1183. doi:10.1016/S0890-6955(98)00092-3

Mahakur, V. K., Bhowmik, S., & Patowari, P. K. (2022). Machining parametric study on the natural fiber reinforced composites: A review. *Proceedings of the Institution of Mechanical Engineers. Part C, Journal of Mechanical Engineering Science*, *236*(11), 6232–6249. doi:10.1177/09544062211063752

Mahdi, M., & Zhang, L. (2001a). A finite element model for the orthogonal cutting of fiber-reinforced composite materials. *Journal of Materials Processing Technology*, *113*(1), 373–377. doi:10.1016/S0924-0136(01)00675-6

Mahdi, M., & Zhang, L. (2001b). An adaptive three-dimensional finite element algorithm for the orthogonal cutting of composite materials. *Journal of Materials Processing Technology*, *113*(1), 368–372. doi:10.1016/S0924-0136(01)00676-8

Makhdum, F., Phadnis, V. A., Roy, A., & Silberschmidt, V. V. (2014). Effect of ultrasonically-assisted drilling on carbon-fibre-reinforced plastics. *Journal of Sound and Vibration*, *333*(23), 5939–5952. doi:10.1016/j.jsv.2014.05.042

Malik, K., Ahmad, F., & Gunister, E. (2022). Drilling performance of natural fiber reinforced polymer composites: A review. *Journal of Natural Fibers*, *19*(12), 4761–4779. doi:10.1080/15440478.2020.1870624

Manjunath Patel, Jagadish, Kumar, & Naidu. (2020). Optimization of Abrasive Water Jet Machining for Green Composites Using Multi-Variant Hybrid Techniques. doi:10.1007/978-3-030-19638-7_6

Mansour, G., Kyratsis, P., Korlos, A., & Tzetzis, D. (2021). Investigation into the Effect of Cutting Conditions in Turning on the Surface Properties of Filament Winding GFRP Pipe Rings. *Machines*, *9*(1), 16. doi:10.3390/machines9010016

Mata, F., Hanafi, I., Beamud, E., Khamlichi, A., & Jabbouri, A. (2012). Modelling of machining force components during turning of PEEK CF30 by TiN coated cutting tools using artificial intelligence. *International Journal of Machining and Machinability of Materials*, *11*(3), 263. doi:10.1504/IJMMM.2012.046885

Mathew, J., Ramakrishnan, N., & Naik, N. K. (1999a). Investigations into the effect of geometry of a trepanning tool on thrust and torque during drilling of GFRP composites. *Journal of Materials Processing Technology*, *91*(1), 1–11. doi:10.1016/S0924-0136(98)00416-6

Mathew, J., Ramakrishnan, N., & Naik, N. K. (1999b). Trepanning on unidirectional composites : Delamination studies. *Composites. Part A, Applied Science and Manufacturing*, *30*(8), 951–959. doi:10.1016/S1359-835X(99)00012-3

Matsumoto, S., Sato, Y., Kamo, M., & Setaka, N. (1982). Vapor Deposition of Diamond Particles from Methane. *Japanese Journal of Applied Physics*, *21*(4), 183–185. doi:10.1143/JJAP.21.L183

Merino-Pérez, J. L., Royer, R., Merson, E., Lockwood, A., Ayvar-Soberanis, S., & Marshall, M. B. (2016). Influence of Workpiece Constituents and Cutting Speed on the Cutting Forces Developed in the Conventional Drilling of CFRP Composites. *Composite Structures*, *140*, 621–629. doi:10.1016/j.compstruct.2016.01.008

Mingzheng, L. I. U. (2023). Analysis of Grinding Mechanics and Improved Grinding Force Model Based on Randomized Grain Geometric Characteristics. *Chinese Journal of Aeronautics*, *36*(7), 160–193. doi:10.1016/j.cja.2022.11.005

Mittal, R. K., Kulkarni, S. S., & Singh, R. K. (2017). Effect of lubrication on machining response and dynamic instability in high-speed micromilling of Ti-6Al-4V. *Journal of Manufacturing Processes*, *28*, 413–421. doi:10.1016/j.jmapro.2017.04.007

Mkaddem, A., Ben Soussia, A., & El Mansori, M. (2013). Wear resistance of CVD and PVD multilayer coatings when dry cutting fiber reinforced polymers (FRP). *Wear*, *302*(1-2), 946–954. doi:10.1016/j.wear.2013.03.017

Mohamed, S. B., Rashid, R. A., Muhamad, M., & Ismail, J. (2019). Composite Materials and Types of Machining. In *SpringerBriefs in Applied Sciences and Technology* (pp. 1–14). Springer Verlag. doi:10.1007/978-981-13-1804-7_1

Mohanavel, V., Raja, T., Yadav, A., Ravichandran, M., & Winczek, J. (2022). Evaluation of Mechanical and Thermal Properties of Jute and Ramie Reinforced Epoxy-Based Hybrid Composites. *Journal of Natural Fibers*, *19*(14), 8022–8032. doi:10.1080/15440478.2021.1958432

Mohan, N., Ramachandra, A., & Kulkarni, S. (2005). Influence of process parameters on cutting force and torque during drilling of glass–fiber polyester reinforced composites. *Composite Structures*, *71*(3–4), 407–413. doi:10.1016/j.compstruct.2005.09.039

Mohanty, A. K., Misra, M., & Drzal, L. T. (2002). Sustainable Bio-Composites from Renewable Resources : Opportunities and Challenges in the Green Materials World. *Journal of Polymers and the Environment*, *10*(1), 19–26. doi:10.1023/A:1021013921916

Mohanty, A. K., Misra, M., & Hinrichsen, G. (2000). Biofibres, biodegradable polymers and biocomposites : An overview. *Macromolecular Materials and Engineering*, *276-277*(1), 1–24. doi:10.1002/(SICI)1439-2054(20000301)276:1<1::AID-MAME1>3.0.CO;2-W

Montesano, J., Fawaz, Z., & Bougherara, H. (2015). Non-Destructive Assessment of the Fatigue Strength and Damage Progression of Satin Woven Fiber Reinforced Polymer Matrix Composites. *Composites. Part B, Engineering*, *71*, 122–130. doi:10.1016/j.compositesb.2014.11.005

Muhamad Khairussaleh, N. K., Che Haron, C. H., & Ghani, J. A. (2016). Study on wear mechanism of solid carbide cutting tool in milling CFRP. *Journal of Materials Research*, *31*(13), 1893–1899. doi:10.1557/jmr.2016.21

Murphy, C., Byrne, G., & Gilchrist, M. D. (2002). The performance of coated tungsten carbide drills when machining carbon fibre-reinforced epoxy composite materials. *Proceedings of the Institution of Mechanical Engineers. Part B, Journal of Engineering Manufacture*, *216*(2), 143–152. doi:10.1243/0954405021519735

Naik, N. K., Asmelash, A., Kavala, V. R., & Ch, V. (2007). Interlaminar Shear Properties of Polymer Matrix Composites: Strain Rate Effect. *Mechanics of Materials*, *39*(12), 1043–1052. doi:10.1016/j.mechmat.2007.05.003

Nasir, A. A., Azmi, A. I., & Khalil, A. N. M. (2015). Parametric Study on the Residual Tensile Strength of Flax Natural Fibre Composites after Drilling Operation. *Procedia Manufacturing*, *2*, 97–101. doi:10.1016/j.promfg.2015.07.017

Negi, R. S., Prasad, L., Yadav, A., & Winczek, J. (2022). Physical and Mechanical Properties of Pinecone Scale Fiber/Vigna Mungo Powder Reinforced Polypropylene Based Hybrid Composites. *Journal of Natural Fibers*, *19*(15), 11458–11468. doi:10.1080/15440478.2022.2025983

Nieminen, I., Paro, J., & Kauppinen, V. (1996). High-speed milling of advanced materials. *Journal of Materials Processing Technology*, *56*(1-4), 24–36. doi:10.1016/0924-0136(95)01817-4

Nurazzi, N. M., Harussani, M. M., Aisyah, H. A., Ilyas, R. A., Norrrahim, M. N. F., Khalina, A., & Abdullah, N. (2021). Treatments of natural fiber as reinforcement in polymer composites—A short review. *Functional Composites and Structures*, *3*(2), 024002. doi:10.1088/2631-6331/abff36

Olonisakin, K., fan, M., Xin-Xiang, Z., Ran, L., Lin, W. S., Zhang, W., & Wenbin, Y. (2022). Key Improvements in Interfacial Adhesion and Dispersion of Fibers/Fillers in Polymer Matrix Composites; Focus on PLA Matrix Composites. *Composite Interfaces, 29*(10), 1071–1120. doi:10.1080/09276440.2021.1878441

Olsen, R. H., Dewes, R. C., & Aspinwall, D. K. (2004). Machining of electrically conductive CVD diamond tool blanks using EDM. *Journal of Materials Processing Technology, 149*(1), 627–632. doi:10.1016/j.jmatprotec.2003.11.052

Ouyang, W., Jiao, J., Xu, Z., Xia, H., Ye, Y., Zou, Q., Tian, R., & Sheng, L. (2021). Experimental study on CFRP drilling with the picosecond laser "double rotation" cutting technique. *Optics & Laser Technology, 142*, 107238. doi:10.1016/j.optlastec.2021.107238

Ozkan, D., Sabri Gok, M., Oge, M., & Cahit Karaoglanli, A. (2019). Milling Behavior Analysis of Carbon Fiber-Reinforced Polymer (CFRP) Composites. *Materials Today: Proceedings, 11*, 526–533. doi:10.1016/j.matpr.2019.01.024

Paige, R. W. (1983). *US Patent for Milling cutter for composite laminates Patent (Patent # 4,810,136 issued March 7, 1989)—Justia Patents Search* (Patent N° 4810136).

Palanikumar, K. (2006). Cutting Parameters Optimization for Surface Roughness in Machining of GFRP Composites using Taguchi's Method. *Journal of Reinforced Plastics and Composites, 25*(16), 1739–1751. doi:10.1177/0731684406068445

Palanikumar, K., Karunamoorthy, L., & Karthikeyan, R. (2006a). Assessment of factors influencing surface roughness on the machining of glass fiber-reinforced polymer composites. *Materials & Design, 27*(10), 862–871. doi:10.1016/j.matdes.2005.03.011

Palanikumar, K., Karunamoorthy, L., & Karthikeyan, R. (2006b). Parametric optimization to minimise the surface roughness on the machining of GFRP composites. *Journal of Materials Science and Technology, 22*(1), 66–72.

Palanikumar, K., Rubio, J. C., Abrão, A., Esteves, A., & Davim, J. P. (2008). Statistical analysis of delamination in drilling glass fiber-reinforced plastics (GFRP). *Journal of Reinforced Plastics and Composites, 27*(15), 1615–1623. doi:10.1177/0731684407083012

Pan, C.T., & Hocheng, H. (1996). The anisotropic heat-affected zone in the laser grooving or fiber-reinforced composite material. *Journal of Materials Processing Technology, 62*(1-3), 54 – 60. doi:10.1016/0924-0136(95)02192-2

Parida, A. K., Das, R., Sahoo, A. K., & Routara, B. C. (2014). Optimization of Cutting Parameters for Surface Roughness in Machining of gfrp Composites with Graphite/fly Ash Filler. *Procedia Materials Science, 6*, 1533–1538. doi:10.1016/j.mspro.2014.07.134

Patel, G. C. M., Chate, G. R., Parappagoudar, M. B., & Gupta, K. (2020). Introduction to Hard Materials and Machining Methods. In *SpringerBriefs in Applied Sciences and Technology* (pp. 1–24). Springer. doi:10.1007/978-3-030-40102-3_1

Patel, J. S., Boddu, V. M., Brenner, M. W., & Kumar, A. (2016). Effect of Fabric Structure and Polymer Matrix on Flexural Strength, Interlaminar Shear Stress, and Energy Dissipation of Glass Fiber-Reinforced Polymer Composites. *Textile Research Journal*, *86*(2), 127–137. doi:10.1177/0040517515586165

Patel, R. V., Yadav, A., & Winczek, J. (2023). Physical, Mechanical, and Thermal Properties of Natural Fiber-Reinforced Epoxy Composites for Construction and Automotive Applications. *Applied Sciences (Basel, Switzerland)*, *13*(8), 5126. doi:10.3390/app13085126

Patel, S. R., & Shaikh, A. A. (2013). A review on machining of fiber reinforced plastic using abrasive waterjet. *International Journal of Innovative Technology & Adaptive Management*, *1*(3).

Patole, P. B., Pol, G. J., Desai, A. A., & Kamble, S. B. (2021). Analysis of Surface Roughness and Cutting Force under MQL Turning Using Nano Fluids. *Materials Today: Proceedings*, *45*, 5684–5688. doi:10.1016/j.matpr.2021.02.501

Petropoulos, G. P., Pandazaras, C. N., Vaxevanidis, N. M., & Antoniadis, A. (2006). Multi-parameter identification and control of turned surface textures. *International Journal of Advanced Manufacturing Technology*, *29*(1), 118–128. doi:10.1007/s00170-004-2483-y

Petropoulos, G., & Pandazaras, C. (2003). Evaluating the real profile length in turning of carbon steels. *Industrial Lubrication and Tribology*, *55*(3), 128–136. doi:10.1108/00368790310470967

Picard, Y. N., Adams, D. P., Vasile, M. J., & Ritchey, M. B. (2003). Focused ion beam-shaped microtools for ultra-precision machining of cylindrical components. *Precision Engineering*, *27*(1), 59–69. doi:10.1016/S0141-6359(02)00188-5

Polini, R. (2006). Chemically vapour deposited diamond coatings on cemented tungsten carbides: Substrate pretreatments, adhesion and cutting performance. *Thin Solid Films*, *515*(1), 4–13. doi:10.1016/j.tsf.2005.12.042

Prakash, M., & Dileep Aditya Dhar, P. V. S. (2018). Investigation on the Effect of Drilling Parameters on the Tool Wear and Delamination of Glass Fibre-Reinforced Polymer Composite Using Vibration Signal Analysis. *Journal of Composite Materials*, *52*(12), 1641–1648. doi:10.1177/0021998317728109

Pramod, R., Basavarajappa, S., Veeresh Kumar, G. B., & Chavali, M. (2022). Drilling Induced Delamination Assessment of Nanoparticles Reinforced Polymer Matrix Composites. *Proceedings of the Institution of Mechanical Engineers. Part C, Journal of Mechanical Engineering Science*, *236*(6), 2931–2948. doi:10.1177/09544062211030967

Prasad, L., Bairwan, R., Yadav, A., Kumar, A., Kumar, V., & Winczek, J. (2022). Evaluation of Physical, Mechanical, and Wear Properties of Jatropha Shell Powder Reinforced Epoxy Glass Fiber Composites. *Journal of Natural Fibers*, *19*(15), 12195–12207. doi:10.1080/15440478.2022.2054892

Prasad, L., Kapri, P., Patel, R. V., Yadav, A., & Winczek, J. (2023). Physical and Mechanical Behavior of Ramie and Glass Fiber Reinforced Epoxy Resin-Based Hybrid Composites. *Journal of Natural Fibers*, *20*(2), 2234080. Advance online publication. doi:10.1080/15440478.2023.2234080

Prasad, L., Kumain, A., Patel, R. V., Yadav, A., & Winczek, J. (2022). Physical and Mechanical Behavior of Hemp and Nettle Fiber-Reinforced Polyester Resin-Based Hybrid Composites. *Journal of Natural Fibers*, *19*(7), 2632–2647. doi:10.1080/15440478.2020.1821284

Prasad, L., Kumar, S., Patel, R. V., Yadav, A., Kumar, V., & Winczek, J. (2020). Physical and Mechanical Behaviour of Sugarcane Bagasse Fibre-Reinforced Epoxy Bio-Composites. *Materials (Basel)*, *13*(23), 5387. doi:10.3390/ma13235387 PMID:33260868

Prasad, L., Singh, G., Yadav, A., Kumar, V., & Kumar, A. (2019). Properties of Functionally Gradient Composites Reinforced with Waste Natural Fillers. *Acta Periodica Technologica*, (50), 250–259. doi:10.2298/APT1950250P

Prasad, L., Singh, V., Patel, R. V., Yadav, A., Kumar, V., & Winczek, J. (2022). Physical and Mechanical Properties of Rambans (Agave) Fiber Reinforced with Polyester Composite Materials. *Journal of Natural Fibers*, *19*(13), 6104–6118. doi:10.1080/15440478.2021.1904481

Prasanth, I. S. N. V. R., Ravishankar, D. V., & Manzoor Hussain, M. (2017). Analysis of Milling Process Parameters and Their Influence on Glass Fiber Reinforced Polymer Composites. *International Journal of Engineering*, *30*(7), 1074–1080. doi:10.5829/ije.2017.30.07a.17

Puw, H. Y., & Hocheng, H. (1993). Machinability Test Of Carbon Fiber-Reinforced Plastics In Milling. *Materials and Manufacturing Processes*, *8*(6), 717–729. doi:10.1080/10426919308934875

Qiu, X., Li, P., & Niu, Q. (2019). New Compound Drill Bit for Damage Reduction in Drilling CFRP. *International Journal of Precision Engineering and Manufacturing – Green Technology*, *6*(1), 75 – 87. . doi:10.1007/s40684-019-00026-3

Rahman, M., Ramakrishna, S., Prakash, J. R. S., & Tan, D. C. G. (1999). Machinability study of carbon fiber reinforced composite. *Journal of Materials Processing Technology*, *89-90*, 292–297. doi:10.1016/S0924-0136(99)00040-0

Rahman, M., Ramakrishna, S., & Thoo, H. C. (1999). Machinability Study of Carbon/Peek Composites. *Machining Science and Technology*, *3*(1), 49–59. doi:10.1080/10940349908945682

Raja, Satheesh, Manisekar, & Manikandan. (2013). Effect of Fly Ash Filler Size on Mechanical Properties of Polymer Matrix Composites. *International Journal of Mining, Metallurgy & Mechanical Engineering*, *1*(1), 34–37.

Rajak, D. K., Pagar, D. D., Kumar, R., & Pruncu, C. I. (2019). Recent progress of reinforcement materials : A comprehensive overview of composite materials. *Journal of Materials Research and Technology*, *8*(6), 6354–6374. doi:10.1016/j.jmrt.2019.09.068

Rajak, D. K., Pagar, D. D., Menezes, P. L., & Linul, E. (2019). Fiber-Reinforced Polymer Composites : Manufacturing, Properties, and Applications. *Polymers*, *11*(10), 10. Advance online publication. doi:10.3390/polym11101667 PMID:31614875

Rajakumar, I., Raguraman, D., Isaac, J. S., Suthan, R., Bhattacharya, S., Seikh, A. H., Khan, S. M. A., & Raghavan, I. K. (2022). Mechanical Properties of Polymer Composites Reinforced with Alkaline-Treated Natural Fibre. *Advances in Polymer Technology*, *2022*, 1–7. doi:10.1155/2022/1458547

Rajamurugan, T., Rajaganapathy, C., Jani, S. P., Gurram, C. S., Allasi, H. L., & Damtew, S. Z. (2022). Analysis of drilling of coir fiber-reinforced polyester composites using multifaceted drill bit. *Advances in Materials Science and Engineering*, *2022*, 1–9. doi:10.1155/2022/9481566

Rajasekaran, T., Palanikumar, K., & Vinayagam, B. K. (2012b). Turning CFRP Composites with Ceramic tool for Surface Roughness Analysis. In R. Rajesh, K. Ganesh, & S. C. L. Koh (Éds.), *International Conference on Modelling Optimization and Computing* (Vol. 38, p. 2922-2929). Elsevier Science Bv. 10.1016/j.proeng.2012.06.341

Rajasekaran, T., Palanikumar, K., & Vinayagam, B. (2012a). Experimental investigation and analysis in turning of CFRP composites. *Journal of Composite Materials*, *46*(7), 809–821. doi:10.1177/0021998311410500

Ramesh, M. V., Seetharamu, K. N., Ganesan, N., & Sivakumar, M. S. (1998). Analysis of machining of FRPs using FEM. *International Journal of Machine Tools & Manufacture*, *38*(12), 1531–1549. doi:10.1016/S0890-6955(98)00008-X

Ramulu, M. (1997). Machining and surface integrity of fibre-reinforced plastic composites. *Sadhana*, *22*(3), 449–472. doi:10.1007/BF02744483

Ramulu, M., Faridnia, M., Garbini, J. L., & Jorgensen, J. E. (1989). Machining of graphite/epoxy materials with polycrystalline diamond (PCD) tools. In M. Ramulu & M. Hashish (Eds.), *Machining Characteristics of Advanced Materials* (Vol. 16, pp. 33–40). ASME.

Ramulu, M., Kim, D., & Choi, G. (2003). Frequency analysis and characterization in orthogonal cutting of glass fiber reinforced composites. *Composites. Part A, Applied Science and Manufacturing*, *34*(10), 949–962. doi:10.1016/S1359-835X(03)00203-3

Ramulu, M., Wern, C. W., & Garbini, J. L. (1993). Effect of fibre direction on surface roughness measurements of machined graphite/epoxy composite. *Composites Manufacturing*, *4*(1), 39–51. doi:10.1016/0956-7143(93)90015-Z

Rance, J., Hall, S., de Bartolomeis, A., & Shokrani, A. (2019). Future Direction of the Sustainable Turning of Difficult-to-Machine Materials. *Sustainable Design and Manufacturing*, *2019*, 111–120. doi:10.1007/978-981-13-9271-9_11

Rao, Y. S., Mohan, N. S., Shetty, N., & Shivamurthy, B. (2019). Drilling and Structural Property Study of Multi-Layered Fiber and Fabric Reinforced Polymer Composite - a Review. *Materials and Manufacturing Processes*, *34*(14), 1549–1579. doi:10.1080/10426914.2019.1686522

Raspall, F., Velu, R., & Vaheed, N. M. (2019). Fabrication of complex 3D composites by fusing automated fiber placement (AFP) and additive manufacturing (AM) technologies. *Advanced Manufacturing: Polymer & Composites Science*, *5*(1), 6–16. doi:10.1080/20550340.2018.1557397

Ravindra, H. V., Srinivasa, Y. G., & Krishnamurthy, R. (1993a). Modelling of tool wear based on cutting forces in turning. *Wear, 169*(1), 25–32. doi:10.1016/0043-1648(93)90387-2

Ravindra, H. V., Srinivasa, Y. G., & Krishnamurthy, R. (1993b). Tool wear monitoring in turning using a pattern recognition technique. *Journal of Materials Processing Technology, 37*(1), 731–740. doi:10.1016/0924-0136(93)90132-P

Reineck, I., Sjöstrand, M. E., Karner, J., & Pedrazzini, M. (1996). Diamond coated cutting tools. *International Journal of Refractory & Hard Metals, 14*(1), 187–193. doi:10.1016/0263-4368(96)83433-0

Ribeiro, M. V., & Coppini, N. L. (1999). An applied database system for the optimization of cutting conditions and tool selection. *Journal of Materials Processing Technology, 92-93*, 371–374. doi:10.1016/S0924-0136(99)00233-2

Ribeiro, P. de O., Krahl, P. A., Carrazedo, R., & Bernardo, L. F. A. (2023). Modeling the Tensile Behavior of Fiber-Reinforced Strain-Hardening Cement-Based Composites : A Review. *Materials (Basel), 16*(9), 9. Advance online publication. doi:10.3390/ma16093365 PMID:37176256

Robinson, D. N., Binienda, W. K., & Ruggles, M. B. (2003). Creep of Polymer Matrix Composites. I: Norton/Bailey Creep Law for Transverse Isotropy. *Journal of Engineering Mechanics, 129*(3), 310–317. doi:10.1061/(ASCE)0733-9399(2003)129:3(310)

Ross, N. S., Sheeba, P. T., & Shibi, C. S. (2023). A novel approach of tool condition monitoring in sustainable machining of Ni alloy with transfer learning models. *Journal of Intelligent Manufacturing*. Advance online publication. doi:10.1007/s10845-023-02074-8

Rusinek, R. (2010). Cutting process of composite materials : An experimental study. *International Journal of Non-linear Mechanics, 45*(4), 458–462. doi:10.1016/j.ijnonlinmec.2010.01.004

Ruszaj, A., Gawlik, J., & Skoczypiec, S. (2016). Electrochemical Machining – Special Equipment and Applications in Aircraft Industry. *Management and Production Engineering Review, 7*(2), 34–41. doi:10.1515/mper-2016-0015

S, S. R., R, B. J., E, C. S., J, P. C. S., G, S. C., A, W., H, W. C. J., & N, D. C. (1998). CVD diamond : A new engineering material for thermal, dielectric and optical applications. *Industrial Diamond Review, 58*(578), 69-71.

Sabet, F. A., Su, F. Y., McKittrick, J., & Jasiuk, I. (2018). Mechanical Properties of Model Two-Phase Composites with Continuous Compared to Discontinuous Phases. *Advanced Engineering Materials, 20*(10), 1800505. doi:10.1002/adem.201800505

Saikrupa, C. H., ChandraMohan Reddy, G., & Venkatesh, S. (2023). *Experimental investigation of cutting conditions in turning of Al-MMC.* Materials Today: Proceedings, 76, 488–494. doi:. matpr.2022.11.154 doi:10. 1016/j

Sakuma, K., & Seto, M. (1983). Tool wear in cutting glass-fiber-reinforced plastics : The relation between fiber orientation and tool wear. *Bulletin of the JSME, 26*(218), 1420–1427. doi:10.1299/jsme1958.26.1420

Sa, M.-W. (2020). Comparison Study on Side Milling of CFRP with AlCrN-based, Diamond-Like-Carbon(DLC), and Diamond-Coated End Mill. *The Korean Society of Manufacturing Process Engineers*, *19*(4), 9–15. doi:10.14775/ksmpe.2020.19.04.009

Santhanakrishnan, G., Krishnamurthy, R., & Malhotra, S. K. (1988). Machinability characteristics of fibre reinforced plastics composites. *Journal of Mechanical Working Technology*, *17*, 195–204. doi:10.1016/0378-3804(88)90021-6

Sarangi, S. K., Chattopadhyay, A., & Chattopadhyay, A. K. (2008). Effect of pretreatment, seeding and interlayer on nucleation and growth of HFCVD diamond films on cemented carbide tools. *International Journal of Refractory & Hard Metals*, *26*(3), 220–231. doi:10.1016/j.ijrmhm.2007.05.002

Sarangi, S. K., Chattopadhyay, A., & Chattopadhyay, A. K. (2012). Influence of process parameters on growth of diamond crystal on cemented carbide substrates by HFCVD system. *International Journal of Refractory & Hard Metals*, *31*, 1–13. doi:10.1016/j.ijrmhm.2011.07.007

Sardiñas, R. Q., Reis, P., & Davim, J. P. (2006). Multi-objective optimization of cutting parameters for drilling laminate composite materials by using genetic algorithms. *Composites Science and Technology*, *66*(15), 3083–3088. doi:10.1016/j.compscitech.2006.05.003

Schindler, S., Zimmermann, M., Aurich, J. C., & Steinmann, P. (2013). Modeling deformations of the workpiece and removal of material when turning. *Procedia CIRP*, *8*, 39–44. doi:10.1016/j.procir.2013.06.062

Schuh, C. A., Hufnagel, T. C., & Ramamurty, U. (2007). Mechanical behavior of amorphous alloys. *Acta Materialia*, *55*(12), 4067–4109. doi:10.1016/j.actamat.2007.01.052

Selvakumar, S., Sreebalaji, V. S., & Ravikumar, K. (2021). Machinability Analysis and Optimization in Micro turning on tool wear for Titanium Alloy. *Materials and Manufacturing Processes*, *36*(7), 792–802. doi:10.1080/10426914.2020.1866198

Seydibeyoğlu, M. Ö., Dogru, A., Wang, J., Rencheck, M., Han, Y., Wang, L., Seydibeyoğlu, E. A., Zhao, X., Ong, K., Shatkin, J. A., Shams Es-haghi, S., Bhandari, S., Ozcan, S., & Gardner, D. J. (2023). Review on Hybrid Reinforced Polymer Matrix Composites with Nanocellulose, Nanomaterials, and Other Fibers. *Polymers*, *15*(4), 4. Advance online publication. doi:10.3390/polym15040984 PMID:36850267

Shanmugam, D. K., Chen, F. L., Siores, E., & Brandt, M. (2002). Comparative study of jetting machining technologies over laser machining technology for cutting composite materials. *Composite Structures*, *57*(1-4), 289–296. doi:10.1016/S0263-8223(02)00096-X

Sharma, S. P., & Lakkad, S. C. (2011). Effect of CNTs Growth on Carbon Fibers on the Tensile Strength of CNTs Grown Carbon Fiber-Reinforced Polymer Matrix Composites. *Composites. Part A, Applied Science and Manufacturing*, *42*(1), 8–15. doi:10.1016/j.compositesa.2010.09.008

Shastri, A., Nargundkar, A., & Kulkarni, A. J. (2021). Optimization of Micro-turning Process. In *Socio-Inspired Optimization Methods for Advanced Manufacturing Processes* (pp. 119–126). Springer. doi:10.1007/978-981-15-7797-0_9

Shaw, M. C. (2005). *Metal Cutting Principles* (2nd ed.). Oxford University Press.

Sheikh-Ahmad, J. Y., Stewart, J. S., & Feld, H. (2003). Failure characteristics of diamond-coated carbides in machining wood-based composites. *Wear*, *255*(7), 1433–1437. doi:10.1016/S0043-1648(03)00179-0

Sheikh-Ahmad, J., & Yadav, R. (2005). Force prediction in milling of carbon fiber reinforced polymers. *American Society of Mechanical Engineers, Materials Division (Publication). MD Medical Newsmagazine*, *100*(MD), 429–436. doi:10.1115/IMECE2005-81909

Sheikh-Ahmed, J. Y. (2009). *Machining of polymer composites*. Springer. doi:10.1007/978-0-387-68619-6

Shyha, I. S., Aspinwall, D. K., Soo, S. L., & Bradley, S. (2009). Drill geometry and operating effects when cutting small diameter holes in CFRP. *International Journal of Machine Tools & Manufacture*, *49*(12-13), 1008–1014. doi:10.1016/j.ijmachtools.2009.05.009

Singh, R., Gupta, M. K., Sarikaya, M., Mia, M., & Garcia-Collado, A. (2021). Evaluation of machinability-based sustainability indicators in the eco-benign turning of Ti3Al2.5V alloy with textured tools. *International Journal of Advanced Manufacturing Technology*, *116*(9), 3051–3061. doi:10.1007/s00170-021-07667-2

Sivakiran, G., Gangwal, Y., Venkatachalam, G., Pandivelan, C., & Ayyappan, S. (2018). Investigations on Machining of Banana Fibre Reinforced Hybrid Polymer Matrix Composite Materials. *Materials Today: Proceedings*, *5*(2), 7908–7914. doi:10.1016/j.matpr.2017.11.472

Slamani, M., Chafai, H., & Chatelain, J. F. (2022). Effect of Milling Parameters on the Surface Quality of a Flax Fiber-Reinforced Polymer Composite. *Proceedings of the Institution of Mechanical Engineers, Part E: Journal of Process Mechanical Engineering*. 10.1177/09544089221126087

Slamani, M., Gauthier, S., & Chatelain, J.-F. (2015). A study of the combined effects of machining parameters on cutting force components during high-speed robotic trimming of CFRPs. *Measurement*, *59*, 268–283. doi:10.1016/j.measurement.2014.09.052

Somireddy, M., & Czekanski, A. (2020). Anisotropic material behavior of 3D printed composite structures – Material extrusion additive manufacturing. *Materials & Design*, *195*, 108953. doi:10.1016/j.matdes.2020.108953

Sommer, M., Haubner, R., & Lux, B. (2000). Diamond deposition on copper treated hardmetal substrates. *Diamond and Related Materials*, *9*(3-6), 351–357. doi:10.1016/S0925-9635(99)00250-2

Song, D.-Y., & Otani, N. (1997). Fatigue Life Prediction of Cross-Ply Composite Laminates. *Materials Science and Engineering A*, *238*(2), 329–335. doi:10.1016/S0921-5093(97)00427-9

Sorrentino, L., Turchetta, S., & Bellini, C. (2017). Analysis of carbon fibre reinforced polymers milling by diamond electroplated tool. *Diamond and Related Materials*, *76*, 184–190. doi:10.1016/j.diamond.2017.05.005

Sousa, V. F. C., & Silva, F. J. G. (2020). Recent Advances in Turning Processes Using Coated Tools—A Comprehensive Review. *Metals*, *10*(2), 2. Advance online publication. doi:10.3390/met10020170

Soutis, C. (2005). Fibre reinforced composites in aircraft construction. *Progress in Aerospace Sciences*, *41*(2), 143–151. doi:10.1016/j.paerosci.2005.02.004

Spencer, A. J. M. (1972). *Deformations of Fibre-reinforced Materials*. Clarendon Press.

Spitsyn, B., Bouilov, L., & Deryagin, B. (1988). Diamond and diamond-like films: Deposition from the vapour phase, structure and properties. *Progress in Crystal Growth and Characterization*, *17*(2), 79–170. doi:10.1016/0146-3535(88)90001-9

Spur, G., & Lachmund, U. (1999). Turning of fiber-reinforced plastics. In Machining of Ceramics and Composites. Marcel Dekker.

Sreejith, P. S., Krishnamurthy, R., & Malhotra, S. K. (2007). Effect of specific cutting pressure and temperature during machining of carbon/phenolic ablative composite using PCBN tools. *Journal of Materials Processing Technology*, *183*(1), 88–95. doi:10.1016/j.jmatprotec.2006.10.003

Sreejith, P. S., Krishnamurthy, R., Malhotra, S. K., & Narayanasamy, K. (2000). Evaluation of PCD tool performance during machining of carbon/phenolic ablative composites. *Journal of Materials Processing Technology*, *104*(1), 53–58. doi:10.1016/S0924-0136(00)00549-5

Stawiszynski, B., Protz, F., & Uhlmann, E. (2021). Selection of diamond coated silicon nitrides for use as ceramic end mills when machining glass and carbon fiber reinforced plastics. *International Journal of Refractory & Hard Metals*, *96*, 105487. doi:10.1016/j.ijrmhm.2021.105487

Su, F., Zheng, L., Sun, F., Wang, Z., Deng, Z., & Qiu, X. (2018). Novel drill bit based on the step-control scheme for reducing the CFRP delamination. *Journal of Materials Processing Technology*, *262*, 157–167. doi:10.1016/j.jmatprotec.2018.06.037

Suh, J. D., Lee, D. G., & Kegg, R. (2002). Composite machine tool structures for high speed milling machines. *CIRP Annals - Manufacturing Technology, 51*(1), 285-288. doi:10.1016/S0007-8506(07)61518-2

Sundi, S. A., Izamshah, R., Kasim, M. S., Jaafar, M. F., & Hassan, M. H. (2021). Milling/Trimming of Carbon Fiber Reinforced Polymers (CFRP): Recent Advances in Tool Geometrical Design. In Machining and Machinability of Fiber Reinforced Polymer Composites (p. 101-128). doi:10.1007/978-981-33-4153-1_4

Sun, Z., Geng, D., Zheng, W., Liu, Y., Liu, L., Ying, E., Jiang, X., & Zhang, D. (2023). An innovative study on high-performance milling of carbon fiber reinforced plastic by combining ultrasonic vibration assistance and optimized tool structures. *Journal of Materials Research and Technology*, *22*, 2131–2146. doi:10.1016/j.jmrt.2022.12.054

Szallies, K., Siebert, N., & Bergmann, J. P. (2017). Low Frequency Oscillated Milling of Carbon Fiber-reinforced Plastics. *Procedia CIRP*, *66*, 153–158. doi:10.1016/j.procir.2017.03.298

Takeyama, H., & Iijima, N. (1988). Machinability of Glassfiber Reinforced Plastics and Application of Ultrasonic Machining. *CIRP Annals*, *37*(1), 93–96. doi:10.1016/S0007-8506(07)61593-5

Tao, C., Qiu, J., Yao, W., & Ji, H. (2016). The Effect of Drilling-Induced Delamination on Tensile Strength and Prediction of Residual Strength of Carbon Fiber-Reinforced Polymer Laminate. *Journal of Composite Materials*, *50*(24), 3373–3384. doi:10.1177/0021998315620292

Teti, R. (2002). Machining of composite materials. *CIRP Annals - Manufacturing Technology*, *51*(2), 611-634. doi:10.1016/S0007-8506(07)61703-X

Tian, J., Wu, F., Zhang, P., Lin, B., Liu, T., & Liu, L. (2019). The coupling effect and damage analysis when drilling GFRP laminates using candlestick drills. *International Journal of Advanced Manufacturing Technology*, *102*(1), 519–531. doi:10.1007/s00170-018-03216-6

Toug, L., Mouritz, A., & Bannister, M. (2002). *3D fiber reinforced polymer composites*. Elsevier.

Trent, E., & Wright, P. (2001). *Metal cutting* (4th ed.). Butterworth – Heinemann.

Tsao, C. C. (2008). Investigation into the effects of drilling parameters on delamination by various step-core drills. *Journal of Materials Processing Technology*, *206*(1-3), 405–411. doi:10.1016/j.jmatprotec.2007.12.057

Tsao, C. C., & Chiu, Y. C. (2011). Evaluation of drilling parameters on thrust force in drilling carbon fiber reinforced plastic (CFRP) composite laminates using compound core-special drills. *International Journal of Machine Tools & Manufacture*, *51*(9), 740–744. doi:10.1016/j.ijmachtools.2011.05.004

Tsao, C. C., & Hocheng, H. (2003). The effect of chisel length and associated pilot hole on delamination when drilling composite materials. *International Journal of Machine Tools & Manufacture*, *43*(11), 1087–1092. doi:10.1016/S0890-6955(03)00127-5

Turchetta, S., Carrino, L., & Polini, W. (2005). CVD diamond insert in stone cutting. *Diamond and Related Materials*, *14*(3), 641–645. doi:10.1016/j.diamond.2004.10.031

Turki, Y., Habak, M., Velasco, R., Aboura, Z., Khellil, K., & Vantomme, P. (2014). Experimental investigation of drilling damage and stitching effects on the mechanical behavior of carbon/epoxy composites. *International Journal of Machine Tools & Manufacture*, *87*, 61–72. doi:10.1016/j.ijmachtools.2014.06.004

Uhlmann, E., & Protz, F. (2019). Ultrasonic assisted peripheral milling of fiber reinforced plastics in consideration of clamping and cutting edge geometry. *Procedia Manufacturing*, *40*, 56–63. doi:10.1016/j.promfg.2020.02.011

Uhlmann, E., Protz, F., Stawiszynski, B., & Heidler, S. (2017). Ultrasonic Assisted Milling of Reinforced Plastics. *Procedia CIRP*, *66*, 164–168. doi:10.1016/j.procir.2017.03.278

Uhlmann, E., Sammler, F., Richarz, S., Reucher, G., Hufschmied, R., Frank, A., Stawiszynski, B., & Protz, F. (2016). Machining of Carbon and Glass Fibre Reinforced Composites. *Procedia CIRP*, *46*, 63–66. doi:10.1016/j.procir.2016.03.197

Ullen, N. B., Hasak, D. S. M. A., & Dirikolu, M. H. (2020). Factors influencing the machinability during turning sinter-hardened Cu-Ni-Mo based steel : Dependency on cutting speed, feed rate and cutting depth. *Journal of Engineering Research*, *8*(4), 4. Advance online publication. doi:10.36909/jer.v8i4.9031

Ullram, S., & Haubner, R. (2006). Temperature pre-treatments of hardmetal substrates to reduce the cobalt content and improve diamond deposition. *Diamond and Related Materials*, *15*(4-8), 994–999. doi:10.1016/j.diamond.2005.12.055

Ullrich, K., von Elling, M., Gutzeit, K., Dix, M., Weigold, M., Aurich, J. C., Wertheim, R., Jawahir, I. S., & Ghadbeigi, H. (2024). AI-Based Optimisation of Total Machining Performance: A Review. *CIRP Journal of Manufacturing Science and Technology*, *50*, 40–54. doi:10.1016/j.cirpj.2024.01.012

UMECO. (2013). *An Introduction to Advanced Composites and Prepreg Technology*. Available from https://www.yumpu.com/en/document/read/9334619/introduction-to-advanced-composites-and-prepreg-technology

Urresti, I., Llanos, I., de Lacalle, L. N. L., & Zelaieta, O. (2022). Study of the surface integrity during CFRP trimming: Tool material and geometry, fiber orientation and tool wear effect analysis. *Procedia CIRP*, *108*, 660–664. doi:10.1016/j.procir.2022.03.102

Vandevelde, T.C.S., Vandierendonck, K., & Van Stappen, M. (1999). Cutting applications of DLC, hard carbon and diamond films. *Surface and Coating Technology, 113*(1 – 2), 80 – 85. doi:10.1016/S0257-8972(98)00831-7

Varatharajan, R., Malhotra, S. K., Vijayaraghavan, L., & Krishnamurthy, R. (2006). Mechanical and machining characteristics of GF/PP and GF/Polyester composites. *Materials Science and Engineering B*, *132*(1), 134–137. doi:10.1016/j.mseb.2006.02.010

Varma, M., Chandran, S., Vijay Kumar, V., Suyambulingam, I., & Siengchin, S. (2024). A Comprehensive Review on the Machining and Joining Characteristics of Natural Fiber-reinforced Polymeric Composites. *Polymer Composites*, *45*(6), 4850–4875. Advance online publication. doi:10.1002/pc.28148

Vigneshwaran, S., Uthayakumar, M., & Arumugaprabu, V. (2018). Abrasive water jet machining of fiber-reinforced composite materials. *Journal of Reinforced Plastics and Composites*, *37*(4), 230–237. doi:10.1177/0731684417740771

Voss, R., Henerichs, M., & Kuster, F. (2016). Comparison of conventional drilling and orbital drilling in machining carbon fibre reinforced plastics (CFRP). *CIRP Annals - Manufacturing Technology, 65*(1), 137-140. doi:10.1016/j.cirp.2016.04.001

Voss, H., & Friedrich, K. (1987). On the wear behaviour of short-fibre-reinforced peek composites. *Wear*, *116*(1), 1–18. doi:10.1016/0043-1648(87)90262-6

Wambua, P., Ivens, J., & Verpoest, I. (2003). Natural fibres : Can they replace glass in fibre reinforced plastics? *Composites Science and Technology*, *63*(9), 1259–1264. doi:10.1016/S0266-3538(03)00096-4

Wang, Z., Ma, Y., Yuan, B., Wu, C., Li, C., & Sun, S. (2023). Development of Laser Processing Carbon-Fiber-Reinforced Plastic. In Sensors (Basel, Switzerland) (Vol. 23, Numéro 7). Sensors (Basel). doi:10.3390/s23073659

Wang, C., Liu, G., An, Q., & Chen, M. (2017). Occurrence and formation mechanism of surface cavity defects during orthogonal milling of CFRP laminates. *Composites. Part B, Engineering*, *109*, 10–22. doi:10.1016/j.compositesb.2016.10.015

Wang, C., Wen, L., Ming, W., An, Q., & Chen, M. (2018). Experimental study on effects of fiber cutting angle in milling of high-strength unidirectional carbon fiber–reinforced polymer laminates. *Proceedings of the Institution of Mechanical Engineers. Part B, Journal of Engineering Manufacture*, *232*(10), 1813–1824. doi:10.1177/0954405416679868

Wang, D. H., Ramulu, M., & Arola, D. (1995a). Orthogonal cutting mechanisms of graphite/epoxy composite. Part I : Unidirectional laminate. *International Journal of Machine Tools & Manufacture*, *35*(12), 1623–1638. doi:10.1016/0890-6955(95)00014-O

Wang, D. H., Ramulu, M., & Arola, D. (1995b). Orthogonal cutting mechanisms of graphite/epoxy composite. Part II : Multi-directional laminate. *International Journal of Machine Tools & Manufacture*, *35*(12), 1639–1648. doi:10.1016/0890-6955(95)00015-P

Wang, F. (2022). Effects of cutting speed and fiber orientation on tool wear and machining quality in milling CFRP with PCD cutter. *Proceedings of the Institution of Mechanical Engineers. Part B, Journal of Engineering Manufacture*.

Wang, F. J., Yan, J. B., Zhao, M., Wang, D., Wang, X. N., & Hao, J. X. (2020). Surface damage reduction of dry milling carbon fiber reinforced plastic/polymer using left–right edge milling tool. *Journal of Reinforced Plastics and Composites*, *39*(11-12), 409–421. doi:10.1177/0731684420912018

Wang, H., Ning, F., Hu, Y., Fernando, P. K. S. C., Pei, Z. J., & Cong, W. (2016). Surface grinding of carbon fiber-reinforced plastic composites using rotary ultrasonic machining : Effects of tool variables. *Advances in Mechanical Engineering*, *8*(9), 1–14. doi:10.1177/1687814016670284

Wang, H., Qin, X., Wu, D., & Song, A. (2018). Optimization of Cutting Parameters in Helical Milling of Carbon Fiber Reinforced Polymer. *Transactions of Tianjin University*, *24*(1), 91–100. doi:10.1007/s12209-017-0079-5

Wang, H., Sun, J., Li, J., Lu, L., & Li, N. (2016). Evaluation of cutting force and cutting temperature in milling carbon fiber-reinforced polymer composites. *International Journal of Advanced Manufacturing Technology*, *82*(9-12), 1517–1525. doi:10.1007/s00170-015-7479-2

Wang, J. (1999). A machinability study of polymer matrix composites using abrasive waterjet cutting technology. *Journal of Materials Processing Technology*, *94*(1), 30–35. doi:10.1016/S0924-0136(98)00443-9

Wang, Q., Wu, Y., Lu, D., Bitoh, T., & Feng, M. (2018). Experimental investigation on the tilt helical milling of carbon fiber reinforced plastics (CFRP). *Key Engineering Materials*, *792*, 173–178. . doi:10.4028/www.scientific.net/KEM.792.173

Wang, X. B., & Zhao, C. Y. (2014). Design of special ultrasonic device in composite materials processing. *Advanced Materials Research*, *1027*, 331–335. . doi:10.4028/www.scientific.net/AMR.1027.331

Wang, X. M., & Zhang, L. C. (2003). An experimental investigation into the orthogonal cutting of unidirectional fibre reinforced plastics. *International Journal of Machine Tools & Manufacture*, *43*(10), 1015–1022. doi:10.1016/S0890-6955(03)00090-7

Wang, X., Kwon, P. Y., Sturtevant, C., Kim, D. D.-W., & Lantrip, J. (2013). Tool wear of coated drills in drilling CFRP. *Journal of Manufacturing Processes*, *15*(1), 127–135. doi:10.1016/j.jmapro.2012.09.019

Wang, X., Shen, X., Yang, G., & Sun, F. (2017). Evaluation of boron-doped- microcrystalline/nanocrystalline diamond composite coatings in drilling of CFRP. *Surface and Coatings Technology*, *330*, 149–162. doi:10.1016/j.surfcoat.2017.10.002

Wang, X., Shen, X., Zeng, C., & Sun, F. (2018). Combined influences of tool shape and as-deposited diamond film on cutting performance of drills for CFRP machining. *Surface and Coatings Technology*, *347*, 390–397. doi:10.1016/j.surfcoat.2018.05.024

Wang, X., Zhang, J., & Sun, F. (2014). Fracture and solid particle erosion of micro-crystalline, nano-crystalline and boron-doped diamond films. *International Journal of Refractory & Hard Metals*, *45*, 31–40. doi:10.1016/j.ijrmhm.2014.02.005

Wan, M., Li, S.-E., Yuan, H., & Zhang, W.-H. (2019). Cutting force modelling in machining of fiber-reinforced polymer matrix composites (PMCs): A review. *Composites. Part A, Applied Science and Manufacturing*, *117*, 34–55. doi:10.1016/j.compositesa.2018.11.003

Wisnom, M. R. (1965). The Role of Delamination in Failure of Fibre-Reinforced Composites. *Philosophical Transactions of the Royal Society of London. Series A, Mathematical and Physical Sciences*, *370*(1965), 1850–1870. doi:10.1098/rsta.2011.0441 PMID:22431760

Xie, Y., Hill, C. A. S., Xiao, Z., Militz, H., & Mai, C. (2010). Silane coupling agents used for natural fiber/polymer composites : A review. *Composites. Part A, Applied Science and Manufacturing*, *41*(7), 806–819. doi:10.1016/j.compositesa.2010.03.005

Xu, J., An, Q., Cai, X., & Chen, M. (2013). Drilling machinability evaluation on new developed high- strength T800S/250F CFRP laminates. *International Journal of Precision Engineering and Manufacturing*, *14*(10), 1687–1696. doi:10.1007/s12541-013-0252-2

Xu, J., An, Q., & Chen, M. (2014). A comparative evaluation of polycrystalline diamond drills in drilling high-strength T800S/250F CFRP. *Composite Structures*, *117*, 71–82. doi:10.1016/j.compstruct.2014.06.034

Xu, J., & El Mansori, M. (2017). Wear characteristics of polycrystalline diamond tools in orthogonal cutting of CFRP/Ti stacks. *Wear*, *376-377*, 91–106. doi:10.1016/j.wear.2016.11.038

Xu, J., Feng, P., Feng, F., Zha, H., & Liang, G. (2021). Subsurface damage and burr improvements of aramid fiber reinforced plastics by using longitudinal–torsional ultrasonic vibration milling. *Journal of Materials Processing Technology*, *297*, 117265. doi:10.1016/j.jmatprotec.2021.117265

Xu, J., Li, C., Mi, S., An, Q., & Chen, M. (2018). Study of drilling induced defects for CFRP composites using new criteria. *Composite Structures*, *201*, 1076–1087. doi:10.1016/j.compstruct.2018.06.051

Xu, W., Zhang, L. C., & Wu, Y. (2014). Elliptic vibration-assisted cutting of fibre-reinforced polymer composites : Understanding the material removal mechanisms. *Composites Science and Technology*, *92*, 103–111. doi:10.1016/j.compscitech.2013.12.011

Yang, G., Park, M., & Park, S. J. (2019). Recent progresses of fabrication and characterization of fibers-reinforced composites : A review. In *Composites Communications* (Vol. 14, pp. 34–42). Elsevier. doi:10.1016/j.coco.2019.05.004

Yang, L., Yan, Y., Liu, Y., & Ran, Z. (2012). Microscopic failure mechanisms of fiber-reinforced polymer composites under transverse tension and compression. *Composites Science and Technology*, *72*(15), 1818–1825. doi:10.1016/j.compscitech.2012.08.001

Yang, W. H., & Tarng, Y. S. (1998). Design optimization of cutting parameters for turning operations based on the Taguchi method. *Journal of Materials Processing Technology*, *84*(1), 122–129. doi:10.1016/S0924-0136(98)00079-X

Yousefi, S., & Zohoor, M. (2019). Effect of Cutting Parameters on the Dimensional Accuracy and Surface Finish in the Hard Turning of MDN250 Steel with Cubic Boron Nitride Tool, for Developing a Knowledged Base Expert System. *International Journal of Mechanical and Materials Engineering*, *14*(1), 1. doi:10.1186/s40712-018-0097-7

Zemann, R., Kain, L., & Bleicher, F. (2014). Vibration assisted machining of carbon fibre reinforced polymers. *Procedia Engineering*, *69*, 536–543. doi:10.1016/j.proeng.2014.03.023

Zhang, B., Lu, S., Rabiey, M., Axinte, D., & Bleicher, F. (2023). Grinding of Composite Materials. *CIRP Annals*, *72*(2), 645–671. doi:10.1016/j.cirp.2023.05.001

Zhang, D., Shao, Z., Geng, D., Jiang, X., Liu, Y., Zhou, Z., & Li, S. (2021). Feasibility study of wave-motion milling of carbon fiber reinforced plastic holes. *International Journal of Extreme Manufacturing*, *3*(1), 010401. doi:10.1088/2631-7990/abc675

Zhang, G., Schlarb, A. K., Tria, S., & Elkedim, O. (2008). Tensile and Tribological Behaviors of PEEK/Nano-SiO2 Composites Compounded Using a Ball Milling Technique. *Composites Science and Technology*, *68*(15–16), 3073–3080. doi:10.1016/j.compscitech.2008.06.027

Zhang, H., Chen, W., Chen, D., & Zhang, L. C. (2001). Assessment of the Exit Defects in Carbon Fibre-Reinforced Plastic Plates Caused by Drilling. *Key Engineering Materials, 196*, 43–52. . doi:10.4028/www.scientific.net/KEM.196.43

Zhang, J., Wang, X., Shen, B., & Sun, F. (2013). Effect of boron and silicon doping on improving the cutting performance of CVD diamond coated cutting tools in machining CFRP. *International Journal of Refractory & Hard Metals, 41*, 285–292. doi:10.1016/j.ijrmhm.2013.04.017

Zhang, X., Zhao, N., & He, C. (2020). The superior mechanical and physical properties of nanocarbon reinforced bulk composites achieved by architecture design – A review. *Progress in Materials Science, 113*, 100672. doi:10.1016/j.pmatsci.2020.100672

Zhang, Y. D., Xiao, H. H., & Teng, Y. (2013). Study of surface roughness in milling carbon fiber reinforced plastics using PCD tool. *Key Engineering Materials, 531-532*, 63–66. . doi:10.4028/www.scientific.net/KEM.531-532.63

Zheng, H., & Liu, K. (2013). Machinability of Engineering Materials. In *Handbook of Manufacturing Engineering and Technology* (pp. 1–34). Springer. doi:10.1007/978-1-4471-4976-7_2-1

Zhi, C., Long, H., & Miao, M. (2017). Influence of Microbond Test Parameters on Interfacial Shear Strength of Fiber Reinforced Polymer-Matrix Composites. *Composites. Part A, Applied Science and Manufacturing, 100*, 55–63. doi:10.1016/j.compositesa.2017.05.004

Zweben, C. (2015). Composite Materials. In *Mechanical Engineers' Handbook* (pp. 1–37). John Wiley & Sons, Ltd. doi:10.1002/9781118985960.meh110

Related References

To continue our tradition of advancing academic research, we have compiled a list of recommended IGI Global readings. These references will provide additional information and guidance to further enrich your knowledge and assist you with your own research and future publications.

Abbasnejad, B., Moeinzadeh, S., Ahankoob, A., & Wong, P. S. (2021). The Role of Collaboration in the Implementation of BIM-Enabled Projects. In J. Underwood & M. Shelbourn (Eds.), *Handbook of Research on Driving Transformational Change in the Digital Built Environment* (pp. 27–62). IGI Global. https://doi.org/10.4018/978-1-7998-6600-8.ch002

Abdulrahman, K. O., Mahamood, R. M., & Akinlabi, E. T. (2022). Additive Manufacturing (AM): Processing Technique for Lightweight Alloys and Composite Material. In K. Kumar, B. Babu, & J. Davim (Ed.), *Handbook of Research on Advancements in the Processing, Characterization, and Application of Lightweight Materials* (pp. 27-48). IGI Global. https://doi.org/10.4018/978-1-7998-7864-3.ch002

Agrawal, R., Sharma, P., & Saxena, A. (2021). A Diamond Cut Leather Substrate Antenna for BAN (Body Area Network) Application. In V. Singh, V. Dubey, A. Saxena, R. Tiwari, & H. Sharma (Eds.), *Emerging Materials and Advanced Designs for Wearable Antennas* (pp. 54–59). IGI Global. https://doi.org/10.4018/978-1-7998-7611-3.ch004

Ahmad, F., Al-Ammar, E. A., & Alsaidan, I. (2022). Battery Swapping Station: A Potential Solution to Address the Limitations of EV Charging Infrastructure. In M. Alam, R. Pillai, & N. Murugesan (Eds.), *Developing Charging Infrastructure and Technologies for Electric Vehicles* (pp. 195–207). IGI Global. doi:10.4018/978-1-7998-6858-3.ch010

Aikhuele, D. (2018). A Study of Product Development Engineering and Design Reliability Concerns. *International Journal of Applied Industrial Engineering*, *5*(1), 79–89. doi:10.4018/IJAIE.2018010105

Al-Khatri, H., & Al-Atrash, F. (2021). Occupants' Habits and Natural Ventilation in a Hot Arid Climate. In R. González-Lezcano (Ed.), *Advancements in Sustainable Architecture and Energy Efficiency* (pp. 146–168). IGI Global. https://doi.org/10.4018/978-1-7998-7023-4.ch007

Al-Shebeeb, O. A., Rangaswamy, S., Gopalakrishan, B., & Devaru, D. G. (2017). Evaluation and Indexing of Process Plans Based on Electrical Demand and Energy Consumption. *International Journal of Manufacturing, Materials, and Mechanical Engineering*, *7*(3), 1–19. doi:10.4018/IJMMME.2017070101

Amuda, M. O., Lawal, T. F., & Akinlabi, E. T. (2017). Research Progress on Rheological Behavior of AA7075 Aluminum Alloy During Hot Deformation. *International Journal of Materials Forming and Machining Processes*, *4*(1), 53–96. doi:10.4018/IJMFMP.2017010104

Amuda, M. O., Lawal, T. F., & Mridha, S. (2021). Microstructure and Mechanical Properties of Silicon Carbide-Treated Ferritic Stainless Steel Welds. In L. Burstein (Ed.), *Handbook of Research on Advancements in Manufacturing, Materials, and Mechanical Engineering* (pp. 395–411). IGI Global. https://doi.org/10.4018/978-1-7998-4939-1.ch019

Anikeev, V., Gasem, K. A., & Fan, M. (2021). Application of Supercritical Technologies in Clean Energy Production: A Review. In L. Chen (Ed.), *Handbook of Research on Advancements in Supercritical Fluids Applications for Sustainable Energy Systems* (pp. 792–821). IGI Global. https://doi.org/10.4018/978-1-7998-5796-9.ch022

Arafat, M. Y., Saleem, I., & Devi, T. P. (2022). Drivers of EV Charging Infrastructure Entrepreneurship in India. In M. Alam, R. Pillai, & N. Murugesan (Eds.), *Developing Charging Infrastructure and Technologies for Electric Vehicles* (pp. 208–219). IGI Global. https://doi.org/10.4018/978-1-7998-6858-3.ch011

Araujo, A., & Manninen, H. (2022). Contribution of Project-Based Learning on Social Skills Development: An Industrial Engineer Perspective. In A. Alves & N. van Hattum-Janssen (Eds.), *Training Engineering Students for Modern Technological Advancement* (pp. 119–145). IGI Global. https://doi.org/10.4018/978-1-7998-8816-1.ch006

Armutlu, H. (2018). Intelligent Biomedical Engineering Operations by Cloud Computing Technologies. In U. Kose, G. Guraksin, & O. Deperlioglu (Eds.), *Nature-Inspired Intelligent Techniques for Solving Biomedical Engineering Problems* (pp. 297–317). Hershey, PA: IGI Global. doi:10.4018/978-1-5225-4769-3.ch015

Atik, M., Sadek, M., & Shahrour, I. (2017). Single-Run Adaptive Pushover Procedure for Shear Wall Structures. In V. Plevris, G. Kremmyda, & Y. Fahjan (Eds.), *Performance-Based Seismic Design of Concrete Structures and Infrastructures* (pp. 59–83). Hershey, PA: IGI Global. doi:10.4018/978-1-5225-2089-4.ch003

Attia, H. (2021). Smart Power Microgrid Impact on Sustainable Building. In R. González-Lezcano (Ed.), *Advancements in Sustainable Architecture and Energy Efficiency* (pp. 169–194). IGI Global. https://doi.org/10.4018/978-1-7998-7023-4.ch008

Aydin, A., Akyol, E., Gungor, M., Kaya, A., & Tasdelen, S. (2018). Geophysical Surveys in Engineering Geology Investigations With Field Examples. In N. Ceryan (Ed.), *Handbook of Research on Trends and Digital Advances in Engineering Geology* (pp. 257–280). Hershey, PA: IGI Global. doi:10.4018/978-1-5225-2709-1.ch007

Ayoobkhan, M. U. D., Y., A., J., Easwaran, B., & R., T. (2021). Smart Connected Digital Products and IoT Platform With the Digital Twin. In P. Vasant, G. Weber, & W. Punurai (Ed.), Research Advancements in Smart Technology, Optimization, and Renewable Energy (pp. 330-350). IGI Global. https://doi.org/ doi:10.4018/978-1-7998-3970-5.ch016

Baeza Moyano, D., & González Lezcano, R. A. (2021). The Importance of Light in Our Lives: Towards New Lighting in Schools. In R. González-Lezcano (Ed.), *Advancements in Sustainable Architecture and Energy Efficiency* (pp. 239–256). IGI Global. https://doi.org/10.4018/978-1-7998-7023-4.ch011

Bagdadee, A. H. (2021). A Brief Assessment of the Energy Sector of Bangladesh. *International Journal of Energy Optimization and Engineering, 10*(1), 36–55. doi:10.4018/IJEOE.2021010103

Baklezos, A. T., & Hadjigeorgiou, N. G. (2021). Magnetic Sensors for Space Applications and Magnetic Cleanliness Considerations. In C. Nikolopoulos (Ed.), *Recent Trends on Electromagnetic Environmental Effects for Aeronautics and Space Applications* (pp. 147–185). IGI Global. https://doi.org/10.4018/978-1-7998-4879-0.ch006

Bas, T. G. (2017). Nutraceutical Industry with the Collaboration of Biotechnology and Nutrigenomics Engineering: The Significance of Intellectual Property in the Entrepreneurship and Scientific Research Ecosystems. In T. Bas & J. Zhao (Eds.), *Comparative Approaches to Biotechnology Development and Use in Developed and Emerging Nations* (pp. 1–17). Hershey, PA: IGI Global. doi:10.4018/978-1-5225-1040-6.ch001

Bazeer Ahamed, B., & Periakaruppan, S. (2021). Taxonomy of Influence Maximization Techniques in Unknown Social Networks. In P. Vasant, G. Weber, & W. Punurai (Eds.), *Research Advancements in Smart Technology, Optimization, and Renewable Energy* (pp. 351-363). IGI Global. https://doi.org/10.4018/978-1-7998-3970-5.ch017

Beale, R., & André, J. (2017). *Design Solutions and Innovations in Temporary Structures*. Hershey, PA: IGI Global. doi:10.4018/978-1-5225-2199-0

Behnam, B. (2017). Simulating Post-Earthquake Fire Loading in Conventional RC Structures. In P. Samui, S. Chakraborty, & D. Kim (Eds.), *Modeling and Simulation Techniques in Structural Engineering* (pp. 425–444). Hershey, PA: IGI Global. doi:10.4018/978-1-5225-0588-4.ch015

Ben Hamida, I., Salah, S. B., Msahli, F., & Mimouni, M. F. (2018). Distribution Network Reconfiguration Using SPEA2 for Power Loss Minimization and Reliability Improvement. *International Journal of Energy Optimization and Engineering*, *7*(1), 50–65. doi:10.4018/IJEOE.2018010103

Bentarzi, H. (2021). Fault Tree-Based Root Cause Analysis Used to Study Mal-Operation of a Protective Relay in a Smart Grid. In A. Recioui & H. Bentarzi (Eds.), *Optimizing and Measuring Smart Grid Operation and Control* (pp. 289–308). IGI Global. https://doi.org/10.4018/978-1-7998-4027-5.ch012

Beysens, D. A., Garrabos, Y., & Zappoli, B. (2021). Thermal Effects in Near-Critical Fluids: Piston Effect and Related Phenomena. In L. Chen (Ed.), *Handbook of Research on Advancements in Supercritical Fluids Applications for Sustainable Energy Systems* (pp. 1–31). IGI Global. https://doi.org/10.4018/978-1-7998-5796-9.ch001

Bhaskar, S. V., & Kudal, H. N. (2017). Effect of TiCN and AlCrN Coating on Tribological Behaviour of Plasma-nitrided AISI 4140 Steel. *International Journal of Surface Engineering and Interdisciplinary Materials Science*, *5*(2), 1–17. doi:10.4018/IJSEIMS.2017070101

Bhuyan, D. (2018). Designing of a Twin Tube Shock Absorber: A Study in Reverse Engineering. In K. Kumar & J. Davim (Eds.), *Design and Optimization of Mechanical Engineering Products* (pp. 83–104). Hershey, PA: IGI Global. doi:10.4018/978-1-5225-3401-3.ch005

Blumberg, G. (2021). Blockchains for Use in Construction and Engineering Projects. In J. Underwood & M. Shelbourn (Eds.), *Handbook of Research on Driving Transformational Change in the Digital Built Environment* (pp. 179–208). IGI Global. https://doi.org/10.4018/978-1-7998-6600-8.ch008

Bolboaca, A. M. (2021). Considerations Regarding the Use of Fuel Cells in Combined Heat and Power for Stationary Applications. In G. Badea, R. Felseghi, & I. Aşchilean (Eds.), *Hydrogen Fuel Cell Technology for Stationary Applications* (pp. 239–275). IGI Global. https://doi.org/10.4018/978-1-7998-4945-2.ch010

Burstein, L. (2021). Simulation Tool for Cable Design. In L. Burstein (Ed.), *Handbook of Research on Advancements in Manufacturing, Materials, and Mechanical Engineering* (pp. 54–74). IGI Global. https://doi.org/10.4018/978-1-7998-4939-1.ch003

Calderon, F. A., Giolo, E. G., Frau, C. D., Rengel, M. G., Rodriguez, H., Tornello, M., ... Gallucci, R. (2018). Seismic Microzonation and Site Effects Detection Through Microtremors Measures: A Review. In N. Ceryan (Ed.), *Handbook of Research on Trends and Digital Advances in Engineering Geology* (pp. 326–349). Hershey, PA: IGI Global. doi:10.4018/978-1-5225-2709-1.ch009

Ceryan, N., & Can, N. K. (2018). Prediction of The Uniaxial Compressive Strength of Rocks Materials. In N. Ceryan (Ed.), *Handbook of Research on Trends and Digital Advances in Engineering Geology* (pp. 31–96). Hershey, PA: IGI Global. doi:10.4018/978-1-5225-2709-1.ch002

Ceryan, S. (2018). Weathering Indices Used in Evaluation of the Weathering State of Rock Material. In N. Ceryan (Ed.), *Handbook of Research on Trends and Digital Advances in Engineering Geology* (pp. 132–186). Hershey, PA: IGI Global. doi:10.4018/978-1-5225-2709-1.ch004

Chen, H., Padilla, R. V., & Besarati, S. (2017). Supercritical Fluids and Their Applications in Power Generation. In L. Chen & Y. Iwamoto (Eds.), *Advanced Applications of Supercritical Fluids in Energy Systems* (pp. 369–402). Hershey, PA: IGI Global. doi:10.4018/978-1-5225-2047-4.ch012

Chen, H., Padilla, R. V., & Besarati, S. (2021). Supercritical Fluids and Their Applications in Power Generation. In L. Chen (Ed.), *Handbook of Research on Advancements in Supercritical Fluids Applications for Sustainable Energy Systems* (pp. 566–599). IGI Global. https://doi.org/10.4018/978-1-7998-5796-9.ch016

Chen, L. (2017). Principles, Experiments, and Numerical Studies of Supercritical Fluid Natural Circulation System. In L. Chen & Y. Iwamoto (Eds.), *Advanced Applications of Supercritical Fluids in Energy Systems* (pp. 136–187). Hershey, PA: IGI Global. doi:10.4018/978-1-5225-2047-4.ch005

Chen, L. (2021). Principles, Experiments, and Numerical Studies of Supercritical Fluid Natural Circulation System. In L. Chen (Ed.), *Handbook of Research on Advancements in Supercritical Fluids Applications for Sustainable Energy Systems* (pp. 219–269). IGI Global. https://doi.org/10.4018/978-1-7998-5796-9.ch007

Chiba, Y., Marif, Y., Henini, N., & Tlemcani, A. (2021). Modeling of Magnetic Refrigeration Device by Using Artificial Neural Networks Approach. *International Journal of Energy Optimization and Engineering*, *10*(4), 68–76. https://doi.org/10.4018/IJEOE.2021100105

Clementi, F., Di Sciascio, G., Di Sciascio, S., & Lenci, S. (2017). Influence of the Shear-Bending Interaction on the Global Capacity of Reinforced Concrete Frames: A Brief Overview of the New Perspectives. In V. Plevris, G. Kremmyda, & Y. Fahjan (Eds.), *Performance-Based Seismic Design of Concrete Structures and Infrastructures* (pp. 84–111). Hershey, PA: IGI Global. doi:10.4018/978-1-5225-2089-4.ch004

Codinhoto, R., Fialho, B. C., Pinti, L., & Fabricio, M. M. (2021). BIM and IoT for Facilities Management: Understanding Key Maintenance Issues. In J. Underwood & M. Shelbourn (Eds.), *Handbook of Research on Driving Transformational Change in the Digital Built Environment* (pp. 209–231). IGI Global. doi:10.4018/978-1-7998-6600-8.ch009

Cortés-Polo, D., Calle-Cancho, J., Carmona-Murillo, J., & González-Sánchez, J. (2017). Future Trends in Mobile-Fixed Integration for Next Generation Networks: Classification and Analysis. *International Journal of Vehicular Telematics and Infotainment Systems*, *1*(1), 33–53. doi:10.4018/IJVTIS.2017010103

Costa, H. G., Sheremetieff, F. H., & Araújo, E. A. (2022). Influence of Game-Based Methods in Developing Engineering Competences. In A. Alves & N. van Hattum-Janssen (Eds.), *Training Engineering Students for Modern Technological Advancement* (pp. 69–88). IGI Global. https://doi.org/10.4018/978-1-7998-8816-1.ch004

Cui, X., Zeng, S., Li, Z., Zheng, Q., Yu, X., & Han, B. (2018). Advanced Composites for Civil Engineering Infrastructures. In K. Kumar & J. Davim (Eds.), *Composites and Advanced Materials for Industrial Applications* (pp. 212–248). Hershey, PA: IGI Global. doi:10.4018/978-1-5225-5216-1.ch010

Dalgıç, S., & Kuşku, İ. (2018). Geological and Geotechnical Investigations in Tunneling. In N. Ceryan (Ed.), *Handbook of Research on Trends and Digital Advances in Engineering Geology* (pp. 482–529). Hershey, PA: IGI Global. doi:10.4018/978-1-5225-2709-1.ch014

Dang, C., & Hihara, E. (2021). Study on Cooling Heat Transfer of Supercritical Carbon Dioxide Applied to Transcritical Carbon Dioxide Heat Pump. In L. Chen (Ed.), *Handbook of Research on Advancements in Supercritical Fluids Applications for Sustainable Energy Systems* (pp. 451–493). IGI Global. https://doi.org/10.4018/978-1-7998-5796-9.ch013

Daus, Y., Kharchenko, V., & Yudaev, I. (2021). Research of Solar Energy Potential of Photovoltaic Installations on Enclosing Structures of Buildings. *International Journal of Energy Optimization and Engineering*, *10*(4), 18–34. https://doi.org/10.4018/IJEOE.2021100102

Daus, Y., Kharchenko, V., & Yudaev, I. (2021). Optimizing Layout of Distributed Generation Sources of Power Supply System of Agricultural Object. *International Journal of Energy Optimization and Engineering*, *10*(3), 70–84. https://doi.org/10.4018/IJEOE.2021070104

de la Varga, D., Soto, M., Arias, C. A., van Oirschot, D., Kilian, R., Pascual, A., & Álvarez, J. A. (2017). Constructed Wetlands for Industrial Wastewater Treatment and Removal of Nutrients. In Á. Val del Río, J. Campos Gómez, & A. Mosquera Corral (Eds.), *Technologies for the Treatment and Recovery of Nutrients from Industrial Wastewater* (pp. 202–230). Hershey, PA: IGI Global. doi:10.4018/978-1-5225-1037-6.ch008

Deb, S., Ammar, E. A., AlRajhi, H., Alsaidan, I., & Shariff, S. M. (2022). V2G Pilot Projects: Review and Lessons Learnt. In M. Alam, R. Pillai, & N. Murugesan (Eds.), *Developing Charging Infrastructure and Technologies for Electric Vehicles* (pp. 252–267). IGI Global. https://doi.org/10.4018/978-1-7998-6858-3.ch014

Dekhandji, F. Z., & Rais, M. C. (2021). A Comparative Study of Power Quality Monitoring Using Various Techniques. In A. Recioui & H. Bentarzi (Eds.), *Optimizing and Measuring Smart Grid Operation and Control* (pp. 259–288). IGI Global. https://doi.org/10.4018/978-1-7998-4027-5.ch011

Deperlioglu, O. (2018). Intelligent Techniques Inspired by Nature and Used in Biomedical Engineering. In U. Kose, G. Guraksin, & O. Deperlioglu (Eds.), *Nature-Inspired Intelligent Techniques for Solving Biomedical Engineering Problems* (pp. 51–77). Hershey, PA: IGI Global. doi:10.4018/978-1-5225-4769-3.ch003

Dhurpate, P. R., & Tang, H. (2021). Quantitative Analysis of the Impact of Inter-Line Conveyor Capacity for Throughput of Manufacturing Systems. *International Journal of Manufacturing, Materials, and Mechanical Engineering, 11*(1), 1–17. https://doi.org/10.4018/IJMMME.2021010101

Dinkar, S., & Deep, K. (2021). A Survey of Recent Variants and Applications of Antlion Optimizer. *International Journal of Energy Optimization and Engineering, 10*(2), 48–73. doi:10.4018/IJEOE.2021040103

Dixit, A. (2018). Application of Silica-Gel-Reinforced Aluminium Composite on the Piston of Internal Combustion Engine: Comparative Study of Silica-Gel-Reinforced Aluminium Composite Piston With Aluminium Alloy Piston. In K. Kumar & J. Davim (Eds.), *Composites and Advanced Materials for Industrial Applications* (pp. 63–98). Hershey, PA: IGI Global. doi:10.4018/978-1-5225-5216-1.ch004

Drabecki, M. P., & Kułak, K. B. (2021). Global Pandemics on European Electrical Energy Markets: Lessons Learned From the COVID-19 Outbreak. *International Journal of Energy Optimization and Engineering, 10*(3), 24–46. https://doi.org/10.4018/IJEOE.2021070102

Dutta, M. M. (2021). Nanomaterials for Food and Agriculture. In M. Bhat, I. Wani, & S. Ashraf (Eds.), *Applications of Nanomaterials in Agriculture, Food Science, and Medicine* (pp. 75–97). IGI Global. doi:10.4018/978-1-7998-5563-7.ch004

Dutta, M. M., & Goswami, M. (2021). Coating Materials: Nano-Materials. In S. Roy & G. Bose (Eds.), *Advanced Surface Coating Techniques for Modern Industrial Applications* (pp. 1–30). IGI Global. doi:10.4018/978-1-7998-4870-7.ch001

Elsayed, A. M., Dakkama, H. J., Mahmoud, S., Al-Dadah, R., & Kaialy, W. (2017). Sustainable Cooling Research Using Activated Carbon Adsorbents and Their Environmental Impact. In T. Kobayashi (Ed.), *Applied Environmental Materials Science for Sustainability* (pp. 186–221). Hershey, PA: IGI Global. doi:10.4018/978-1-5225-1971-3.ch009

Ercanoglu, M., & Sonmez, H. (2018). General Trends and New Perspectives on Landslide Mapping and Assessment Methods. In N. Ceryan (Ed.), *Handbook of Research on Trends and Digital Advances in Engineering Geology* (pp. 350–379). Hershey, PA: IGI Global. doi:10.4018/978-1-5225-2709-1.ch010

Faroz, S. A., Pujari, N. N., Rastogi, R., & Ghosh, S. (2017). Risk Analysis of Structural Engineering Systems Using Bayesian Inference. In P. Samui, S. Chakraborty, & D. Kim (Eds.), *Modeling and Simulation Techniques in Structural Engineering* (pp. 390–424). Hershey, PA: IGI Global. doi:10.4018/978-1-5225-0588-4.ch014

Fekik, A., Hamida, M. L., Denoun, H., Azar, A. T., Kamal, N. A., Vaidyanathan, S., Bousbaine, A., & Benamrouche, N. (2022). Multilevel Inverter for Hybrid Fuel Cell/ PV Energy Conversion System. In A. Fekik & N. Benamrouche (Eds.), *Modeling and Control of Static Converters for Hybrid Storage Systems* (pp. 233–270). IGI Global. https://doi.org/10.4018/978-1-7998-7447-8.ch009

Fekik, A., Hamida, M. L., Houassine, H., Azar, A. T., Kamal, N. A., Denoun, H., Vaidyanathan, S., & Sambas, A. (2022). Power Quality Improvement for Grid-Connected Photovoltaic Panels Using Direct Power Control. In A. Fekik & N. Benamrouche (Eds.), *Modeling and Control of Static Converters for Hybrid Storage Systems* (pp. 107–142). IGI Global. https://doi.org/10.4018/978-1-7998-7447-8.ch005

Fernando, P. R., Hamigah, T., Disne, S., Wickramasingha, G. G., & Sutharshan, A. (2018). The Evaluation of Engineering Properties of Low Cost Concrete Blocks by Partial Doping of Sand with Sawdust: Low Cost Sawdust Concrete Block. *International Journal of Strategic Engineering*, *1*(2), 26–42. doi:10.4018/IJoSE.2018070103

Ferro, G., Minciardi, R., Parodi, L., & Robba, M. (2022). Optimal Charging Management of Microgrid-Integrated Electric Vehicles. In M. Alam, R. Pillai, & N. Murugesan (Eds.), *Developing Charging Infrastructure and Technologies for Electric Vehicles* (pp. 133–155). IGI Global. https://doi.org/10.4018/978-1-7998-6858-3.ch007

Flumerfelt, S., & Green, C. (2022). Graduate Lean Leadership Education: A Case Study of a Program. In A. Alves & N. van Hattum-Janssen (Eds.), *Training Engineering Students for Modern Technological Advancement* (pp. 202–224). IGI Global. https://doi.org/10.4018/978-1-7998-8816-1.ch010

Galli, B. J. (2021). Implications of Economic Decision Making to the Project Manager. *International Journal of Strategic Engineering*, *4*(1), 19–32. https://doi.org/10.4018/IJoSE.2021010102

Gento, A. M., Pimentel, C., & Pascual, J. A. (2022). Teaching Circular Economy and Lean Management in a Learning Factory. In A. Alves & N. van Hattum-Janssen (Eds.), *Training Engineering Students for Modern Technological Advancement* (pp. 183–201). IGI Global. https://doi.org/10.4018/978-1-7998-8816-1.ch009

Ghosh, S., Mitra, S., Ghosh, S., & Chakraborty, S. (2017). Seismic Reliability Analysis in the Framework of Metamodelling Based Monte Carlo Simulation. In P. Samui, S. Chakraborty, & D. Kim (Eds.), *Modeling and Simulation Techniques in Structural Engineering* (pp. 192–208). Hershey, PA: IGI Global. doi:10.4018/978-1-5225-0588-4.ch006

Gil, M., & Otero, B. (2017). Learning Engineering Skills through Creativity and Collaboration: A Game-Based Proposal. In R. Alexandre Peixoto de Queirós & M. Pinto (Eds.), *Gamification-Based E-Learning Strategies for Computer Programming Education* (pp. 14–29). Hershey, PA: IGI Global. doi:10.4018/978-1-5225-1034-5.ch002

Gill, J., Ayre, M., & Mills, J. (2017). Revisioning the Engineering Profession: How to Make It Happen! In M. Gray & K. Thomas (Eds.), *Strategies for Increasing Diversity in Engineering Majors and Careers* (pp. 156–175). Hershey, PA: IGI Global. doi:10.4018/978-1-5225-2212-6.ch008

Godzhaev, Z., Senkevich, S., Kuzmin, V., & Melikov, I. (2021). Use of the Neural Network Controller of Sprung Mass to Reduce Vibrations From Road Irregularities. In P. Vasant, G. Weber, & W. Punurai (Ed.), *Research Advancements in Smart Technology, Optimization, and Renewable Energy* (pp. 69-87). IGI Global. https://doi.org/10.4018/978-1-7998-3970-5.ch005

Gomes de Gusmão, C. M. (2022). Digital Competencies and Transformation in Higher Education: Upskilling With Extension Actions. In A. Alves & N. van Hattum-Janssen (Eds.), *Training Engineering Students for Modern Technological Advancement* (pp. 313–328). IGI Global. https://doi.org/10.4018/978-1-7998-8816-1.ch015A

Goyal, N., Ram, M., & Kumar, P. (2017). Welding Process under Fault Coverage Approach for Reliability and MTTF. In M. Ram & J. Davim (Eds.), *Mathematical Concepts and Applications in Mechanical Engineering and Mechatronics* (pp. 222–245). Hershey, PA: IGI Global. doi:10.4018/978-1-5225-1639-2.ch011

Gray, M., & Lundy, C. (2017). Engineering Study Abroad: High Impact Strategy for Increasing Access. In M. Gray & K. Thomas (Eds.), *Strategies for Increasing Diversity in Engineering Majors and Careers* (pp. 42–59). Hershey, PA: IGI Global. doi:10.4018/978-1-5225-2212-6.ch003

Güler, O., & Varol, T. (2021). Fabrication of Functionally Graded Metal and Ceramic Powders Synthesized by Electroless Deposition. In S. Roy & G. Bose (Eds.), *Advanced Surface Coating Techniques for Modern Industrial Applications* (pp. 150–187). IGI Global. https://doi.org/10.4018/978-1-7998-4870-7.ch007

Guraksin, G. E. (2018). Internet of Things and Nature-Inspired Intelligent Techniques for the Future of Biomedical Engineering. In U. Kose, G. Guraksin, & O. Deperlioglu (Eds.), *Nature-Inspired Intelligent Techniques for Solving Biomedical Engineering Problems* (pp. 263–282). Hershey, PA: IGI Global. doi:10.4018/978-1-5225-4769-3.ch013

Hamida, M. L., Fekik, A., Denoun, H., Ardjal, A., & Bokhtache, A. A. (2022). Flying Capacitor Inverter Integration in a Renewable Energy System. In A. Fekik & N. Benamrouche (Eds.), *Modeling and Control of Static Converters for Hybrid Storage Systems* (pp. 287–306). IGI Global. https://doi.org/10.4018/978-1-7998-7447-8.ch011

Hasegawa, N., & Takahashi, Y. (2021). Control of Soap Bubble Ejection Robot Using Facial Expressions. *International Journal of Manufacturing, Materials, and Mechanical Engineering, 11*(2), 1–16. https://doi.org/10.4018/IJMMME.2021040101

Hejazi, T., & Akbari, L. (2017). A Multiresponse Optimization Model for Statistical Design of Processes with Discrete Variables. In M. Ram & J. Davim (Eds.), *Mathematical Concepts and Applications in Mechanical Engineering and Mechatronics* (pp. 17–37). Hershey, PA: IGI Global. doi:10.4018/978-1-5225-1639-2.ch002

Hejazi, T., & Hejazi, A. (2017). Monte Carlo Simulation for Reliability-Based Design of Automotive Complex Subsystems. In M. Ram & J. Davim (Eds.), *Mathematical Concepts and Applications in Mechanical Engineering and Mechatronics* (pp. 177–200). Hershey, PA: IGI Global. doi:10.4018/978-1-5225-1639-2.ch009

Hejazi, T., & Poursabbagh, H. (2017). Reliability Analysis of Engineering Systems: An Accelerated Life Testing for Boiler Tubes. In M. Ram & J. Davim (Eds.), *Mathematical Concepts and Applications in Mechanical Engineering and Mechatronics* (pp. 154–176). Hershey, PA: IGI Global. doi:10.4018/978-1-5225-1639-2.ch008

Henao, J., Poblano-Salas, C. A., Vargas, F., Giraldo-Betancur, A. L., Corona-Castuera, J., & Sotelo-Mazón, O. (2021). Principles and Applications of Thermal Spray Coatings. In S. Roy & G. Bose (Eds.), *Advanced Surface Coating Techniques for Modern Industrial Applications* (pp. 31–70). IGI Global. https://doi.org/10.4018/978-1-7998-4870-7.ch002

Henao, J., & Sotelo, O. (2018). Surface Engineering at High Temperature: Thermal Cycling and Corrosion Resistance. In A. Pakseresht (Ed.), *Production, Properties, and Applications of High Temperature Coatings* (pp. 131–159). Hershey, PA: IGI Global. doi:10.4018/978-1-5225-4194-3.ch006

Hrnčič, M. K., Cör, D., & Knez, Ž. (2021). Supercritical Fluids as a Tool for Green Energy and Chemicals. In L. Chen (Ed.), *Handbook of Research on Advancements in Supercritical Fluids Applications for Sustainable Energy Systems* (pp. 761–791). IGI Global. doi:10.4018/978-1-7998-5796-9.ch021

Ibrahim, O., Erdem, S., & Gurbuz, E. (2021). Studying Physical and Chemical Properties of Graphene Oxide and Reduced Graphene Oxide and Their Applications in Sustainable Building Materials. In R. González-Lezcano (Ed.), *Advancements in Sustainable Architecture and Energy Efficiency* (pp. 221–238). IGI Global. https://doi.org/10.4018/978-1-7998-7023-4.ch010

Ihianle, I. K., Islam, S., Naeem, U., & Ebenuwa, S. H. (2021). Exploiting Patterns of Object Use for Human Activity Recognition. In A. Nwajana & I. Ihianle (Eds.), *Handbook of Research on 5G Networks and Advancements in Computing, Electronics, and Electrical Engineering* (pp. 382–401). IGI Global. https://doi.org/10.4018/978-1-7998-6992-4.ch015

Ijemaru, G. K., Ngharamike, E. T., Oleka, E. U., & Nwajana, A. O. (2021). An Energy-Efficient Model for Opportunistic Data Collection in IoV-Enabled SC Waste Management. In A. Nwajana & I. Ihianle (Eds.), *Handbook of Research on 5G Networks and Advancements in Computing, Electronics, and Electrical Engineering* (pp. 1–19). IGI Global. https://doi.org/10.4018/978-1-7998-6992-4.ch001

Ilori, O. O., Adetan, D. A., & Umoru, L. E. (2017). Effect of Cutting Parameters on the Surface Residual Stress of Face-Milled Pearlitic Ductile Iron. *International Journal of Materials Forming and Machining Processes*, *4*(1), 38–52. doi:10.4018/IJMFMP.2017010103

Imam, M. H., Tasadduq, I. A., Ahmad, A., Aldosari, F., & Khan, H. (2017). Automated Generation of Course Improvement Plans Using Expert System. *International Journal of Quality Assurance in Engineering and Technology Education*, *6*(1), 1–12. doi:10.4018/IJQAETE.2017010101

Injeti, S. K., & Kumar, T. V. (2018). A WDO Framework for Optimal Deployment of DGs and DSCs in a Radial Distribution System Under Daily Load Pattern to Improve Techno-Economic Benefits. *International Journal of Energy Optimization and Engineering*, *7*(2), 1–38. doi:10.4018/IJEOE.2018040101

Ishii, N., Anami, K., & Knisely, C. W. (2018). *Dynamic Stability of Hydraulic Gates and Engineering for Flood Prevention*. Hershey, PA: IGI Global. doi:10.4018/978-1-5225-3079-4

Iwamoto, Y., & Yamaguchi, H. (2021). Application of Supercritical Carbon Dioxide for Solar Water Heater. In L. Chen (Ed.), *Handbook of Research on Advancements in Supercritical Fluids Applications for Sustainable Energy Systems* (pp. 370–387). IGI Global. https://doi.org/10.4018/978-1-7998-5796-9.ch010

Jayapalan, S. (2018). A Review of Chemical Treatments on Natural Fibers-Based Hybrid Composites for Engineering Applications. In K. Kumar & J. Davim (Eds.), *Composites and Advanced Materials for Industrial Applications* (pp. 16–37). Hershey, PA: IGI Global. doi:10.4018/978-1-5225-5216-1.ch002

Kapetanakis, T. N., Vardiambasis, I. O., Ioannidou, M. P., & Konstantaras, A. I. (2021). Modeling Antenna Radiation Using Artificial Intelligence Techniques: The Case of a Circular Loop Antenna. In C. Nikolopoulos (Ed.), *Recent Trends on Electromagnetic Environmental Effects for Aeronautics and Space Applications* (pp. 186–225). IGI Global. https://doi.org/10.4018/978-1-7998-4879-0.ch007

Karkalos, N. E., Markopoulos, A. P., & Dossis, M. F. (2017). Optimal Model Parameters of Inverse Kinematics Solution of a 3R Robotic Manipulator Using ANN Models. *International Journal of Manufacturing, Materials, and Mechanical Engineering*, *7*(3), 20–40. doi:10.4018/IJMMME.2017070102

Kelly, M., Costello, M., Nicholson, G., & O'Connor, J. (2021). The Evolving Integration of BIM Into Built Environment Programmes in a Higher Education Institute. In J. Underwood & M. Shelbourn (Eds.), *Handbook of Research on Driving Transformational Change in the Digital Built Environment* (pp. 294–326). IGI Global. https://doi.org/10.4018/978-1-7998-6600-8.ch012

Kesimal, A., Karaman, K., Cihangir, F., & Ercikdi, B. (2018). Excavatability Assessment of Rock Masses for Geotechnical Studies. In N. Ceryan (Ed.), *Handbook of Research on Trends and Digital Advances in Engineering Geology* (pp. 231–256). Hershey, PA: IGI Global. doi:10.4018/978-1-5225-2709-1.ch006

Knoflacher, H. (2017). The Role of Engineers and Their Tools in the Transport Sector after Paradigm Change: From Assumptions and Extrapolations to Science. In H. Knoflacher & E. Ocalir-Akunal (Eds.), *Engineering Tools and Solutions for Sustainable Transportation Planning* (pp. 1–29). Hershey, PA: IGI Global. doi:10.4018/978-1-5225-2116-7.ch001

Kose, U. (2018). Towards an Intelligent Biomedical Engineering With Nature-Inspired Artificial Intelligence Techniques. In U. Kose, G. Guraksin, & O. Deperlioglu (Eds.), *Nature-Inspired Intelligent Techniques for Solving Biomedical Engineering Problems* (pp. 1–26). Hershey, PA: IGI Global. doi:10.4018/978-1-5225-4769-3.ch001

Kostić, S. (2018). A Review on Enhanced Stability Analyses of Soil Slopes Using Statistical Design. In N. Ceryan (Ed.), *Handbook of Research on Trends and Digital Advances in Engineering Geology* (pp. 446–481). Hershey, PA: IGI Global. doi:10.4018/978-1-5225-2709-1.ch013

Kumar, A., Patil, P. P., & Prajapati, Y. K. (2018). *Advanced Numerical Simulations in Mechanical Engineering*. Hershey, PA: IGI Global. doi:10.4018/978-1-5225-3722-9

Kumar, G. R., Rajyalakshmi, G., & Manupati, V. K. (2017). Surface Micro Patterning of Aluminium Reinforced Composite through Laser Peening. *International Journal of Manufacturing, Materials, and Mechanical Engineering*, *7*(4), 15–27. doi:10.4018/IJMMME.2017100102

Kumar, N., Basu, D. N., & Chen, L. (2021). Effect of Flow Acceleration and Buoyancy on Thermalhydraulics of sCO2 in Mini/Micro-Channel. In L. Chen (Ed.), *Handbook of Research on Advancements in Supercritical Fluids Applications for Sustainable Energy Systems* (pp. 161–182). IGI Global. doi:10.4018/978-1-7998-5796-9.ch005

Kumari, N., & Kumar, K. (2018). Fabrication of Orthotic Calipers With Epoxy-Based Green Composite. In K. Kumar & J. Davim (Eds.), *Composites and Advanced Materials for Industrial Applications* (pp. 157–176). Hershey, PA: IGI Global. doi:10.4018/978-1-5225-5216-1.ch008

Kuppusamy, R. R. (2018). Development of Aerospace Composite Structures Through Vacuum-Enhanced Resin Transfer Moulding Technology (VERTMTy): Vacuum-Enhanced Resin Transfer Moulding. In K. Kumar & J. Davim (Eds.), *Composites and Advanced Materials for Industrial Applications* (pp. 99–111). Hershey, PA: IGI Global. doi:10.4018/978-1-5225-5216-1.ch005

Kurganov, V. A., Zeigarnik, Y. A., & Maslakova, I. V. (2021). Normal and Deteriorated Heat Transfer Under Heating Turbulent Supercritical Pressure Coolants Flows in Round Tubes. In L. Chen (Ed.), *Handbook of Research on Advancements in Supercritical Fluids Applications for Sustainable Energy Systems* (pp. 494–532). IGI Global. https://doi.org/10.4018/978-1-7998-5796-9.ch014

Li, H., & Zhang, Y. (2021). Heat Transfer and Fluid Flow Modeling for Supercritical Fluids in Advanced Energy Systems. In L. Chen (Ed.), *Handbook of Research on Advancements in Supercritical Fluids Applications for Sustainable Energy Systems* (pp. 388–422). IGI Global. https://doi.org/10.4018/978-1-7998-5796-9.ch011

Loy, J., Howell, S., & Cooper, R. (2017). Engineering Teams: Supporting Diversity in Engineering Education. In M. Gray & K. Thomas (Eds.), *Strategies for Increasing Diversity in Engineering Majors and Careers* (pp. 106–129). Hershey, PA: IGI Global. doi:10.4018/978-1-5225-2212-6.ch006

Macher, G., Armengaud, E., Kreiner, C., Brenner, E., Schmittner, C., Ma, Z., ... Krammer, M. (2018). Integration of Security in the Development Lifecycle of Dependable Automotive CPS. In N. Druml, A. Genser, A. Krieg, M. Menghin, & A. Hoeller (Eds.), *Solutions for Cyber-Physical Systems Ubiquity* (pp. 383–423). Hershey, PA: IGI Global. doi:10.4018/978-1-5225-2845-6.ch015

Madhu, M. N., Singh, J. G., Mohan, V., & Ongsakul, W. (2021). Transmission Risk Optimization in Interconnected Systems: Risk-Adjusted Available Transfer Capability. In P. Vasant, G. Weber, & W. Punurai (Ed.), *Research Advancements in Smart Technology, Optimization, and Renewable Energy* (pp. 183-199). IGI Global. https://doi.org/10.4018/978-1-7998-3970-5.ch010

Mahendramani, G., & Lakshmana Swamy, N. (2018). Effect of Weld Groove Area on Distortion of Butt Welded Joints in Submerged Arc Welding. *International Journal of Manufacturing, Materials, and Mechanical Engineering*, *8*(2), 33–44. doi:10.4018/IJMMME.2018040103

Makropoulos, G., Koumaras, H., Setaki, F., Filis, K., Lutz, T., Montowtt, P., Tomaszewski, L., Dybiec, P., & Järvet, T. (2021). 5G and Unmanned Aerial Vehicles (UAVs) Use Cases: Analysis of the Ecosystem, Architecture, and Applications. In A. Nwajana & I. Ihianle (Eds.), *Handbook of Research on 5G Networks and Advancements in Computing, Electronics, and Electrical Engineering* (pp. 36–69). IGI Global. https://doi.org/10.4018/978-1-7998-6992-4.ch003

Meric, E. M., Erdem, S., & Gurbuz, E. (2021). Application of Phase Change Materials in Construction Materials for Thermal Energy Storage Systems in Buildings. In R. González-Lezcano (Ed.), *Advancements in Sustainable Architecture and Energy Efficiency* (pp. 1–20). IGI Global. https://doi.org/10.4018/978-1-7998-7023-4.ch001

Mihret, E. T., & Yitayih, K. A. (2021). Operation of VANET Communications: The Convergence of UAV System With LTE/4G and WAVE Technologies. *International Journal of Smart Vehicles and Smart Transportation*, *4*(1), 29–51. https://doi.org/10.4018/IJSVST.2021010103

Mir, M. A., Bhat, B. A., Sheikh, B. A., Rather, G. A., Mehraj, S., & Mir, W. R. (2021). Nanomedicine in Human Health Therapeutics and Drug Delivery: Nanobiotechnology and Nanobiomedicine. In M. Bhat, I. Wani, & S. Ashraf (Eds.), *Applications of Nanomaterials in Agriculture, Food Science, and Medicine* (pp. 229–251). IGI Global. doi:10.4018/978-1-7998-5563-7.ch013

Mohammadzadeh, S., & Kim, Y. (2017). Nonlinear System Identification of Smart Buildings. In P. Samui, S. Chakraborty, & D. Kim (Eds.), *Modeling and Simulation Techniques in Structural Engineering* (pp. 328–347). Hershey, PA: IGI Global. doi:10.4018/978-1-5225-0588-4.ch011

Molina, G. J., Aktaruzzaman, F., Soloiu, V., & Rahman, M. (2017). Design and Testing of a Jet-Impingement Instrument to Study Surface-Modification Effects by Nanofluids. *International Journal of Surface Engineering and Interdisciplinary Materials Science*, *5*(2), 43–61. doi:10.4018/IJSEIMS.2017070104

Moreno-Rangel, A., & Carrillo, G. (2021). Energy-Efficient Homes: A Heaven for Respiratory Illnesses. In R. González-Lezcano (Ed.), *Advancements in Sustainable Architecture and Energy Efficiency* (pp. 49–71). IGI Global. https://doi.org/10.4018/978-1-7998-7023-4.ch003

Msomi, V., & Jantjies, B. T. (2021). Correlative Analysis Between Tensile Properties and Tool Rotational Speeds of Friction Stir Welded Similar Aluminium Alloy Joints. *International Journal of Surface Engineering and Interdisciplinary Materials Science*, *9*(2), 58–78. https://doi.org/10.4018/IJSEIMS.2021070104

Muigai, M. N., Mwema, F. M., Akinlabi, E. T., & Obiko, J. O. (2021). Surface Engineering of Materials Through Weld-Based Technologies: An Overview. In S. Roy & G. Bose (Eds.), *Advanced Surface Coating Techniques for Modern Industrial Applications* (pp. 247–260). IGI Global. doi:10.4018/978-1-7998-4870-7.ch011

Mukherjee, A., Saeed, R. A., Dutta, S., & Naskar, M. K. (2017). Fault Tracking Framework for Software-Defined Networking (SDN). In C. Singhal & S. De (Eds.), *Resource Allocation in Next-Generation Broadband Wireless Access Networks* (pp. 247–272). Hershey, PA: IGI Global. doi:10.4018/978-1-5225-2023-8.ch011

Mukhopadhyay, A., Barman, T. K., & Sahoo, P. (2018). Electroless Nickel Coatings for High Temperature Applications. In K. Kumar & J. Davim (Eds.), *Composites and Advanced Materials for Industrial Applications* (pp. 297–331). Hershey, PA: IGI Global. doi:10.4018/978-1-5225-5216-1.ch013

Mwema, F. M., & Wambua, J. M. (2022). Machining of Poly Methyl Methacrylate (PMMA) and Other Olymeric Materials: A Review. In K. Kumar, B. Babu, & J. Davim (Eds.), *Handbook of Research on Advancements in the Processing, Characterization, and Application of Lightweight Materials* (pp. 363–379). IGI Global. https://doi.org/10.4018/978-1-7998-7864-3.ch016

Mykhailyshyn, R., Savkiv, V., Boyko, I., Prada, E., & Virgala, I. (2021). Substantiation of Parameters of Friction Elements of Bernoulli Grippers With a Cylindrical Nozzle. *International Journal of Manufacturing, Materials, and Mechanical Engineering, 11*(2), 17–39. https://doi.org/10.4018/IJMMME.2021040102

Náprstek, J., & Fischer, C. (2017). Dynamic Stability and Post-Critical Processes of Slender Auto-Parametric Systems. In V. Plevris, G. Kremmyda, & Y. Fahjan (Eds.), *Performance-Based Seismic Design of Concrete Structures and Infrastructures* (pp. 128–171). Hershey, PA: IGI Global. doi:10.4018/978-1-5225-2089-4.ch006

Nautiyal, L., Shivach, P., & Ram, M. (2018). Optimal Designs by Means of Genetic Algorithms. In M. Ram & J. Davim (Eds.), *Soft Computing Techniques and Applications in Mechanical Engineering* (pp. 151–161). Hershey, PA: IGI Global. doi:10.4018/978-1-5225-3035-0.ch007

Nazir, R. (2017). Advanced Nanomaterials for Water Engineering and Treatment: Nano-Metal Oxides and Their Nanocomposites. In T. Saleh (Ed.), *Advanced Nanomaterials for Water Engineering, Treatment, and Hydraulics* (pp. 84–126). Hershey, PA: IGI Global. doi:10.4018/978-1-5225-2136-5.ch005

Nikolopoulos, C. D. (2021). Recent Advances on Measuring and Modeling ELF-Radiated Emissions for Space Applications. In C. Nikolopoulos (Ed.), *Recent Trends on Electromagnetic Environmental Effects for Aeronautics and Space Applications* (pp. 1–38). IGI Global. https://doi.org/10.4018/978-1-7998-4879-0.ch001

Nogueira, A. F., Ribeiro, J. C., Fernández de Vega, F., & Zenha-Rela, M. A. (2018). Evolutionary Approaches to Test Data Generation for Object-Oriented Software: Overview of Techniques and Tools. In M. Khosrow-Pour, D.B.A. (Ed.), Incorporating Nature-Inspired Paradigms in Computational Applications (pp. 162-194). Hershey, PA: IGI Global. https://doi.org/ doi:10.4018/978-1-5225-5020-4.ch006

Nwajana, A. O., Obi, E. R., Ijemaru, G. K., Oleka, E. U., & Anthony, D. C. (2021). Fundamentals of RF/Microwave Bandpass Filter Design. In A. Nwajana & I. Ihianle (Eds.), *Handbook of Research on 5G Networks and Advancements in Computing, Electronics, and Electrical Engineering* (pp. 149–164). IGI Global. https://doi.org/10.4018/978-1-7998-6992-4.ch005

Ogbodo, E. A. (2021). Comparative Study of Transmission Line Junction vs. Asynchronously Coupled Junction Diplexers. In A. Nwajana & I. Ihianle (Eds.), *Handbook of Research on 5G Networks and Advancements in Computing, Electronics, and Electrical Engineering* (pp. 326–336). IGI Global. https://doi.org/10.4018/978-1-7998-6992-4.ch013

Orosa, J. A., Vergara, D., Fraguela, F., & Masdías-Bonome, A. (2021). Statistical Understanding and Optimization of Building Energy Consumption and Climate Change Consequences. In R. González-Lezcano (Ed.), *Advancements in Sustainable Architecture and Energy Efficiency* (pp. 195–220). IGI Global. https://doi.org/10.4018/978-1-7998-7023-4.ch009

Osho, M. B. (2018). Industrial Enzyme Technology: Potential Applications. In S. Bharati & P. Chaurasia (Eds.), *Research Advancements in Pharmaceutical, Nutritional, and Industrial Enzymology* (pp. 375–394). Hershey, PA: IGI Global. doi:10.4018/978-1-5225-5237-6.ch017

Ouadi, A., & Zitouni, A. (2021). Phasor Measurement Improvement Using Digital Filter in a Smart Grid. In A. Recioui & H. Bentarzi (Eds.), *Optimizing and Measuring Smart Grid Operation and Control* (pp. 100–117). IGI Global. https://doi.org/10.4018/978-1-7998-4027-5.ch005

Padmaja, P., & Marutheswar, G. (2017). Certain Investigation on Secured Data Transmission in Wireless Sensor Networks. *International Journal of Mobile Computing and Multimedia Communications*, *8*(1), 48–61. doi:10.4018/IJMCMC.2017010104

Palmer, S., & Hall, W. (2017). An Evaluation of Group Work in First-Year Engineering Design Education. In R. Tucker (Ed.), *Collaboration and Student Engagement in Design Education* (pp. 145–168). Hershey, PA: IGI Global. doi:10.4018/978-1-5225-0726-0.ch007

Panchenko, V. (2021). Prospects for Energy Supply of the Arctic Zone Objects of Russia Using Frost-Resistant Solar Modules. In P. Vasant, G. Weber, & W. Punurai (Eds.), *Research Advancements in Smart Technology, Optimization, and Renewable Energy* (pp. 149-169). IGI Global. https://doi.org/10.4018/978-1-7998-3970-5.ch008

Panchenko, V. (2021). Photovoltaic Thermal Module With Paraboloid Type Solar Concentrators. *International Journal of Energy Optimization and Engineering*, *10*(2), 1–23. https://doi.org/10.4018/IJEOE.2021040101

Pandey, K., & Datta, S. (2021). Dry Machining of Inconel 825 Superalloys: Performance of Tool Inserts (Carbide, Cermet, and SiAlON). *International Journal of Manufacturing, Materials, and Mechanical Engineering*, *11*(4), 26–39. doi:10.4018/IJMMME.2021100102

Panneer, R. (2017). Effect of Composition of Fibers on Properties of Hybrid Composites. *International Journal of Manufacturing, Materials, and Mechanical Engineering*, *7*(4), 28–43. doi:10.4018/IJMMME.2017100103

Pany, C. (2021). Estimation of Correct Long-Seam Mismatch Using FEA to Compare the Measured Strain in a Non-Destructive Testing of a Pressurant Tank: A Reverse Problem. *International Journal of Smart Vehicles and Smart Transportation*, *4*(1), 16–28. doi:10.4018/IJSVST.2021010102

Paul, S., & Roy, P. (2018). Optimal Design of Power System Stabilizer Using a Novel Evolutionary Algorithm. *International Journal of Energy Optimization and Engineering*, *7*(3), 24–46. doi:10.4018/IJEOE.2018070102

Paul, S., & Roy, P. K. (2021). Oppositional Differential Search Algorithm for the Optimal Tuning of Both Single Input and Dual Input Power System Stabilizer. In P. Vasant, G. Weber, & W. Punurai (Eds.), *Research Advancements in Smart Technology, Optimization, and Renewable Energy* (pp. 256-282). IGI Global. https://doi.org/10.4018/978-1-7998-3970-5.ch013

Pavaloiu, A. (2018). Artificial Intelligence Ethics in Biomedical-Engineering-Oriented Problems. In U. Kose, G. Guraksin, & O. Deperlioglu (Eds.), *Nature-Inspired Intelligent Techniques for Solving Biomedical Engineering Problems* (pp. 219–231). Hershey, PA: IGI Global. doi:10.4018/978-1-5225-4769-3.ch010

Pioro, I., Mahdi, M., & Popov, R. (2017). Application of Supercritical Pressures in Power Engineering. In L. Chen & Y. Iwamoto (Eds.), *Advanced Applications of Supercritical Fluids in Energy Systems* (pp. 404–457). Hershey, PA: IGI Global. doi:10.4018/978-1-5225-2047-4.ch013

Plaksina, T., & Gildin, E. (2017). Rigorous Integrated Evolutionary Workflow for Optimal Exploitation of Unconventional Gas Assets. *International Journal of Energy Optimization and Engineering*, *6*(1), 101–122. doi:10.4018/IJEOE.2017010106

Popat, J., Kakadiya, H., Tak, L., Singh, N. K., Majeed, M. A., & Mahajan, V. (2021). Reliability of Smart Grid Including Cyber Impact: A Case Study. In R. Singh, A. Singh, A. Dwivedi, & P. Nagabhushan (Eds.), *Computational Methodologies for Electrical and Electronics Engineers* (pp. 163–174). IGI Global. https://doi.org/10.4018/978-1-7998-3327-7.ch013

Quiza, R., La Fé-Perdomo, I., Rivas, M., & Ramtahalsing, V. (2021). Triple Bottom Line-Focused Optimization of Oblique Turning Processes Based on Hybrid Modeling: A Study Case on AISI 1045 Steel Turning. In L. Burstein (Ed.), *Handbook of Research on Advancements in Manufacturing, Materials, and Mechanical Engineering* (pp. 215–241). IGI Global. https://doi.org/10.4018/978-1-7998-4939-1.ch010

Rahmani, M. K. (2022). Blockchain Technology: Principles and Algorithms. In S. Khan, M. Syed, R. Hammad, & A. Bushager (Eds.), *Blockchain Technology and Computational Excellence for Society 5.0* (pp. 16–27). IGI Global. https://doi.org/10.4018/978-1-7998-8382-1.ch002

Ramdani, N., & Azibi, M. (2018). Polymer Composite Materials for Microelectronics Packaging Applications: Composites for Microelectronics Packaging. In K. Kumar & J. Davim (Eds.), *Composites and Advanced Materials for Industrial Applications* (pp. 177–211). Hershey, PA: IGI Global. doi:10.4018/978-1-5225-5216-1.ch009

Ramesh, M., Garg, R., & Subrahmanyam, G. V. (2017). Investigation of Influence of Quenching and Annealing on the Plane Fracture Toughness and Brittle to Ductile Transition Temperature of the Zinc Coated Structural Steel Materials. *International Journal of Surface Engineering and Interdisciplinary Materials Science*, *5*(2), 33–42. doi:10.4018/IJSEIMS.2017070103

Robinson, J., & Beneroso, D. (2022). Project-Based Learning in Chemical Engineering: Curriculum and Assessment, Culture and Learning Spaces. In A. Alves & N. van Hattum-Janssen (Eds.), *Training Engineering Students for Modern Technological Advancement* (pp. 1–19). IGI Global. https://doi.org/10.4018/978-1-7998-8816-1.ch001

Rondon, B. (2021). Experimental Characterization of Admittance Meter With Crude Oil Emulsions. *International Journal of Electronics, Communications, and Measurement Engineering*, *10*(2), 51–59. https://doi.org/10.4018/IJECME.2021070104

Rudolf, S., Biryuk, V. V., & Volov, V. (2018). Vortex Effect, Vortex Power: Technology of Vortex Power Engineering. In V. Kharchenko & P. Vasant (Eds.), *Handbook of Research on Renewable Energy and Electric Resources for Sustainable Rural Development* (pp. 500–533). Hershey, PA: IGI Global. doi:10.4018/978-1-5225-3867-7.ch021

Sah, A., Bhadula, S. J., Dumka, A., & Rawat, S. (2018). A Software Engineering Perspective for Development of Enterprise Applications. In A. Elçi (Ed.), *Handbook of Research on Contemporary Perspectives on Web-Based Systems* (pp. 1–23). Hershey, PA: IGI Global. doi:10.4018/978-1-5225-5384-7.ch001

Sahli, Y., Zitouni, B., & Hocine, B. M. (2021). Three-Dimensional Numerical Study of Overheating of Two Intermediate Temperature P-AS-SOFC Geometrical Configurations. In G. Badea, R. Felseghi, & I. Aşchilean (Eds.), *Hydrogen Fuel Cell Technology for Stationary Applications* (pp. 186–222). IGI Global. https://doi.org/10.4018/978-1-7998-4945-2.ch008

Sahoo, P., & Roy, S. (2017). Tribological Behavior of Electroless Ni-P, Ni-P-W and Ni-P-Cu Coatings: A Comparison. *International Journal of Surface Engineering and Interdisciplinary Materials Science, 5*(1), 1–15. doi:10.4018/IJSEIMS.2017010101

Sahoo, S. (2018). Laminated Composite Hypar Shells as Roofing Units: Static and Dynamic Behavior. In K. Kumar & J. Davim (Eds.), *Composites and Advanced Materials for Industrial Applications* (pp. 249–269). Hershey, PA: IGI Global. doi:10.4018/978-1-5225-5216-1.ch011

Sahu, H., & Hungyo, M. (2018). Introduction to SDN and NFV. In A. Dumka (Ed.), *Innovations in Software-Defined Networking and Network Functions Virtualization* (pp. 1–25). Hershey, PA: IGI Global. doi:10.4018/978-1-5225-3640-6.ch001

Salem, A. M., & Shmelova, T. (2018). Intelligent Expert Decision Support Systems: Methodologies, Applications, and Challenges. In T. Shmelova, Y. Sikirda, N. Rizun, A. Salem, & Y. Kovalyov (Eds.), *Socio-Technical Decision Support in Air Navigation Systems: Emerging Research and Opportunities* (pp. 215–242). Hershey, PA: IGI Global. doi:10.4018/978-1-5225-3108-1.ch007

Samal, M. (2017). FE Analysis and Experimental Investigation of Cracked and Un-Cracked Thin-Walled Tubular Components to Evaluate Mechanical and Fracture Properties. In P. Samui, S. Chakraborty, & D. Kim (Eds.), *Modeling and Simulation Techniques in Structural Engineering* (pp. 266–293). Hershey, PA: IGI Global. doi:10.4018/978-1-5225-0588-4.ch009

Samal, M., & Balakrishnan, K. (2017). Experiments on a Ring Tension Setup and FE Analysis to Evaluate Transverse Mechanical Properties of Tubular Components. In P. Samui, S. Chakraborty, & D. Kim (Eds.), *Modeling and Simulation Techniques in Structural Engineering* (pp. 91–115). Hershey, PA: IGI Global. doi:10.4018/978-1-5225-0588-4.ch004

Samarasinghe, D. A., & Wood, E. (2021). Innovative Digital Technologies. In J. Underwood & M. Shelbourn (Eds.), *Handbook of Research on Driving Transformational Change in the Digital Built Environment* (pp. 142–163). IGI Global. https://doi.org/10.4018/978-1-7998-6600-8.ch006

Sawant, S. (2018). Deep Learning and Biomedical Engineering. In U. Kose, G. Guraksin, & O. Deperlioglu (Eds.), *Nature-Inspired Intelligent Techniques for Solving Biomedical Engineering Problems* (pp. 283–296). Hershey, PA: IGI Global. doi:10.4018/978-1-5225-4769-3.ch014

Schulenberg, T. (2021). Energy Conversion Using the Supercritical Steam Cycle. In L. Chen (Ed.), *Handbook of Research on Advancements in Supercritical Fluids Applications for Sustainable Energy Systems* (pp. 659–681). IGI Global. doi:10.4018/978-1-7998-5796-9.ch018

Sezgin, H., & Berkalp, O. B. (2018). Textile-Reinforced Composites for the Automotive Industry. In K. Kumar & J. Davim (Eds.), *Composites and Advanced Materials for Industrial Applications* (pp. 129–156). Hershey, PA: IGI Global. doi:10.4018/978-1-5225-5216-1.ch007

Shaaban, A. A., & Shehata, O. M. (2021). Combining Response Surface Method and Metaheuristic Algorithms for Optimizing SPIF Process. *International Journal of Manufacturing, Materials, and Mechanical Engineering*, *11*(4), 1–25. https://doi.org/10.4018/IJMMME.2021100101

Shafaati Shemami, M., & Sefid, M. (2022). Implementation and Demonstration of Electric Vehicle-to-Home (V2H) Application: A Case Study. In M. Alam, R. Pillai, & N. Murugesan (Eds.), *Developing Charging Infrastructure and Technologies for Electric Vehicles* (pp. 268–293). IGI Global. https://doi.org/10.4018/978-1-7998-6858-3.ch015

Shah, M. Z., Gazder, U., Bhatti, M. S., & Hussain, M. (2018). Comparative Performance Evaluation of Effects of Modifier in Asphaltic Concrete Mix. *International Journal of Strategic Engineering*, *1*(2), 13–25. doi:10.4018/IJoSE.2018070102

Sharma, N., & Kumar, K. (2018). Fabrication of Porous NiTi Alloy Using Organic Binders. In K. Kumar & J. Davim (Eds.), *Composites and Advanced Materials for Industrial Applications* (pp. 38–62). Hershey, PA: IGI Global. doi:10.4018/978-1-5225-5216-1.ch003

Shivach, P., Nautiyal, L., & Ram, M. (2018). Applying Multi-Objective Optimization Algorithms to Mechanical Engineering. In M. Ram & J. Davim (Eds.), *Soft Computing Techniques and Applications in Mechanical Engineering* (pp. 287–301). Hershey, PA: IGI Global. doi:10.4018/978-1-5225-3035-0.ch014

Shmelova, T. (2018). Stochastic Methods for Estimation and Problem Solving in Engineering: Stochastic Methods of Decision Making in Aviation. In S. Kadry (Ed.), *Stochastic Methods for Estimation and Problem Solving in Engineering* (pp. 139–160). Hershey, PA: IGI Global. doi:10.4018/978-1-5225-5045-7.ch006

Siero González, L. R., & Romo Vázquez, A. (2017). Didactic Sequences Teaching Mathematics for Engineers With Focus on Differential Equations. In M. Ramírez-Montoya (Ed.), *Handbook of Research on Driving STEM Learning With Educational Technologies* (pp. 129–151). Hershey, PA: IGI Global. doi:10.4018/978-1-5225-2026-9.ch007

Sim, M. S., You, K. Y., Esa, F., & Chan, Y. L. (2021). Nanostructured Electromagnetic Metamaterials for Sensing Applications. In M. Bhat, I. Wani, & S. Ashraf (Eds.), *Applications of Nanomaterials in Agriculture, Food Science, and Medicine* (pp. 141–164). IGI Global. https://doi.org/10.4018/978-1-7998-5563-7.ch009

Singh, R., & Dutta, S. (2018). Visible Light Active Nanocomposites for Photocatalytic Applications. In K. Kumar & J. Davim (Eds.), *Composites and Advanced Materials for Industrial Applications* (pp. 270–296). Hershey, PA: IGI Global. doi:10.4018/978-1-5225-5216-1.ch012

Skripov, P. V., Yampol'skiy, A. D., & Rutin, S. B. (2021). High-Power Heat Transfer in Supercritical Fluids: Microscale Times and Sizes. In L. Chen (Ed.), *Handbook of Research on Advancements in Supercritical Fluids Applications for Sustainable Energy Systems* (pp. 424–450). IGI Global. https://doi.org/10.4018/978-1-7998-5796-9.ch012

Sözbilir, H., Özkaymak, Ç., Uzel, B., & Sümer, Ö. (2018). Criteria for Surface Rupture Microzonation of Active Faults for Earthquake Hazards in Urban Areas. In N. Ceryan (Ed.), *Handbook of Research on Trends and Digital Advances in Engineering Geology* (pp. 187–230). Hershey, PA: IGI Global. doi:10.4018/978-1-5225-2709-1.ch005

Stanciu, I. (2018). Stochastic Methods in Microsystems Engineering. In S. Kadry (Ed.), *Stochastic Methods for Estimation and Problem Solving in Engineering* (pp. 161–176). Hershey, PA: IGI Global. doi:10.4018/978-1-5225-5045-7.ch007

Strebkov, D., Nekrasov, A., Trubnikov, V., & Nekrasov, A. (2018). Single-Wire Resonant Electric Power Systems for Renewable-Based Electric Grid. In V. Kharchenko & P. Vasant (Eds.), *Handbook of Research on Renewable Energy and Electric Resources for Sustainable Rural Development* (pp. 449–474). Hershey, PA: IGI Global. doi:10.4018/978-1-5225-3867-7.ch019

Sukhyy, K., Belyanovskaya, E., & Sukhyy, M. (2021). *Basic Principles for Substantiation of Working Pair Choice*. IGI Global. doi:10.4018/978-1-7998-4432-7.ch002

Suri, M. S., & Kaliyaperumal, D. (2022). Extension of Aspiration Level Model for Optimal Planning of Fast Charging Stations. In A. Fekik & N. Benamrouche (Eds.), *Modeling and Control of Static Converters for Hybrid Storage Systems* (pp. 91–106). IGI Global. https://doi.org/10.4018/978-1-7998-7447-8.ch004

Tallet, E., Gledson, B., Rogage, K., Thompson, A., & Wiggett, D. (2021). Digitally-Enabled Design Management. In J. Underwood & M. Shelbourn (Eds.), *Handbook of Research on Driving Transformational Change in the Digital Built Environment* (pp. 63–89). IGI Global. https://doi.org/10.4018/978-1-7998-6600-8.ch003

Terki, A., & Boubertakh, H. (2021). A New Hybrid Binary-Real Coded Cuckoo Search and Tabu Search Algorithm for Solving the Unit-Commitment Problem. *International Journal of Energy Optimization and Engineering, 10*(2), 104–119. https://doi.org/10.4018/IJEOE.2021040105

Tüdeş, Ş., Kumlu, K. B., & Ceryan, S. (2018). Integration Between Urban Planning and Natural Hazards For Resilient City. In N. Ceryan (Ed.), *Handbook of Research on Trends and Digital Advances in Engineering Geology* (pp. 591–630). Hershey, PA: IGI Global. doi:10.4018/978-1-5225-2709-1.ch017

Ulamis, K. (2018). Soil Liquefaction Assessment by Anisotropic Cyclic Triaxial Test. In N. Ceryan (Ed.), *Handbook of Research on Trends and Digital Advances in Engineering Geology* (pp. 631–664). Hershey, PA: IGI Global. doi:10.4018/978-1-5225-2709-1.ch018

Valente, M., & Milani, G. (2017). Seismic Assessment and Retrofitting of an Under-Designed RC Frame Through a Displacement-Based Approach. In V. Plevris, G. Kremmyda, & Y. Fahjan (Eds.), *Performance-Based Seismic Design of Concrete Structures and Infrastructures* (pp. 36–58). Hershey, PA: IGI Global. doi:10.4018/978-1-5225-2089-4.ch002

Vargas-Bernal, R. (2021). Advances in Electromagnetic Environmental Shielding for Aeronautics and Space Applications. In C. Nikolopoulos (Ed.), *Recent Trends on Electromagnetic Environmental Effects for Aeronautics and Space Applications* (pp. 80–96). IGI Global. https://doi.org/10.4018/978-1-7998-4879-0.ch003

Vasant, P. (2018). A General Medical Diagnosis System Formed by Artificial Neural Networks and Swarm Intelligence Techniques. In U. Kose, G. Guraksin, & O. Deperlioglu (Eds.), *Nature-Inspired Intelligent Techniques for Solving Biomedical Engineering Problems* (pp. 130–145). Hershey, PA: IGI Global. doi:10.4018/978-1-5225-4769-3.ch006

Verner, C. M., & Sarwar, D. (2021). Avoiding Project Failure and Achieving Project Success in NHS IT System Projects in the United Kingdom. *International Journal of Strategic Engineering*, *4*(1), 33–54. https://doi.org/10.4018/IJoSE.2021010103

Verrollot, J., Tolonen, A., Harkonen, J., & Haapasalo, H. J. (2018). Challenges and Enablers for Rapid Product Development. *International Journal of Applied Industrial Engineering*, *5*(1), 25–49. doi:10.4018/IJAIE.2018010102

Wan, A. C., Zulu, S. L., & Khosrow-Shahi, F. (2021). Industry Views on BIM for Site Safety in Hong Kong. In J. Underwood & M. Shelbourn (Eds.), *Handbook of Research on Driving Transformational Change in the Digital Built Environment* (pp. 120–140). IGI Global. https://doi.org/10.4018/978-1-7998-6600-8.ch005

Yardimci, A. G., & Karpuz, C. (2018). Fuzzy Rock Mass Rating: Soft-Computing-Aided Preliminary Stability Analysis of Weak Rock Slopes. In N. Ceryan (Ed.), *Handbook of Research on Trends and Digital Advances in Engineering Geology* (pp. 97–131). Hershey, PA: IGI Global. doi:10.4018/978-1-5225-2709-1.ch003

You, K. Y. (2021). Development Electronic Design Automation for RF/Microwave Antenna Using MATLAB GUI. In A. Nwajana & I. Ihianle (Eds.), *Handbook of Research on 5G Networks and Advancements in Computing, Electronics, and Electrical Engineering* (pp. 70–148). IGI Global. https://doi.org/10.4018/978-1-7998-6992-4.ch004

Yousefi, Y., Gratton, P., & Sarwar, D. (2021). Investigating the Opportunities to Improve the Thermal Performance of a Case Study Building in London. *International Journal of Strategic Engineering*, *4*(1), 1–18. https://doi.org/10.4018/IJoSE.2021010101

Zindani, D., & Kumar, K. (2018). Industrial Applications of Polymer Composite Materials. In K. Kumar & J. Davim (Eds.), *Composites and Advanced Materials for Industrial Applications* (pp. 1–15). Hershey, PA: IGI Global. doi:10.4018/978-1-5225-5216-1.ch001

Zindani, D., Maity, S. R., & Bhowmik, S. (2018). A Decision-Making Approach for Material Selection of Polymeric Composite Bumper Beam. In K. Kumar & J. Davim (Eds.), *Composites and Advanced Materials for Industrial Applications* (pp. 112–128). Hershey, PA: IGI Global. doi:10.4018/978-1-5225-5216-1.ch006

About the Contributors

Francisco Mata Cabrera has over 200 publications, including books, book chapters, journal articles, and conference communications. Participation as principal or collaborator researcher in 20 international, national, regional, and company collaboration projects. Deputy Director of the Almadén School of Engineering from 2004 to 2012. Director of the Almadén School of Mining and Industrial Engineering from 2012 to 2022. 2000-2005 and 2006-2011 evaluation periods favorably evaluated by the National Commission for Research Activity Evaluation. Secondary and vocational education professor from 1997 to 2000. Laboratory technician from 2000 to 2001. Associate Professor, Collaborating Professor, and Contracted Doctor Professor. ANECA accreditation as University Full Professor since 2015. Head of the research group "Technologies Applied to Design". Head of the "Technology" section of the UCLM Applied Geology Institute. Network of collaborators in Morocco, Portugal, Brazil, Mexico, India, France, Greece, Turkey, and Germany. Member of the editorial committee of several international journals and of the scientific committee of conferences. Technical advisor and editorial writer for EDEBE. Training consultant.

Issam Hanafi is a professor at National School of Applied Sciences - Al Huceima (ENSAH), Abdelmalek Essaâdi University, Morocco. He holds a Ph.D. in mechanical engineering from the University of Abdelmalek Essaâdi, Morocco. His main interest field is devoted to the development of novel machining processes for composite materials. His graduate degrees include a master's degree in civil engineering from Faculty of Sciences and Techniques of Tangier, Morocco, and a Master of advanced studies (DESA) from Faculty of Sciences at Tétouan, Morocco. His present research interests are in the areas of quality engineering, intelligent control as it relates to neural networks and fuzzy control, also genetic algorithms in industrial applications.

* * *

Chiara Borsellino was born in Palermo March 24, 1971. In July 1989 she obtained a Diploma in Industrial Chemistry Technician, address "Environmental Chemistry", at the Istituto Tecnico Industriale E. Majorana Palermo with a score of 60/60. In July of 1994 she graduated in Industrial Engineering and Management, with the score: 110/110 cum laude with honors at a thesis entitled "Prediction of the formation of ductile fracture during plastic deformation processes", at the Faculty of Engineering University of Palermo. Subsequently, in November of the same year he qualified to practice. In November 1994 she won a scholarship to the attainment of the degree of Doctor of Philosophy in Production Engineering - Cycle X, at the University of Palermo. The title of PhD was obtained in May 1998 with a thesis entitled: "Models of ductile fracture and their application to some cold forming processes." As part of the PhD she spent six months of study and research (February-July 1995), at the Université de France-Comte, Besancon-Cedex, France. Since December 1998 she has received a research grant at the University of Palermo, for research on the topic "study of the hydroforming process of sheet metal", until February 2001. In March 2001 she became researcher in the ING-IND 16, at the Faculty of Engineering of Messina. Since May 2005, she is Associate Professor of Mechanical Technology (SSD ING-IND16-TECHNOLOGIES AND SYSTEMS) at the Department 1 Engineering, University of Messina.

Juan Carlos Campos Rubio was born in Chile, he is a Full Professor of Mechanical Engineering at Federal University of Minas Gerais (Belo Horizonte, Brazil). He obtained his BS in Mechanical Engineering from the Federal University of Santa Maria (Brazil), in 1988 and his MSc in Mechanical Engineering (electromechanics and control) from the Federal University of Santa Catarina (Brazil) in 1992. He obtained his PhD in Mechanical Engineering (precision engineering) from the University of São Paulo (Brazil) in 2000. His research interests include precision engineering, composite materials, and machining process. He is the Head of Materials Innovation and Technology Group (GiTeM).

Khadija Haboubi has dedicated her career to the fields of Environmental Engineering and Materials Science. After completing her undergraduate studies in Chemical Engineering at the University of Abdelmalek Essaâdi, she pursued a Master's degree in Materials Science and Engineering at the University of Mohamed V in Rabat. She then went on to earn a PhD in Environmental Engineering from the Facultu of sciences at Tetouan. Upon completing her studies, Dr. Khadija Haboubi joined the National School of Applied Sciences in Al-Hoceima, Morocco, where she currently serves as a Professor and Researcher in the field of Environmental Engineering and Materials Science. Over the course of her career, she has become an expert in the development of green technologies for waste management, specifically in the area of solid waste management. In addition to her academic and research responsibilities, Dr. Dr. Khadija Haboubi is also the President of the Moroccan Association for Scientific Research and Development, and the Coordinator of the Water Management and Environmental Sciences program at the National School of Applied Sciences in Al-Hoceima. She also leads the research team in charge of Materials Science, Energy, and Environment. Dr. Khadija Haboubi has made significant contributions to the field of environmental engineering through her work on the use of green technologies for waste management. Her research has focused on the development of new methods and technologies for the collection, sorting, treatment, disposal, and recycling of solid waste. Her research has been recognized internationally and has been published in numerous peer-reviewed journals.

Kaushik Kumar holds a Ph.D. in Engineering from Jadavpur University, India, an MBA in Marketing Management from Indira Gandhi National Open University, India and a Bachelor of Technology from Regional Engineering College (Now National Institute of Technology), Warangal, India. For 11 years, he worked in a manufacturing unit of Global repute. He is currently working as an Associate Professor in the Department of Mechanical Engineering, Birla Institute of Technology, Mesra, Ranchi, India. He has 22 Years of Teaching and Research Experience. His research interests include Composites, Optimization, Non-Conventional Machining, CAD / CAM, Rapid Prototyping and Quality Management Systems towards product development for societal and industrial usage and has received 27 patents for them. He has published 55+ Books (including 31 Edited Book Volume) (They are referred as Text Books and Reference Books by 40+ Universities / Institutes in their academic curriculum), 80 Book Chapters and 200+ Research Papers in peer-reviewed reputed national and international journals. Kaushik has also served as Editor–in–Chief, Series Editor, Guest Editor, Editor, Editorial Board Member and Reviewers for International and National Journals. He has been felicitated with many awards and honours including DISTINGUISHED ALUMNUS AWARD for PROFESSIONAL EXECELLENCE 2023 under Academic and Research from his Alma Matar National Institute of Technology, Warangal, India. He has also received Sponsored Research and Consultancy Projects of more than 1 Crore from Govt. of India and abroad. Kaushik has delivered expert lectures as Keynote Speaker at International & National Conferences, Resource Person at Various Workshops, FDP's and Short-Term Courses. He has guided many students of Doctoral, Masters and Under Graduate programmes of his home and other institutions in India and abroad. He also has served as Reviewer and Examiner of Doctoral and Masters dissertation for institutes in India and abroad.

Chikesh Ranjan is a highly accomplished Mechanical Engineer, PhD in Mechanical Engineering at the Department of Mechanical Engineering, Birla Institute of Technology Mesra, Ranchi, Jharkhand. He holds a Master of Engineering (M.E) degree in Mechanical Engineering with a specialization in Design of Mechanical Equipment from the same institution. He also holds a Bachelor of Engineering (B.E) in Mechanical Engineering from Marathwara Institute of Technology, Aurangabad, Maharashtra, achieving First Class with distinction upon graduation. Presently, he is serving as a Project Engineer at the National Institute of Technology Rourkela, Sandergarh, Odisha, where his innovative contributions have been recognized through the publication of 15 journal papers, 10 authored books, and 5 book chapters. He has also been actively involved in various workshops, conferences, and webinars, organizing 7 workshops and 8 events, and coordinating 13 events. Additionally, he has participated in 31 training/FDP programs and attended 34 national/international workshops, seminars, and webinars. He has been granted 10 design patents and 5 foreign patents, demonstrating his dedication to advancing mechanical design and technology. With 8 years of Teaching & Research experience, his areas of teaching and research interest are Composites, Non-conventional machining, CAD / CAM Drone Technology, and Robotics. He is a proud member of the Institution of Engineers, India, and has received 7 prestigious awards and honors for his remarkable contributions to the field of Mechanical Engineering.

Ritesh Kumar Singh is an Associate Professor in the Department of Production & Industrial Engineering, BIT, Mesra, Ranchi, India. He has 23 years of teaching and research experience. His PhD dissertation was concerned with waste minimisation strategies in manufacturing industries. He has done post-doctoral research at INRIA Nancy, France. His research interests include continuous improvement philosophies such as lean manufacturing, TOC, Six Sigma, Reconfigurable Manufacturing Systems (RMS), supply chain management, e-sourcing, machining etc.. He has authored nearly 60 technical papers in peer-reviewed journals and at various international conferences. His contributions have appeared in various reputable journals such as Advance Engineering Informatics, Engineering Application of Artificial Intelligence, IJPR, IJAMT, PPC, JEM, etc.

Index

S

T

IGI Global
Publishing Tomorrow's Research Today
www.igi-global.com

Publishing Tomorrow's Research Today

Uncover Current Insights and Future Trends in
Scientific, Technical, & Medical (STM)
with IGI Global's Cutting-Edge Recommended Books

Print Only, E-Book Only, or Print + E-Book.
Order direct through IGI Global's Online Bookstore at **www.igi-global.com** or through your preferred provider.

ISBN: 9798369303689
© 2024; 299 pp.
List Price: US$ **300**

ISBN: 9798369314791
© 2024; 287 pp.
List Price: US$ **330**

ISBN: 9798369300442
© 2023; 542 pp.
List Price: US$ **270**

ISBN: 9798369308929
© 2024; 426 pp.
List Price: US$ **265**

ISBN: 9781668489383
© 2023; 299 pp.
List Price: US$ **325**

ISBN: 9781668465776
© 2023; 454 pp.
List Price: US$ **325**

Do you want to stay current on the latest research trends, product announcements, news, and special offers?
Join IGI Global's mailing list to receive customized recommendations, exclusive discounts, and more.
Sign up at: www.igi-global.com/newsletters.

Scan the QR Code here to view more related titles in STM.

www.igi-global.com Sign up at www.igi-global.com/newsletters facebook.com/igiglobal twitter.com/igiglobal linkedin.com/igiglobal

www.ingramcontent.com/pod-product-compliance
Lightning Source LLC
LaVergne TN
LVHW061545290926
849162LV00004B/911